EVOLUTION OF
THE BRAIN
AND INTELLIGENCE

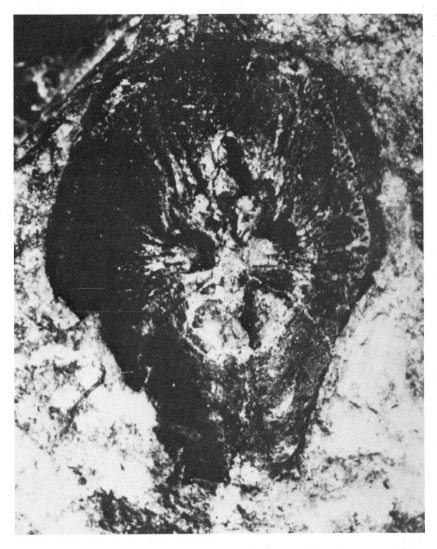

One of the earliest known vertebrate brains is that of the 400-million-year-old fossil fish *Kiaeraspis auchenaspidoides* (described by Stensiö, 1963). This ventral view of the fossilized head armor shows the pits, or depressions, that provide evidence of the brain. The pits contained the brain, eyes, and vestibular system (see pp. 84–85).

EVOLUTION OF THE BRAIN AND INTELLIGENCE

HARRY J. JERISON
DEPARTMENTS OF PSYCHIATRY AND PSYCHOLOGY
AND THE NEUROPSYCHIATRIC INSTITUTE
UNIVERSITY OF CALIFORNIA
LOS ANGELES, CALIFORNIA

ACADEMIC PRESS   New York and London   1973
A Subsidiary of Harcourt Brace Jovanovich, Publishers

QP
376
.J45
1973

COPYRIGHT © 1973, BY ACADEMIC PRESS, INC.
ALL RIGHTS RESERVED.
NO PART OF THIS PUBLICATION MAY BE REPRODUCED OR
TRANSMITTED IN ANY FORM OR BY ANY MEANS, ELECTRONIC
OR MECHANICAL, INCLUDING PHOTOCOPY, RECORDING, OR ANY
INFORMATION STORAGE AND RETRIEVAL SYSTEM, WITHOUT
PERMISSION IN WRITING FROM THE PUBLISHER.

ACADEMIC PRESS, INC.
111 Fifth Avenue, New York, New York 10003

United Kingdom Edition published by
ACADEMIC PRESS, INC. (LONDON) LTD.
24/28 Oval Road, London NW1

LIBRARY
LOS ANGELES COUNTY MUSEUM OF NATURAL HISTORY

Library of Congress Cataloging in Publication Data

Jerison, Harry J
    Evolution of the brain and intelligence.

    Bibliography: p.
    1.   Brain.   2.   Psychology, Comparative.
3.   Intellect.   I.   Title.
QP376.J45   596'.01'88   73-2062
ISBN 0–12–385250–1

PRINTED IN THE UNITED STATES OF AMERICA

To Tilly Edinger
1897–1967

Contents

Preface, xi
Acknowledgments, xiii

PART I. THE APPROACH

CHAPTER 1. *Brain, Behavior, and the Evolution of Mind* 3

    The Evolutionary Approach, 5
    Principles of Brain Function and Evolution, 7
    Biological Intelligence, 16

CHAPTER 2. *Evidence, Background, and Methods* 26

    Fossil Brains, 26
    Vertebrate History, 33
    Measuring the Evolution of the Brain, 41

CHAPTER 3. *Gross Brain Indices and the Meaning of Brain Size* 55

    Gross Brain Indices, 56
    The Meaning of Brain Size, 63
    Gross Brain Size and Brain Functions, 74

CHAPTER 4. *Beginnings: Habits and Brains* 82

    The First Vertebrates, 82
    The Generalized Vertebrate Brain: A Primer for Endocasts, 88

PART II. THE BASIC VERTEBRATE RADIATION

CHAPTER 5. *The Lower Vertebrates: Fish*   99

    Brains and Bodies of Fish, 100
    Brain Evolution in Living Classes of Fish, 108
    Potential for Life on Land, 115
    Conclusions, 122

CHAPTER 6. *Invasion of the Land: The First Tetrapods*   124

    Adaptive Radiation, 124
    Brains and Bodies of Amphibians, 130
    Conclusions, 135

CHAPTER 7. *The Radiation of the Reptiles*   137

    Evolutionary History, 138
    Relative Brain Size in Dinosaurs, 141
    The Mammallike Reptiles, 149
    Conclusions, 154

CHAPTER 8. *Flying Reptiles and Aerial Niches*   156

    The Control of Flight, 156
    Flying Reptiles, 161
    Conclusions, 171

PART III. BRAIN ENLARGEMENT AND THE BASIC VERTEBRATE RADIATION

CHAPTER 9. *Evolution of the Brain in Birds*   177

    Evolutionary Background, 178
    Brain and Body in *Archaeopteryx*, 182
    Cretaceous and Cenozoic Birds, 192
    Brains for Aerial Niches, 197

CHAPTER 10. *Mammalian Brains for Mesozoic Niches*   200

    Relative Brain Size, 206
    Evolutionary Implications, 213

Contents ix

CHAPTER 11. *Archaic Tertiary Mammals and Their Brains* 225
      Evolutionary Background, 226
      Relative Brain Size, 235
      Forebrain, Hindbrain, and Olfactory Bulbs, 245
      Conclusions, 253

CHAPTER 12. *Basic Selection Pressures for Enlarged Brains* 256
      Mammalian Trends and Nocturnal Adaptive Zones, 261
      Avian Trends and Adaptive Zones, 275

PART IV. PROGRESSIVE EVOLUTION OF THE BRAIN

CHAPTER 13. *Progressive Tertiary Evolution: Ungulates and Carnivores* 287
      Evolutionary History, 289
      Brain Morphology, 293
      Relative Brain Size, 303
      Conclusions, 318

CHAPTER 14. *Neotropical Herbivores: An Evolutionary Experiment* 320
      Evolutionary Background, 321
      The Experiment, 323
      Experimental Conclusions, 335

CHAPTER 15. *Special Topics* 340
      Evolutionary Trends, 341
      Lessons from the Pleistocene, 355

CHAPTER 16. *The Primates and Man* 363
      Evolutionary Background, 364
      Relative Brain Size in Prosimians, 373
      Relative Brain Size in Anthropoids, 387
      Enlarged Brains in Primates: Selection Pressures, 402

CHAPTER 17. *The Significance of the Progressive Enlargement of the Brain*   406

    Theoretical Background, 407
    Selection Pressures toward Enlarged Brains, 411
    Hominids and the Human Brain, 420
    The Work of the Hominid Brain, 424
    Conclusions, 432

BIBLIOGRAPHY   435

APPENDIX I. *Wirz's Analysis of Relative Size of Parts of the Brain*   457

APPENDIX II. *Statistical Tests on Mammalian Data*   462

APPENDIX III. *Foramen Magnum, the Size Factor, and Brain Size*   466

INDEX   471

Preface

This book may be unique in that general principles of behavior and brain function are derived from the actual record of the evolution of the vertebrate brain. Known from the endocasts of fossil animals, the history of the brain is analyzed here, quantitatively, in a serious effort to reconstruct the evolutionary forces that shaped that remarkable organ. I emphasize the implications of the history of the brain for the evolution of behavior in vertebrates, and, despite the uncertainties about the details of brain:behavior relations, I do not avoid speculations about the behavioral demands that were met by the evolving brain.

The brain is the "organ" of mind, and the nature of mind, of human consciousness and experience, is one of those paradoxical issues in which wonder, ignorance, and familiarity combine. An intuitive sense of its nature is not quite sufficient for the purpose of this book, because we will be concerned with "mind" in nonhuman animals, and for that one should be aware of studies of animal behavior and their implications about the nature of the animal's world (see Chapter 1). The idea that mind evolved as a result of organic evolution is a concept still foreign to many. But it is hardly arguable if we appreciate the relationship between brain and mind, or, more positively, between brain and behavior, and recognize that, after all, behavior is a natural function of natural organisms.

This book is intended for students of the neurosciences, whether concerned with behavior, organs, tissues, or cells, who would broaden their perspective on the functions of the brain and the evolutionary principles that can account for these functions. The book will be of special interest to anthropologists, who have developed much of the science of paleoanthropology, and who may seek to view the evolution of the human brain, manifested most dramatically by its increase in size, from the more general perspective of the history of the brain in vertebrates. The evolutionary analysis of perception and cognition should interest behavioral scientists, including those not especially biologically oriented, because it offers a different perspective on difficult, yet central, issues in psychology. Finally, it should be stimulating to zoologists and paleontologists, because it is an exercise in methodology, providing a quantitative analysis of the entire range of the evolution of one organ system, and because they join other evolutionists who are puzzled by the role of the brain in the evolution of life. This role has raised difficult and occasionally controversial issues, as I have indicated in my introductory chapter (Chapter 1), because even a superficial view of the evidence does not confirm the popular wisdom that large brains are necessarily useful.

Some of the discussion presented in the body of the text uses simple mathematical arguments that assume some knowledge of high school algebra. Statistical inferences that are discussed in the text are supported by the overall statistical analysis in Appendix II. The discussion of these inferences assumes a fair background in applied statistics, although one can simply accept the inferences as stated if one lacks the background. The statistics are meant to buttress discussions, the basic aspects of which are made obvious by the graphs and figures.

Recognizing that many readers may be interested in a particular topic without going through the book sequentially, I have tried to compromise between the repetitiveness necessary to make each chapter completely self-contained and the very tight organization that forces one to read through much irrelevant material to follow a particular discussion. Three chapters are devoted to speculations and integrative efforts that emphasize broad trends in evolution and their implications for the evolution of intelligence. These are Chapter 1, which is an introductory chapter, and Chapters 12 and 17. Chapters 2 to 4 emphasize methodological issues and the meanings of our measures. The remainder of the book is devoted to substantive matter on the brains, bodies, and associated mechanisms of behavior of vertebrates, emphasizing evolution "above the species level" (Rensch, 1959).

Acknowledgments

Among the individuals whose help and support I would like to acknowledge, I must name, first, Tilly Edinger, to whom this book is dedicated. Her frequent letters, sharing with me, an experimental psychologist, the advances and retreats of vertebrate paleontology as it was concerned with the evolution of the brain, were major contributions to my education and important introductions to the data of this book. Hofer (1969) has reviewed her career. When she died shortly before I began writing, my anticipated pleasure in the work diminished because I could no longer look forward to her reactions. These, I am sure, would have combined pleasure in having data about endocasts used in unusual ways with bewilderment at some of the details of that use—as she put it to me, she never could understand logarithms and other magic (cf. Edinger, 1962, p. 98).

She was my main guide to the endocasts at Harvard's Museum of Comparative Zoology. After her death I was helped by Stephen J. Gould and others, particularly Bryan Patterson and Gail Brown.

Valuable information, material, and facilities were provided by many gracious hosts at their respective institutions. I cannot possibly name all but would like to tender my thanks especially to C. L. Gazin and Nicholas Hotton III, the U.S. National Museum; Lawrence G. Barnes, Theodore Downs, James R. Macdonald, and David P. Whistler, the Los Angeles County Museum; S. P. Welles, the University of California at Berkeley; Malcolm McKenna, G. G. Simpson, and B. E. Taylor, the American Museum of Natural History; William D. Turnbull, the Field Museum of Natural History; Glen L. Jepsen and Donald Baird, Princeton University; E. L. Simons, the Yale Peabody Museum; A. J. Sutcliffe, H. W. Ball, A. J. Charig, and K. P. Oakley, the British Museum (Natural History); D. E. Russell, Denise Sigogneau, and Colette Dechaseaux, The Muséum d'Histoire Naturelle in Paris; and H. O. Hofer, the Delta Regional Primate Center in Louisiana.

I wish to thank several colleagues for criticisms and corrections of parts of the manuscript. These include J. A. Hopson, R. Molnar, L. B. Radinsky, A. S. Romer, G. G. Simpson, E. Stensiö, and S. P. Welles.

W. I. Welker of the University of Wisconsin helped me over the years in matters of mammalian behavior, neuroanatomy, and neurophysiology; and, more recently, I have also benefited from discussions with R. B. Masterton, S. Bernstein, R. Molnar, and E. C. Olson. I thank Dr. A. Towe for useful discussions of the final draft of the manuscript.

My wife, Irene, sons Jon and Andrew, and daughter Elizabeth made many suggestions for improving the manuscript and were invaluable in helping with measurements and data analysis.

For technical advice on statistical analysis I thank A. Affifi and A. Forsyth of UCLA. Computations were provided by the Health Sciences Computing Facility, UCLA, sponsored by the NIH Special Resources Grant RR-3.

Many of the finer drawings and graphs were prepared by Patricia Blake, and the photographs by Don Bedard, through the Visual Aids facility of the Mental Retardation Center, UCLA.

I have received financial support from the National Institutes of Mental Health; the National Institute of Child Health and Human Development, through Fellowship Award No. 1 F3 HD-38, 118-01; the Mental Retardation Research Center, UCLA, under Center Grant HD-04612; the Research Corporation through a grant to Antioch College; and the Department of Mental Hygiene of the state of California in which I held an appointment as a research specialist. Dr. George Tarjan, Director of the Mental Retardation Program in the Neuropsychiatric Institute at UCLA, and the late Dr. Stanley Wright, Associate Program Director for Research, generously agreed to my devoting time to complete the research for this book.

It is a special pleasure to acknowledge, finally, the unique institutional support that I received as a fellow at the Center for Advanced Study in the Behavioral Sciences during the 1967–1968 academic year. In particular I am indebted to the Center's Director, Meredith Wilson, and Associate Director, Preston Cutler; staff members, Irene Bickenbeck and David Pizer; past fellows, Herman Chernoff and Richard Savage, and the fellows who were my colleagues during that memorable year. They helped me get this work underway, and I thank them for the pleasure of their fellowship.

Part I

The Approach

The facts in this book are about the evolution of the vertebrate brain, and the explanations are about the evolution of behavior, in particular the evolution of intelligence. Among the facts of brain evolution none are more certain than those obtained from endocasts (casts of the endocranial cavities) of fossil animals, although like all facts these must be interpreted with care. The analysis of the evolution of intelligence is, of course, much more speculative.

Some of the interpretations of the evolution of the brain are very simple and involve no more than labeling the endocasts. The labels are of two types, taxonomic and chronological. There is little question about their validity in their present applications, but underlying the labels are assumptions, made by taxonomists, geochronologists, and other scientists, that are sometimes vigorously debated by specialists. Other facts about the endocasts may involve more difficult and occasionally controversial interpretations. There is always a temptation to treat an endocast as if it were a "fossil brain," no matter how often one repeats the caveat that it is at most an impression of a brain on the skull. It is almost impossible to avoid this identification of an endocast with a brain when one analyzes endocasts for information about the evolution of the brain, but this rarely leads to serious problems in actual work with the endocasts.

The most controversial facts in this book will probably be associated with the extensive use of the evidence of relative brain size. It should be emphasized that there is no debate about the validity of the measures of size, and enough information is provided about the methods of measurement to reassure the most skeptical reader that these are repeatable and reliable. There is a debate, however, about their meaningfulness. I believe that the debate has resulted in an overly skeptical view of measures of relative brain size. I indicate my reasons in several of the chapters in this section, especially Chapter 3, and the entire book can serve as evidence for the utility of simple brain:body relations when they are carefully analyzed.

The discussion of the evolution of intelligence and related behavior will suggest explanations for the course of the evolution of the brain. Although it forms important parts of Chapters 1 and 3, most of that discussion is presented in later parts of this book, when the selection pressures that were effective in the evolution of enlarged brains are analyzed. That discussion is really an analysis of the problems faced by vertebrate species when they invaded and coped with certain niches and adaptive zones.

Of the four chapters in this section the first is a general introductory chapter for the entire book. The second chapter describes the methods of analysis of brains, endocasts, and body size and presents other background material that will be particularly helpful in interpreting and understanding the data. The third chapter is a discussion of the meaning of brain size, both from the point of view of the relevance of this measure for other measures of the brain and its relevance for understanding the work of the brain. Although Chapters 2 and 3 emphasize technical and methodological issues, they also present something of a philosophy about how to think about those issues. The final chapter of this section introduces the fossil evidence on the brains of the first vertebrates. In the process it also interweaves the evidence of the brain with conjectures and reconstructions of the niches within which these vertebrates evolved, an approach followed throughout the rest of the book. That chapter concludes with a primer for the external morphology of the brain, emphasizing those parts visible in endocasts. Although I do not emphasize this kind of morphological analysis, the primer enables one to perform such analyses on some of the material presented later in the book by correlating the sketches and photographs of the many illustrated endocasts with the tables and illustrations in the primer.

Chapter 1

Brain, Behavior, and the Evolution of Mind

The "persistent problems in the evolution of mind," aptly labeled and discussed by Karl Lashley (1949), were the challenge that inspired this book. To solve these problems we should understand evolution and understand mind. Evolution can be approached directly, through the study of fossil records, and this book usually follows the direct evolutionary approach. But the mind's evolution is an obscure scientific issue which is generally approached indirectly by asking how perception and cognition, or intelligence, may have evolved. I devote many pages to these questions and their possible answers, but I also make a direct approach with the help of a simple tactic: I will often equate "mind" with "brain," in particular a measure of the brain that Lashley identified as especially relevant for the evolution of intelligence or behavioral capacity.

> The only neurological character for which a correlation with behavioral capacity in different animals is supported by significant evidence is the total mass of tissue, or rather, the index of cephalization, measured by the ratio of brain to body weight, which seems to represent the amount of brain tissue in excess of that required for transmitting impulses to and from the integrative centers [Lashley, 1949, p. 33].

In a later chapter (Chapter 3) I review the evidence that helps to explain why gross brain size is an effective basis for the index of cephalization, and I define the index more precisely. The evidence leads to the conclusion that brain size is a natural biological statistic for estimating more subtle and fundamental characteristics of the brain. It can be used to estimate the total number of neurons and glial cells in the brains of living mammals and perhaps also the complexity of the neural interconnections.

That the gross brain size should have these characteristics is particularly fortunate for anyone interested in the evolution of mind or intelligence.

It is possible to measure brain sizes from endocranial casts (endocasts) and body sizes from skeletal reconstructions, and this has been done in many fossil animals. It is, therefore, possible to determine quantitatively the neurological correlate of "behavioral capacity," as Lashley put it, and to see how that capacity has changed during the history of the vertebrates. This is as close as one is likely to come to a direct analysis of the evolution of mind.[1]

When a psychologist of Lashley's stature used the term "behavioral capacity" rather than "mind," he was exercising sensible discretion, recognizing that definitions of mind can trap the best of us. The safest approach to such definitions is to avoid them and to let a sense of one's meaning be developed as part of a simpler theme. I was tempted to be equally discrete and to avoid terms such as "mind," "consciousness," or "intelligence." But it was impossible to maintain the pose of innocence. Certain issues had demanded the use of these terms throughout the preparation of this book, and I had to discuss them in analyzing the selection pressures toward the enlargement of the vertebrate brain. The definitions, or hints at them, are present in the discussions of almost every chapter. They have been presented most explicitly in this chapter and throughout the final chapter, in particular in the final conclusions.

In a few words, I regard the mind and conscious experience as constructions of nervous systems to handle the overwhelming amount of information that they process. Intelligence, in a "between-species" sense, is a measure of the capacity for such constructions.[2] To the extent that it is a valid evolutionary concept, intelligence would be reflected in the variations among species in their capacities to integrate sensory information from various sense modalities and to construct "perceptual invariants" or "ob-

---

[1] A paleoneurological analysis of the external gross anatomy of the brain, which is all that can be observed in an endocranial cast, can be supplemented by selected evidence from the extensive literature on living brains. Only the small fraction of that literature that is helpful for understanding the significance of gross brain size is discussed in this book. There are many additional evolutionary implications in the data on the comparative neuroanatomy and neurophysiology of living animals; an outstanding example is the article by Diamond and Hall (1969). The symposia edited by Hassler and Stephan (1966) and by Petras and Noback (1969) provide timely and generally excellent introductions to the comparative literature considered in an evolutionary framework. Soviet literature on quantitative comparative neuroanatomy was summarized by Blinkov and Glezer (1968) and more recently by Shevchenko (1971).

[2] "Within-species" and "between-species" are used in a more or less statistical sense to refer, respectively, to variations among individuals within a species and among "typical" representatives of different species. Measures of these kinds of variation are, in principle, distinguishable from one another and can be used for further numerical analysis.

jects," that is, real objects of the real world. These are perceived as constant (unchanging in time) although they may result from changing patterns of stimulation at the sense organ or even from remembered images. This capacity to construct perceptual invariants eventually evolved into the human capacity for elaborate imagery, language, and culture.

Each defining word in the last paragraph demands more definition—"perception," "imagery," "language," "culture"—these hardly qualify as terms to simplify a concept. The brain, for all the mystery that surrounds its functions, is a tangible structure with many measurable features, which can be used for the analysis of the evolution of mind.

THE EVOLUTIONARY APPROACH

By treating relative brain size as a measure of intelligence at the species level, a between-species measure, it is possible to develop a coherent story about the probable history of intelligence as a biological phenomenon. Like other biological processes, intelligence must have evolved under the influence of natural selection. I will try to reconstruct the story of that evolution with the goal of establishing a framework for understanding the place of intelligence in the history of life and reaching a better insight into the nature of intelligence.

The story can be genuinely historical. The brains, as well as the bodies, of many species from critical periods in vertebrate history are analyzed, and I also describe the selection pressures under which these species evolved as they entered their environmental niches.

Let me anticipate a few results of this analysis to illustrate its nature. We will see that selection pressures toward enlargement of the brain beyond the requirements of larger bodies (pressures toward "intelligence") were probably rare until the birds and mammals diverged from their different reptilian ancestral stocks. Even in these "higher" vertebrate classes, selection for enlarged brains did not continue in all orders.

Also, the initial enlargement of the brain in mammals, as they evolved from their reptilian ancestors, in fact occurred in the earliest known forms more than 150 million years (150 m.y.) ago. This was probably related to the development of new sensory capacities for life in nocturnal niches rather than to the evolution of intelligence. As I see it, it was only many millions of years later, in the Tertiary period, about 50 m.y. ago, that there was an expansion of the mammalian brain that was correlated with intelligence. The procedure used here will enable us to perform a retrospective evolutionary "experiment" in which the effect of a critical selection pressure (the presence of progressive predators) on relative brain size will be examined. We will examine some unique aspects of the evolution of the human brain in biological perspective by comparing it with the evolution of the brain in other lineages.

This direct evolutionary approach can be contrasted with the comparative study of living species performed in an evolutionary framework. Although the comparative method is often the only one available, in particular for behavioral functions, its evolutionary significance depends partly on correlations with paleontological data. A major purpose of the analysis in this work is to provide the right kind of paleontological data to be related to differences in intelligent behavior among living species. The result of such an analysis will be to identify some living species as "primitive," or "relicts," and others as "progressive," with respect to traits or characters that are relevant for behavior. In that way a true phylogeny is approximated by a sample of living species. Let us be clear about the significance, assumptions, and limitations of such a phylogeny.

Living Approximations to True Phylogenies

It is not enough to use information from formal studies of systematics and evolution to decide that certain species are appropriate "relicts" because they seem relatively unchanged compared to ancestral forms. Different characters have evolved at different rates, and a species that is primitive in most respects may have diverged considerably from its ancestral stock in those characters involved in the evolution of intelligence. The human species may be the best example of such uneven rates of evolution because in many aspects of skeletal morphology (e.g., dentition, number of digits on the hands and feet, persistence of the clavicle) we are more like ancestral generalized mammals than are rats, cats, sheep, or horses. We are progressive with respect to our brain, but primitive in many other ways.

Comparably uneven rates of evolution of different characters can be found in other lineages. The well-studied evolution of horses is an excellent example of differential rates of evolution of the brain versus other parts of the body. A major and rapid expansion of the horse brain occurred in middle to late Eocene times, relatively early in the history of the equids, perhaps 50 m.y. ago. At that time body size and other skeletal features were relatively stable. During the later evolution of the horse lineage, when the equids became more diversified skeletally, there was only moderate modification of the brain. There was an increase in its size, and the expected increase in convolutedness associated with size, in those species that evolved larger bodies. The later increases in brain size were approximately those expected for the greater body sizes (Edinger, 1948; Simpson, 1951).

Among the more enigmatic groups of living mammals, the monotremes have recently been the subjects of an important revival of interest (Allison and Van Twyver, 1970; Lende, 1964). They warrant that interest because of their evolutionary importance. Among the monotremes, however, the echidna, in particular, illustrates another pitfall in the search for "relicts" among living animals. In the echidna the overall skeletal characteristics do suggest a living species that might represent the ancestral mammalian condition. Assuming that Hopson and Crompton (1969) are correct to consider the monotremes as related to Mesozoic fossil mammals (100 m.y. ago or more), or the comparable argument that the monotremes represent surviving therapsid reptiles (Van Valen, 1960), they are a paradoxical group of species that are primitive in most skeletal aspects, and actually reptilian in their shoulder girdles, but are well in advance of didelphids like the opossum or insectivores like the hedgehog in relative brain size and brain differentiation (Lende, 1963, 1964; Lende and Sadler, 1967). To the extent that intelligence is correlated with the mass of tissue in the brain, the monotremes are best considered to be at almost the same level as living progressive placental mammals; they presumably have reached that level by parallel evolution.

PRINCIPLES OF BRAIN FUNCTION AND EVOLUTION

Relatively few of the principles of brain function that are relevant for evolutionary analyses were developed within an evolutionary framework, and some were actually discovered before Darwin's great synthesis of evolutionary theory. Early in the nineteenth century the basic correlation between structure and function in nervous systems was beginning to be appreciated from the discovery of the sensory role of the dorsal horn and the motor role of the ventral horn of the spinal cord. At the level of the brain, the doctrine of localization of function was proposed by Broca and the phrenologists and opposed by Flourens' version of the mass action hypothesis (see Brazier, 1959). The correlation of structure and function is important for evolutionary thinking about the brain because functional or behavioral capacities can be inferred from the visible brain structures or from the total mass of the brain as determined from fossil endocasts.

Other principles, developed independently of evolutionary points of view, and after the publication of "The Origin of Species," included the neuron doctrine, which established a possible basic unit of analysis of neural activity (see Brazier, 1959). Recent work on "miniature nervous systems" may have disclosed an even more appropriate unit for the analysis of neural activity as it occurs in the brain (Bullock, 1967), one that could serve in the role of cell assemblies as discussed by Hebb (1949). In addition there are new statements of doctrines of localization of function (Woolsey, 1958) and of mass action (Lashley, 1950). Less widely appreciated, but equally important, are the quantitative studies of relations between microscopic and gross structures in the brains of different animals (Bok, 1959; Harman, 1957; Stephan et al., 1970).

The only basically evolutionary principle of neural activity, that of encephalization, or corticalization, of function, states that as advances on the phylogenetic scale took place more rostral parts of the nervous system took over specific functions. This principle is certainly wrong unless one accepts a vaguely teleological definition of "function" and even then its validity is uncertain (Weiskrantz, 1961).

The direct analysis of the evolution of the brain from the data of fossil endocasts yielded a number of important structural principles, the most important of which appeared a few years after Darwin's publication. Lartet (1868) discovered that relative brain size increased with the passage of geological time in at least some lineages of the mammals. This discovery was incorporated among Marsh's (1874) "laws," which are discussed later in this section. The disproof or proof and quantitative restatements of several of these "laws" are a major theme of this book.

LOCALIZATION OF FUNCTION AND THE PRINCIPLE OF PROPER MASS

According to the localization doctrine, neuropsychological functions are controlled or determined by localized structures in the nervous system. For example, if a localized region of a mammal's retina is stimulated with light, it can be shown that neural units respond in localized regions of the retina, optic nerve, optic tract, superior colliculi, lateral geniculate bodies, and a specifiable region of the cerebral cortex. The procedures for determining the limits of this projection system are complex, and the specification itself is far from complete at this time. Thus, one may characterize a portion of the cortex as "visual cortex" by the latency of responses, by anatomical studies of pathways between thalamus and cortex, and by other methods, while recognizing that neural activity in other regions of the brain may also be altered by the presentation of a simple stimulus to the peripheral sense cells of the retina. As a first approximation it is appropriate to think of a restricted region of the cortex as receiving the primary projections from the retina, and that would be an instance of "cortical localization."

In the analysis of the evolution of the brain and its role in the adaptive radiation of animals, the doctrine of localization is the basis for an important principle that we will call the **principle of proper mass**: *The mass of neural tissue controlling a particular function is appropriate to the amount of information processing involved in performing the function.* This implies that in comparisons among species the importance of a function in the life of each species will be reflected by the absolute amount of neural tissue for that function in each species. It also implies that, within a species, the relative masses of neural tissue associated with different functions are related to the relative importance of the functions in the species. Among the mammals, "visual" species have enlarged superior colliculi and an enlarged visual cortex, and "auditory" species have enlarged inferior colliculi and an enlarged auditory cortex.

An example that is presented in more detail in a later chapter (Chapter 11) illustrates the application of the principle of proper mass. It is often stated, probably with some justice, that primitive mammals tended to have enlarged olfactory bulbs and were, therefore, "olfactory" animals. One proposed measure of the progressiveness of a species of mammals has been whether there was replacement of olfaction by other senses with a relative reduction of the size of the neural olfactory system (e.g., Tilney, 1931). A quantitative analysis, however, reveals that the absolute size of the olfactory bulbs in mammals has increased throughout evolution (pp. 221, 251). It is only when the olfactory bulbs are compared with other parts of the same brain that they may appear relatively smaller. Thus, it may be surmised that living mammals are generally more efficient in their ability to use olfactory information or in the amount of olfactory information that they use compared to their fossil ancestors. However, the elaboration of other systems in the brain has been even greater in the evolution of the mammals, so much so that the olfactory system has been overshadowed in size.

The doctrine presented in opposition to localization of function is that of mass action, generally associated with Lashley's early work (1929). As currently conceived, that doctrine might be stated somewhat as follows: complex behavioral functions tend to be governed by an extensive and diffuse network of neural structures; hence, the degree of incapacitation with respect to those functions following brain damage will tend to be related to the amount of tissue destroyed, more or less independently of the locus of the destroyed tissue. It was probably because of this principle that Lashley was able to recognize the role of gross brain size as an anatomical correlate of intelligence or "behavioral capacity."

NEURONS AND MINIATURE NERVOUS SYSTEMS

The evolutionary significance of the neuron arises from its near universality as an information processing unit in the metazoan nervous system. The same near universality is probably true of the "miniature nervous system" considered as a unit, although such systems have only recently come under intensive study. Their widespread occurrence reflects a fundamental evolutionary adaptation that must have occurred before the major multicellular phyla became differentiated from one another, because miniature systems have been found in many species from different phyla, including mollusks (Kandel, 1967), arthropods (Kennedy, 1967; Wilson, 1968), and vertebrates (Rovainen, 1967). At present it seems likely that the miniature system should be regarded as the fundamental unit, or building block, with which more complex or elaborate neural systems are constructed. Its role could be as a "prewired cell assembly," in Hebb's sense (1949), in which specific nerve cells act together to carry out fairly elaborate actions. This point is important for this general introductory chapter —it suggests that some superficially innovative evolutionary experimentation, in particular in the evolution of learning, but perhaps also in the evolution of attentive behavior, may actually represent the use of very ancient behavior mechanisms inherent in the structure of synaptic nervous systems, organized as in miniature nervous systems.

To clarify the point, let us first recall the nature of "miniature" nervous systems. These are systems of a few neurons in which each neuron has a specified physiological role. One neuron may always be excited or always inhibited when a second neuron is stimulated, or the pattern of excitation and inhibition may change in consistent ways in response to repeated stimulation of the second neuron. The network of neurons can be labeled with respect to the location as well as the function of each cell, and the labeling holds for all individuals of a species. As the basic system under selection pressure, the miniature nervous system would be the link between the behaviors and the genetic system of the phenotype. The major "adaptive radiation" of miniature nervous systems, evident from the variations on this adaptive theme in many metazoan phyla, indicates that the basic evolutionary innovation arose in the common ancestor of those metazoans.

Behavioral functions that occur in miniature nervous systems cannot be considered as newly evolved functions in the vertebrates. Thus, simple sensorimotor integrations must be considered as a common feature of the nervous systems of most multicellular animals and of all vertebrate species. The evidence with respect to miniature nervous systems also indicates that simple kinds of learning, habituation in particular, but perhaps also conditioning, occur in these systems (Bullock, 1967). In other words, learning as well as "lower" functions may actually be a property of the elementary units (miniature systems) that go into the building of a vertebrate brain.

A less obvious implication of this point is that it would be best to improve the definition of "function." Is "learning" really a "function"? For our purpose it may only be confusing to answer affirmatively. There are many kinds of learning, and there are many different neural structures associated with them. To learn visual habits and perform visual tasks will involve

Principles of Brain Function and Evolution

structures in the visual system; auditory habits will involve the auditory system. There are probably common systems (e.g., for activation or arousal) for both auditory and visual habits, but the learning of these habits cannot be separated from the localized structures associated with them. Damage to specific structures affects specific habits, visual brain for visual habits and auditory brain for auditory habits.

Some potential confusion may be avoided if we identify the "function" that is localized as the mathematical concept of function and think in terms of specific equations relating measures of stimuli and responses. Such a definition covers all the valid usages of "function" for this book, except as used in the statement of the principle of proper mass, when a broader definition (as a system of functions, such as the "functions" of hearing) was intended.

With this more precise definition of function, a result such as Lashley's famous discovery that the disruption of learned behavior in rats was correlated with the amount of cortex excised leads to the view that learning by Lashley's rats was based on many behavioral functions. It is not learning as a general behavioral capacity that should be discussed to describe Lashley's result. Rather, one should ask what was learned; how many "functions" were involved in the learning, retention, and execution of the habit; and how these interacted to produce the measured performance (errors and time in moving through a maze).

A similar attitude may be taken toward Bitterman's (1965) work on reversal learning in fish and rats, and the same kind of question may be asked. What are the actual dimensions of reversal learning, and how are functions for these dimensions carried out by brains when they process the information? The issue for the evolution of "behavioral capacity" then is not whether learning is possible at a particular phylogenetic stage. Elementary learning functions may be demonstrated in simple miniature systems. The issue may be, rather, how numerous and how complex are the things that can be learned and retained. Only in that sense can learning data provide a basis for defining animal intelligence. I will suggest later in this chapter that dimensions other than learning may also be considered as the basis of "intelligence" in animals, in particular, dimensions of perception, imagery, and consciousness.

Encephalization and Corticalization of Function

The problem of encephalization must be discussed because, as indicated earlier, this is the one evolutionary principle that has been proposed for comparative neuroanatomy and neurophysiology. It is based on several kinds of evidence, in particular, the effects of brain damage in different species at different levels of the nervous system. Thus, for many years it was considered that lesions of the visual cortex in man produced complete blindness, whereas comparable lesions in monkeys produced almost complete blindness, with the capacity to discriminate luminous flux remaining. In rats and cats they produced only the loss of habits based on brightness discrimination; furthermore, these habits could be relearned, indicating that the visual capacity was not completely lost. Weiskrantz (1961) rejected the interpretation of the data as evidence of encephalization (corticalization) and, essentially, reached the conclusion presented in the next paragraphs. Corticalization and encephalization of somatosensory functions (Marquis, 1934), of motor functions (Ruch, 1935), of auditory functions (Ades, 1959), and of primate functions as a whole (Noback and Moskowitz, 1962) have also been described. All of these analyses are subject to the kind of criticism presented by Weiskrantz.

As an evolutionary argument, the preceding description of encephalization can be dismissed. The facts as stated imply no more than that several species of mammals have evolved different sets of functions for processing certain information, and damage to cortical and subcortical brain structures has different effects in these species. There is nothing in the data relevant to an evolutionary succession because there is no reason to consider rats, cats, and monkeys as more progressive or less progressive than one another or as representing any kind of phylogenetic series.

On the basis of the other principles discussed earlier, a simple statement of encephalization or corticalization of function makes little sense. It is much more likely that nervous tissue is conservatively organized in the sense that a particular input–output relationship (function) once established in particular networks will probably be performed by homologous networks in descendant species in an evolutionary series (cf. Stebbins, 1969). It would be an unnecessary burden for the evolutionary process to have to evolve new systems to solve a problem already solved by an existing system. The doctrine of corticalization would require that, for example, a function established in a midbrain network in a primitive species be transferred to a cortical network in the descendant species. Furthermore, this kind of transfer would have to have occurred independently for many functions in many different evolutionary lineages. Encephalization is, therefore, an unlikely anatomical or physiological "fact," and it is even more far-fetched to assume it as a common process in different evolving genetic systems. It is much more likely—and would make good evolutionary sense—that sets of related visual functions were elaborated differently in different species when they entered their varied niches, for example, that form vision is different in rat, cat, and monkey and hence differently organized, rather than that the same form vision is handled by different neural systems.

A comparative sketch of the auditory system may clarify this point.

Auditory information used by primitive land vertebrates could have been analyzed in terms of frequency, phase, and amplitude. Peripheral auditory systems near the sense cells and in the medulla probably had only a few hundred or perhaps a few thousand neurons at most, but that number was probably sufficient for simple forms of that kind of information processing (Wever, 1965). With the evolution of the much more elaborate auditory apparatus in mammals, other kinds of auditory information began to be used, such as information for echolocation by bats in which tectal (midbrain) systems are emphasized (Grinnell, 1970). If all auditory functions were grouped together as the imprecisely defined "function" in the old sense, it could be correctly noted that "auditory function" is much more disrupted in bats than in lizards as a result of lesions in the midbrain. Audition is clearly more encephalized in bats but only because of the broad use of the term "audition" to cover the full range of functions in which the auditory system is involved. It makes better sense simply to recognize that in the bat there are functions of hearing that do not exist in the lizard and that these functions are localized in more rostral structures than are the functions that are common to bat and lizard.

It may be possible to accept a restricted view of encephalization as referring to the likelihood that the evolution of new functions involved the elaboration of new neural structures, which tended to be more rostral in the central nervous system. That principle is also questionable. The elaboration of hindbrain structures such as the cerebellum was highly correlated with the elaboration of the forebrain in mammals. The most nearly correct statement is probably that the functions that first appeared in the evolution of mammals and birds were governed by forebrain structures more than by other parts of the brain. These were, generally, sensorimotor functions and coordinating or integrative functions related to information processed by distance-sense modalities. As a result, the evolution of the brain in birds and mammals has been characterized, generally, by greater enlargement of the forebrain than of the rest of the brain (save possibly the cerebellum in some species).

Whole:Part Relations in the Brain

The structure of the brain is probably sufficiently orderly at the microscopic level to enable one to use the gross brain sizes as a natural biological statistic that estimates the characteristics of the parts. This orderliness gives some meaning to gross brain weight or volume, and since many of the data that are discussed are derived from these gross measures, much of Chapter 3 is devoted to this issue.

The orderliness of the relationship between microscopic and gross structures of the brain does not imply complete geometric simplicity. There is occasional simplicity, such as the fact that the surface area and thickness of the cerebral cortex in progressive mammals appear to be related simply to gross brain size (Elias and Schwartz, 1969; Harman, 1957). The orderliness is statistical in that there are often close relationships between gross brain size and many numerical measures that can be taken from large samples of neurons, glia, and fiber networks.

The use of gross brain size as a "statistic" leads us to emphasize variables ("parameters") that are often neglected in neuroanatomy and neurophysiology. The major present direction of research in these fields is to determine the "wiring diagram" of the nervous system. Specific functions can then be correlated with the activity of specific networks of fibers. When gross brain size is recognized as a meaningful biological statistic, this directs attention to parameters of the nervous system that are estimated by that statistic. These are not directly related to wiring diagrams. Some of the parameters are neuron density, absolute number of neurons, number of glial cells, size of cell bodies, length of the dendrite trees, and neurochemical correlates of these morphological factors (Tower and Young, 1973). The significance of these variables can only be studied by the comparative method in which the functions of neurons and systems of neurons are correlated with the values of the parameters (or of brain size) in different species. Questions of this general type are rarely asked, however, for the scientifically trivial but technically overwhelming reason that it is a major undertaking to develop expertise with more than a small number of species as laboratory animals. The demonstrations in this book of orderly evolutionary changes in absolute and relative brain size should direct attention to the probable importance of neural parameters associated with brain size.

"Laws" of Brain Evolution

Brain casts (endocasts) were discovered in the earliest days of scientific paleontology, at the end of the eighteenth century (Cuvier, 1835), and the meaning of brain size in relation to body size had been discussed a half-century earlier (von Haller, 1762). It was, therefore, almost inevitable that brain:body relationships would be considered for fossil animals. The first published statement about the increase in relative brain size in evolution was probably Lartet's (1868).

> The further back that mammals went into geological time, the more was the volume of their brain reduced in relation to the volume of their head and to the overall dimensions of their body [Lartet, 1868, p. 1120].

Perhaps independently, but certainly soon afterward, Marsh (1874) recognized the same phenomenon in his fossil materials collected on expeditions to the great fossil deposits of the North American West. Marsh's work has been discussed, perhaps too critically, by Edinger (1962). It is the natural precursor to the present analysis, and it is worth repeating Marsh's "laws" of brain evolution as he wrote them, both for a flavor of the science of his time and as background for our analysis.

> 1. All Tertiary mammals had small brains.
> 2. There was a gradual increase in the size of the brain during this period.
> 3. This increase was confined mainly to the cerebral hemispheres, or higher portion of the brain.
> 4. In some groups, the convolutions of the brain have gradually become more complex.
> 5. In some, the cerebellum and the olfactory lobes have even diminished in size.
> 6. There is some evidence that the same general law of brain growth holds good for Birds and Reptiles from the Cretaceous to the present time.
>
> The author [Marsh] has since continued this line of investigation, and has ascertained that the same general law of brain growth is true for Birds and Reptiles from the Jurassic to the present time.
>
> To this general law of brain growth two additions may now be made, which briefly stated are as follows:
>
> 1. The brain of a mammal belonging to a vigorous race, fitted for a long survival, is larger than the average brain, of that period, in the same group.
> 2. The brain of a mammal of a declining race is smaller than the average of its co[n]temporaries of the same group [Marsh, 1886, pp. 58–59].

Marsh was wrong in most respects, but he was certainly right in verifying Lartet's observation. In general, the brains of mammals did increase in relative size as they evolved, although this is properly a statistical rather than a deterministic "law." A few of his observations are completely wrong insofar as they can be checked by measurements; thus, the evolutionary enlargement of the brain was apparently limited to birds and mammals and was not a major feature of the evolution of reptiles or fish (Chapters 6–8; Jerison, 1969).

Most of this book is devoted to clarifying, modifying, quantifying, and generally modernizing Lartet's and Marsh's statements. I would replace Marsh's laws by a simple general principle (Jerison, 1970b; 1971a):

> The brains of all animals have evolved in ways appropriate to life in their niches or adaptive zones, in accordance with principles such as enunciated earlier that describe the relationship to behavior of the structure of the brain as an organ of the body.

The reason for the increase in brain size in birds and mammals was that they had invaded new niches in which there was an adaptive advantage for enlarged brains (Chapter 12; Jerison, 1971a).

The principle of proper mass implies that, other things being equal, brain size will be related to body size. This follows directly from the fact that most cells are about the same size in all vertebrates, and differences in body size are associated mainly with differences in the number of cells in the body. Neural control of the body is generally by a fairly fixed ratio of neurons to sensory and motor cells. Hence, a larger body means that there are more cells to be controlled and more nerve cells to do the controlling. In short, larger animals, which have generally larger organs such as livers or hearts, have to have larger brains for essentially the same reason. The nerve cells have more body to control and service.

BIOLOGICAL INTELLIGENCE

Although not my initial central concern, intelligence was an important determinant of the direction of this work. Some discussion of its nature, beyond that presented in the first paragraphs, will provide useful background for understanding the specific approaches that I have adopted. It should be recognized that the statements in this section represent a personal view and are not in all instances the dominant approach of ethologists or comparative psychologists.

At the outset I had noted that, as a characteristic of a living species, intelligence must have had an evolutionary history. It could conceivably have been the history of a trait in a limited succession of species such as the hominid lineage. It is likely, however, that a history of biological intelligence can be traced throughout the vertebrates, and I will present evidence in later chapters indicating that its significant development probably occurred when mammals evolved. Let me, first, try to develop an intuitive sense of the dimensions of behavior (and experience) that were under selection pressures when intelligence was evolving. I will then present a short sketch of the evolutionary history of intelligence as I see it.

INTELLIGENCE, *Umwelt*, AND CONSCIOUSNESS

Psychologists are more comfortable with "intelligence" as an individual trait rather than as a characteristic of a species (e.g., Butcher, 1968, p. 28), but it is only as a species characteristic under genetic control that the evolution of biological intelligence can be understood. Although experts differ with regard to the nature of human intelligence, a more or less common ground is that it is a dimension of cognitive behavior—the way one knows the world and uses that knowledge when adapting to changing situations. It is, therefore, appropriate to define biological intelligence in relation to what animals "know," that is, the cognitive processes that may enter into animal behavior and consciousness.

Biological intelligence may be nothing more (or less) than the capacity to construct a perceptual world. For man this is the real world of which he is conscious. Animals, too, may have "real" worlds, according to this view, but these worlds will differ depending on how the animal's brain does the work of integrating sensory and motor events. This neo-Kantian notion of different worlds formed by different kinds of brains has been made familiar by von Uexküll (1934) in his vivid descriptions of *Umwelt* as a species-specific perceptual world.

A more modern way of thinking about the perceptual world, of which either man or other animals are conscious, is as a construction of the nervous system designed to explain the sensory and motor information processed by the brain. Such a view, derived from Craik's (1943) discussion of the "nature of explanation," would have consciousness as a simplifying device, a model of a possible reality. Adaptive behavior in a changing environment would be represented in such a model (conscious experience) as actions of objects on a panoramic stage. The construction of the model, which is consciousness, is the work of the brain, and its work for man involves the resolution of thousands of millions of changing events in its neural networks. If a significant number of these events can be recoded as "objects" in "space" and "time" (subroutines, perhaps, to the computer technologist), the work of the brain as it processes its information will obviously be easier.

We may then assume that vertebrate species with less elaborate brains transform sensory information into motor neural information with little or no intervention of the kind of modeling implied by consciousness and the construction of perceptual worlds. Their worlds of prepotent stimulus complexes (the innate releaser mechanisms of the ethologists) are tightly coupled to appropriate response processes. This kind of coupling may have reached its apex in avian evolution.

It was only with the evolution of mammals, and perhaps birds to a lesser extent, that there evolved a significant capacity (or necessity) to transform input into output by the intervention of a conscious perceptual world. My reason for suggesting such a development in information processing is based on certain aspects of the history of mammals as presented in Chapters 12 and 17. I will anticipate some of the conclusions from that account by summarizing its implications for the evolution of consciously perceived worlds.

Evolution of Perceptual Worlds: A Hypothesis

Our awareness of the real world is based on information we receive from our distance senses, mainly vision and, to a lesser extent, hearing and smell. This "distance information" can then be combined with proximal information from touch and taste and information from internal systems that tells us where our head and limbs are and something about many other internal states. Much of this information is not really conscious, but when we speak of conscious experience, we generally mean a perceived panoramic view of things with oneself as the point of reference. Let us try to reconstruct the evolutionary history of the consciousness of such perceptual worlds. As mentioned above, most of the detailed evidence for this argument must be deferred to later chapters (especially Chapter 12). We will restrict ourselves for the present to an evolutionary sketch—a hypothesis about the evolution of sensorimotor systems and the consciously perceived worlds to which they led.

For every history there must be a prehistory, and the prehistory of consciousness, I believe, occurred in the late Paleozoic era and the beginning of the Mesozoic era or "Age of Reptiles." At that time, more than 200 m.y. ago, several major groups of reptiles radiated adaptively to fill many new niches for land-dwelling animals. It is clearly advantageous for such animals to have information about events at a distance. From the remarkable structure of the reptilian eye (Polyak, 1957), it is evident that early reptiles lived in visual worlds, but I suggest that they had not achieved the truly visual perceptual world in the sense of the previous paragraph. Without detailing the evidence, it seems as likely that the reptiles' worlds were more nearly concatenations of specific stimulus patterns and their associated response patterns, systems of reflexes as it were, although the extensive neural information-processing machinery of their retinas would have enabled them to respond selectively to many different stimuli.

During this prehistory of consciousness, behavior was extremely stereotyped, and perhaps the best model for it is in the response of living frogs to moving stimuli such as flies. If, by surgical intervention, its eye is permanently rotated in its socket to produce "inverted" images, the frog never adjusts to the strangely inverted "visual space." Instead, it persistently strikes out at a point in space appropriate to the corresponding point of stimulation as it would be mirrored on the retina of an intact eye (Sperry, 1951). One result of the evolution of perceptual worlds would have been to reduce such stereotypy and enable an animal to modify its behavior appropriately to its experience with its world. The corresponding human experiment (Kohler, 1964) indicates that when perceptual space is inverted by the use of special lenses it eventually rights itself after one is persistently exposed to it.

The first stage in the history of perceptual worlds occurred, according to this hypothesis, when the auditory modality became a significant source of information. I suggest that this occurred because the mammals first evolved as the nocturnal "reptiles" of the late Triassic period, about 200 m.y. ago (Chapter 12; Jerison, 1971a). Audition then replaced vision as the fine distance sense. To perform the functions of such a sense, neural circuits had to translate temporally encoded patterns of auditory nerve impulses into the equivalent of the spatial "maps" that had been generated more directly by the spatially distributed sensory elements of their reptilian ancestors' retinas.

To appreciate the additional demands on an information-processing system when spatial information has to be set in a temporal code, we can consider the situation in which temporal information is set in a spatial code. An example that many of us can share is the problem of reading music and imagining thematic and harmonic structures from the visual appearance of a musical score. When an appropriate level of expertise is achieved, one can successfully "hear" a written score, but this is no ordinary accomplishment. (Yet, the direct hearing and "knowing" the sound of any simple melody is accomplished with essentially no prior experience, upon being exposed to the proper sequence of sounds.) If we try to picture a vista from a set of brief exposures to successive fractions of the total scene, with a second or more between exposures, we see the problem of using temporal integration when spatial integration is the more "natural" approach.

In these examples I have considered the capabilities of our own very highly elaborated nervous system. But we should imagine a much simpler neurosensory system faced with this kind of task. If the simpler system already had a spatial analyzer built into its sensing equipment, in which a specific point on its sense organ was stimulated when a particular point in space was energized, then upon stimulation from a specific point in space it need only encode the point that responded on the sense organ. That code can be inherent in the system's structure. If the simpler system could only take spatial information in at a single point on its sense organ, then it would have to scan an area with the sense organ, recording when different amounts of stimulation occurred. Such systems can and have been built, of course, but it is clear that when they are biological systems, the encoding of "when"—that is, the encoding of time—adds a requirement and a load on the information-processing capacity. The simplest vertebrate visual systems are spatially coded at the retina, whereas the most complex auditory systems do not encode spatial information (localizing sound in space) unless information from the two ears is used and integrated at least at the level of the midbrain (Goldberg and Greenwood, 1966).

The unusual use of hearing hypothesized for the earliest mammals was not without some benefits. These can be appreciated retrospectively. The new role of hearing insured a neural representation of a time dimension with time intervals at least of the order of seconds available for "time binding" temporally disparate events into a unitary stimulus for action. For example, a sequence of twig snappings could be translated as the stimulus: "prey (or predator) moving left."

The second stage in the history of perceptual worlds may have been essentially simultaneous with the first, if the following reconstruction is correct. When the earliest Mesozoic mammals successfully invaded nocturnal niches, their distance information, though primarily auditory, probably continued to be at least partially visual. The reptilian daylight visual system would have evolved into the typically mammalian rod system, which is useful in twilight and moonlight as a crude distance sense for position and an excellent one for the response to motion. (This remains the typical visual system of living mammals, most of which are nocturnal.)

We must now recognize that distance information was coming from at least two significant sense modalities, both well endowed with a neural apparatus. The early reptilian visual system, including its neural components, was packaged to a major extent in the retina, and the mammalian auditory system, to expand sufficiently to perform distance sensing, had to be packaged in the brain. (There was no space for integrative circuits near the auditory sense organ analogous to the neural networks of the retina.) In many instances the early mammals could get reasonably good distance information from the same source of stimuli and through two modalities. It would obviously have been efficient for an information-processing system to identify the fact that the source was the same, that is, to "integrate" the information. We, therefore, expect that identification to be represented, or encoded, and this is what I mean when I speak of "integrative" functions of the brain. It is this process that changes patterns of stimulation into "objects" that maintain "constancy" under transformation in space. The objects have duration as well as location, and the complete set of objects in visual and auditory space would necessarily have been placed against some "map" or background. Thus, the elements of perceptual space appear as adaptively warranted in order to coordinate information from several sense modalities.

The third stage in the evolution of perceptual worlds, according to this sketch, occurred after the great extinction of reptiles at the end of the Mesozoic era (70 m.y. ago), and the successful mammalian invasion of the niches that then became open to diurnal land animals. This is the history of the "Age of Mammals" from about 60 m.y. ago to the present. As one may note, this really represents only the final third of the history of the mammals, but it was during that period that the great adaptive radiation of mammals occurred.

Probably from the very beginning of this great Tertiary radiation of the mammals, the mammalian orders (especially Primates) responded to the selection pressures to use photic information and reevolved a visual distance sense. But their visual sense was not the same as the ancient reptilian vision. It was now represented by much more extensive neural networks in the brain itself. Just as hearing as a distance sense had been modeled after the natural reptilian sense of vision, so it may be assumed that the newly reevolved mammalian diurnal vision was modeled after the, by then, natural distance sense of mammals, the auditory sense, which had been evolving for more than 100 m.y. of the previous history of the mammals. The assumption that a new system is modeled after a preexisting system is one of my basic general assumptions.

This means that, in addition to its normal role in spatial mapping, mammalian vision would be time binding, analogous to the way hearing can unify a complex sequence of tones into melodies. Like hearing, this new mammalian vision would be highly cephalized rather than based almost entirely on the neural systems of the retina. Visual images in such a system could be stored in some form for the order of seconds or longer and could maintain "constancy" under transformations in time and space. They would be images of "objects" viewable from several points in space and at different times, which nevertheless are cognitively the same or perceptually "constant" because of the constancy required of the objects as "real" objects in space.

During the evolution of the mammals an elaborate system developed for fine vision and color vision: the newly evolved cone retinas of some mammals, which were probably derived from the rod system rather than the reptilian cones (Walls, 1942). In other orders of mammals, perhaps less fine but still useful diurnal visual systems evolved, based on panoramic rod vision adapted to daylight effectiveness, as in the large ungulates, rabbits, and other species (Marler and Hamilton, 1966). Parallel to these visual systems, but also giving finely localized information about particular "objects" at a distance, the elaborately cephalized auditory system of the more archaic mammals would surely have persisted for distance sensing.

An enormous amount of information in the form of nerve impulses would be arriving at the central nervous system at every instant of wakefulness from these several distance-sensing systems, and there had to be some organization and simplification of that information. One would expect it to be organized into clumps of some sorts (the subroutines of the computer technologist mentioned earlier are obvious models). The point of this discussion is to suggest that the clumps are the neural basis for the "objects" in perceived "space" and "time" and the true meaning of our "real" world of everyday experience. It is in this sense that reality can be thought of as a creation of the brain.

HUMAN AND ANIMAL PERCEPTUAL WORLDS

Let us now consider the special case of man and our personal perceptual worlds. These are so intuitively "real" that the discovery of their constructed nature is a titillating stage in one's development and often the beginning of personal philosophies that recognize the solipsistic dilemma (for example, Mark Twain's "The Mysterious Stranger"). Our uniqueness is in our use of language, and it is important to realize that perceptual worlds are affected by language. One need not retreat to philosophical texts or anthropological treatises on cultural differences, as they affect our experience, for evidence. When we move into a new neighborhood, there is a familiar sequence of stages as we learn more and more about the homes and their inhabitants. A walk in a forest is obviously different after we have learned to identify the leaves and grasses and rock formations.

From the perspective of this hypothesis, human perceptual worlds involve some new developments beyond those of other animals, and these developments are logical extensions of those previously evolved. The history of the previous developments was the introduction of novel sensory systems that did work previously done by other well-adapted systems. To summarize this view: audition in mammals was elaborated as a nocturnal distance sense when distance information could no longer be handled by the normal adaptive system, the diurnally functional retinally organized visual system. Much later in evolutionary history, mammalian vision appeared as a novel adaptation of a newly organized visual system (largely in the brain, proper, rather than in the retina) for processing distance information for the once nocturnal but newly diurnal mammals of the Cenozoic era. Elaborate perceptual worlds could then be constructed by the brain, which integrated time-coded information that was both auditory and visual, to provide analysis of events at a distance by encoding the neural data as objects in space and time.

I suggest that the next stage in the evolution of perceptual systems occurred with the evolution of man, especially during the last million years or so, by further evolution of auditory space. Without considering the special selection pressures that would have acted to produce further elaboration of the auditory system, let us consider the possible directions of evolution of the ancient mammalian auditory system, which had served as a distance-sensing system for the mammals throughout their history in the

Mesozoic and Cenozoic eras. According to the method of the analysis thus far, one would expect a still more elaborately evolved auditory system to be modeled after the corticalized visual system of advanced mammals. The auditory system would then have the property of converting sounds and sound patterns into objects and producing yet further elaborations of the perceptual world, in particular with respect to time. We would hardly predict the perceptual structures of such a system a priori. But we may recognize, after the fact, that subsequent evolution was the auditory equivalent of elaborate pictures of the temporally extended visual world.

The types of structures that were actually evolved by the elaborated auditory systems included the motor as well as the sensory systems for language and the visual imagery produced by language. That the structures were only to a limited extent "prewired" is secondary to this argument, since we are concerned with the information-processing capacity and the activity of the system, not with the way in which that capacity was attained. Given that the sensorimotor structures for producing and receiving sounds and the neural auditory system evolved according to the model of the fully corticalized visual system, we would expect extensive capacities for encoding information about perceptual worlds in significantly more ways than are available in simple mappings. We can only suggest the richness of the new codes by recognizing the similarity among the perceptual experiences that are derived from written and oral messages and from the direct experience of the events described in such messages. We have all enjoyed entering into the lives of the characters in a well-written novel, and our entries enable us to participate in the existence of these characters.

A key element in these expanding perceptual worlds is the role of temporal integration. We should marvel (though it is too common a direct experience) at the persistence of our temporal world—that changing experience is interpreted as personal movement or changing mood rather than a changing world observed by a fixed observer. Yet, this is the essential point of the "constancy" principle as discussed by the Gestalt psychologists (Koffka, 1935). When "time" became a central element in the system of analysis for distance information, the threshold was reached for the evolution of perceptual worlds, the constancy or inconstancy of which was to become a fundamental issue to psychology and natural philosophy.

The Evolution of Intelligence

It is a small step from the previous discussion to the recognition that, if the nature of the perceptual world defines a dimension of intelligence, the evolution of intelligence is to be sought in the changes among species with respect to their perceptual worlds. One should be wary of species-specific traits, peculiarly evolved for life in a specialized niche, when defining intelligence. For example, effective color vision has evolved in many animals and in several groups of "intelligent" mammals. It is shared by some rodents, tree shrews, and most primates. Is it a dimension of intelligence?

The definition of intelligence for between-species comparisons might place color-receptive species above those that lack that dimension of discrimination, but only if the use of color information is integrated with other senses. The point is to define intelligence among species as a behavioral variable involved in the richness of the characteristic perceptual world of the species. In these terms, human color blindness is a within-species effect that involves a dimension of human variation and cannot be correlated with the unitary and invariant value that would be assigned to human intelligence for between-species measures. As a characteristic of the human species the occasional appearance of color blindness is entirely trivial from the point of view of comparing man with other primates or other animals on a dimension of intelligence. Can intelligence be defined as an evolutionary trait?

The present discussion suggests two important dimensions of intelligence and excludes the extent to which specific sensory capacities have been developed as specific to adaptive niches occupied by a species. The first dimension is the extent to which sensory capacities have been integrated with one another, and the second dimension is the extent to which behavior in response to sensory information is flexible and adjustable to inconsistencies in that information. Some of the familiar categories of experimental comparative psychology are consistent with this definition. The use of performance on reversal learning tasks as developed by Bitterman (1965) is relevant for a dimension of "flexibility." There is little research on intersensory effects, but some on learning sets (Meyer and Harlow, 1949) is relevant to this criterion. For example, when an animal learns to choose the odd object, the fact of oddity may be transferred from a visual to a tactile dimension, and learning sets can be shown to be different in species of different "intelligence" (Warren, 1965).

Brain and Biological Intelligence

It is easier to appreciate now why gross measures of the brain should be most closely related to biological intelligence. The number of neurons and the complexity of their interconnections should reflect the degree to which sensory systems have become elaborated and interconnected; it might make little difference from the point of view of intelligence which systems have actually been emphasized in a particular species. The facts of reptilian vision indicate that relatively small neuronal networks (of the order of perhaps a million elements) may be sufficient to process information within a single sensory channel. The reptilian auditory system, which is reasonably effective for responding to simple auditory stimuli, has an even smaller number of elements, numbering in the hundreds at the periphery and perhaps a few thousand centrally (Wever, 1965).

Greater masses of neural units in the brain, when they occur in the tens of millions or more, are to be identified with the integration of sensory information either within or among sensory modalities. Implications of this argument may be better appreciated if we return to the problem of defining function. There will be many different "auditory" functions handled by the auditory system, and these, too, must be integrated with one another. The integration of separate pure tones into melodies seems to be a peculiarly cortical function in cats and monkeys (Diamond, 1967), and comparable integrations of pattern information in the visual system may be a function of so-called primary visual cortex.

The implication is that the integrative functions of the brain, which will define intelligence for us, are limited by the amount of brain that is typical for an animal of a particular species. As shown in Chapter 3, the amount of neural material available for information-processing and integration is related to the gross brain weight or volume. It is for this reason that Lashley's surprising statement, quoted at the outset of this chapter, can be true. And it is for this reason that I approach the evolution of intelligence by analyzing the evolution of the brain, in particular, the evolution of its size.

Chapter 2

Evidence, Background, and Methods

The endocasts of fossil vertebrates provide unequivocal and direct evidence about the evolution of the brain and brain size. In mammals these endocasts look so much like brains that they almost have to be called "fossil brains." Endocasts are also brainlike in birds, extinct flying reptiles, and a few small fossil fish. Among other vertebrates, both living and fossil, the endocast mirrors the brain less adequately because the cranial cavity fits less closely about the brain and acts as a poorer mold. But endocasts, more and less adequate as "fossil brains," are known from almost the entire evolutionary history of the vertebrates. The first part of this chapter is devoted to a general discussion of fossil endocasts.

In order to see the history of the brain in perspective, the second part of this chapter is devoted to a review of selected topics in the history of the vertebrates. Romer (1966) is our usual authority on taxonomic and paleontological issues, except in cases where more recently proposed names have become generally accepted (see Romer, 1968, pp. 4–9).

The final part of this chapter introduces the problems of measurement of the evolution of the brain. Since our most significant quantitative measure is derived from the relationship of brain size to body size, some space is devoted to methods of estimating these sizes. Some normative data on brain:body relations in living vertebrates are also presented to serve as the standard for the analysis of much of the fossil material. It will be possible to demonstrate the simplest of the methods of analysis in this chapter, the "method of minimum convex polygons." In Chapter 3 other more elaborate, though not necessarily more valid, analytic tools will be introduced.

FOSSIL BRAINS

When an individual animal dies and suffers a normal burial, it will normally disappear entirely as a palpable structure through the action of various scavengers and finally through the action of microorganisms that consume the skeleton. The process in a buried animal may take several months, and it is quite complete. In order to be preserved as a fossil, the animal must usually be entombed under circumstances that will permit its skeleton to become mineralized, with all the original organic material replaced by minerals. The process is discussed in the introductions to most texts on paleontology (Romer, 1966). During the period of fossilization the soft tissues in the cavities of the body, including the brain and its supporting tissue in the endocranial cavity, will decay and may be replaced by a "matrix" of sands, mud, clays, and pebbles that can be washed or blown into open holes. If these remains are buried for long enough a period, the matrix will become hardened sedimentary rock. A skull that is uncrushed or not too badly crushed may then contain a natural "endocast," which can be a nearly perfect "positive" for which the endocranial cavity, as the mold, is the "negative."

We may imagine such a process as being completed in a few million years, during which the fossil remains buried as a result of sedimentation. Subsequent land movements can expose the fossil bones to the air, and, if found in time, the skeleton, and in particular the skull, might be retrieved as a petrified mineralized fossil with the cranial cavity filled with hard material. The cranium can then be sectioned, the matrix removed, and an essentially clean endocranial surface would be available as the mold for a rubber or latex cast of the cavity. In the literature of paleoneurology such casts are referred to as "artificial," in the sense that they are man-made casts of a cleaned out endocranial cavity. They can then be copied in plaster, and these copies are present in most museum collections of fossil vertebrates (Radinsky, 1968a).

Some of the most beautiful endocasts have been found as "natural" endocasts. These occur under a special conjunction of lucky events. First, the original fossil must be in an uncrushed state when buried and fossilized. Second, the matrix that packs its endocranial cavity must fill it completely and become mineralized to a harder form than the mineralization of the skull itself. Third, when the fossil is exposed to air as a result of land movements, wind, erosion, and so forth (millions of years after the fossilization of the specimen), it must remain on the surface for a period long enough for the fossilized bone of the skull to weather away but not long enough for the fossilized endocast to suffer the same fate. The period of exposure of the endocast might be only a few months or years, and somebody has to walk by during that brief period, notice the endocast, and recognize it as something worth picking up (which may not really be too difficult, since it would be an odd-shaped rock if it were a mammalian endocast). When this occurs, we have on our hands a fossil "brain" with no other visible remains of the skeleton.

It is more common to find a partially exposed endocast in a skull that is still intact in many places. These are the most useful finds because they permit accurate identification of the animal and estimation of its age at death (from the condition of the teeth, which are very likely to be preserved under such circumstances), and, if associated bones are recovered, it is possible to reconstruct the animal's appearance in life and to estimate the animal's body size as well as brain size.

BRAIN VERSUS ENDOCAST IN "LOWER" VERTEBRATES

The "lower" vertebrates from the point of view of the evolution of brain size include the reptiles as well as the amphibians and the several classes of fish. In living lower vertebrates most bodily organs continue to grow throughout life, although growth occurs at different rates for different organs. The brain grows more slowly than the cartilaginous and bony structures of the cranium around it. In very young or very small specimens an endocast can be reasonably brainlike, whereas in older or larger specimens there is sometimes almost no relationship between endocast and brain with respect to either size or appearance. This fact makes for obvious difficulties in the analysis of absolute brain size and brain size in relation to body size in fish, amphibians, and reptiles. There is no convenient benchmark for the measurement of the kind available in birds and mammals when the analysis can be restricted to adult specimens. Despite these difficulties it is possible to determine relationships of brain to body in lower vertebrates that are orderly and consistent with other information about such relationships.

There is some difficulty in deciding exactly what an endocranial cavity is in many living lower vertebrates because the structures of the cranium usually include cartilage as well as bone and may have other mesenteric tissue supporting the brain in a much larger cavity (de Beer, 1937). Specific examples will be described in later chapters on these forms, but it may be helpful to mention a few brain:endocast relations in living fish and reptiles. The most extreme case that I have seen described is in *Latimeria,* the "living fossil" coelacanth, in which the brain weighed less than 3 g and was in an endocranial cavity of over 300 ml (Fig. 5.10).

In reptiles differences are usually less dramatic. In several specimens dissected for illustrations in this book (Fig. 2.1), the endocasts were about twice as voluminous as the brains, and this was also reported by Dendy (1911) for *Sphenodon*. The details of the comparisons are instructive because one might legitimately speak of a "chondrosteoendocast" and an "osteoendocast" in many living reptiles and presumably in the fossil forms as well. The former would refer to a cast of the endocranial cavity in which bone and cartilage are left intact when molding the endocast by the

Fossil Brains

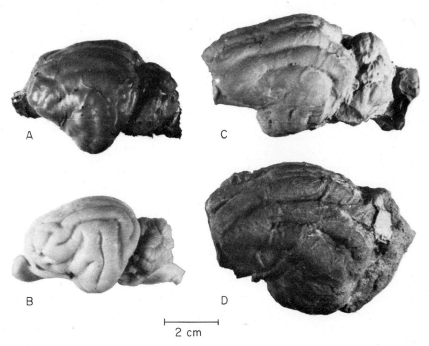

Fig. 2.2. Brain and endocasts of living and fossil felids. Domestic cat 30-ml endocast (A) and 29.1-g brain (B) from same specimen. Oligocene sabretooth *Hoplophoneus*, USNM 22538, endocast (C), and AMNH 11320 [? *Hoplophoneus* sp.], endocast (D), smaller and larger species of about 35 m.y. ago. See Fig. 13.5D for another *Hoplophoneus* endocast and Fig. 15.5 for comparisons of the brain and endocast in insectivores. "Brains" in lateral orientation, facing left. From Jerison (1963).

These and other methodological questions have, at best, arbitrary answers, and the answers that are chosen may result in changes of as much as 20% in the measurements (Ariëns Kappers, 1929; Hrdlička, 1906; Mettler, 1956; Stephan, 1960). One should probably consider any measurement of brain weight as involving an inherent error of the order of at least 5 or 10%, and possibly more, because of vagueness in the definition of the weight. Measurement of the volume of the brain, which is done by water displacement, is also affected by the adequacy of the dissection and the extent to which the brain has shrunk during fixation or become distorted in shape when weighed fresh.

Estimates of volumes of endocasts can also have significant errors. A natural fossil endocast that results from the replacement of soft tissues during the period of fossilization is only rarely retrieved in perfect condition, although it may be nearly perfect as in the case of the *Hoplophoneus* cast shown in Fig. 2.2C. But even in that specimen the anterior parts of the olfactory bulbs are missing. Artificial endocasts may have other errors related to the skill of the preparator who clears out the matrix from the endocranial cavity. A notorious example of an error in this regard that persisted in the paleontological literature for many years occurred in a cast described by Marsh (1876) of the endocranial cavity of *Coryphodon*, a large ungulate of some 55 m.y. or so ago (Chapter 11). In this specimen the entire auditory bullae were apparently removed when preparing the skull for an endocranial cast, and Marsh mistook that region of the endocast for an extraordinary cerebellum—most of which was created by the preparator during his excavation of the endocranial cavity! The error was finally pointed out by Edinger (1929) and by Tilney (1931), more than 50 years later.

The most recent analysis of this issue in mammals involved a very careful study of brain:endocast relations in living insectivores (Bauchot and Stephan, 1967). According to their methods of preparing the brain, they found its weight in grams to be 1.05 times the volume of the endocast in milliliters (compare the data presented earlier, in which the volume of the endocast in the cat was 3% greater than the weight of its brain). One should probably avoid making adjustments in endocast volume when estimating the brain weight, since the difference is of the order of magnitude of the errors of measurement and is much affected by methods of preparing the brain. Whenever I deal with mammalian or avian endocasts, in which the brain fills the cranial cavity, I will assume that the brain has a specific gravity of 1.0, and consider the volume of the endocast in milliliters to be the same as the weight of the brain in grams.

Making inferences from surface features of endocasts, in particular for the analysis of the evolution of fissural and gyral systems in the brain, is a difficult and somewhat hazardous enterprise. Connolly (1950), for example, has pointed out that the Sylvian fissure, one of the outstanding landmarks in the primate brain, may be considerably displaced in the brain relative to its position on an endocast from the same individual. Extensive analyses of the fissural patterns of early fossil hominids were reported by Schepers (1946). These were severely criticized by later writers, in particular by von Bonin (1963), whose criticisms were perhaps unnecessarily extended to all studies of fossil endocasts.

Careful studies of the evolution of gyri are presently being performed by Radinsky (1968b, 1969, 1970), and some of his results are presented in later chapters. In my view these excellent studies are of special importance in determining the modification of the brain to fill a particular niche. They are most clearly relevant for the evolution of species-specific functions but are somewhat tangential, though still important, to the analysis of the evolution of intelligence as considered in the previous chapter.

VERTEBRATE HISTORY

Evidence on brain evolution implies evidence about the brain, some of which has just been discussed and more of which is presented throughout this text. It also implies evidence about evolution. It is tempting to present the evidence of evolution by citing some of the many outstanding texts that are now available, and I will not entirely resist this temptation.

Some Reading Matter

If Simpson's (1967) "The Meaning of Evolution" is an unfamiliar title to the reader, he would do well to devote a few pleasant hours to that very readable account of the history of life, in which the evidence of the fossil record is emphasized. Among other paleontologically oriented and relatively elementary texts I have profited particularly from the following (in alphabetical order): Carter (1967), Coon (1962), LeGros Clark (1962), Olson (1965, 1970), and some more technical texts by Simpson (1945, 1953). Romer's (1966) basic text in paleontology has already been mentioned and is an essential resource for intuitions and insights about fossil animals; his companion volume of commentary (Romer, 1968) should not be overlooked as adding depth to those insights.

At a somewhat different level Scott's (1937) text—which Scott considered a popularization but which may be quite forbidding for the untutored in vertebrate taxonomy—is a useful reference because of its many illustrated reconstructions of fossil animals. In the same vein, the Time-Life, Inc., series of texts, in particular "The World We Live in" (Barnett, 1955) and "The Wonders of Life on Earth" (Barnett, 1960), were particularly useful for their accurate reconstructions of many unfamiliar forms. Those superb and genuinely popular texts relied heavily on Scott's work and on that of his artist, Bruce Horsfall.

More than any other text, Rensch's (1959) "Evolution Above the Species Level" overlaps this book in philosophy and orientation. Rensch has also been concerned with the problem of size and the role of size in evolution. His text is a mine of examples illustrating the role of relative size and of the change in size in evolution, with evidence from an amazing range of biological sources. It is, unfortunately, a somewhat idiosyncratic book, with the evidence not always clear; it is especially weak in its occasional excursions into the analysis of behavior. For evolutionary concepts and terminology I found myself relying most on Simpson's (1953) monograph cited earlier, which emphasizes rates of evolution and the achievement of levels or "grades" of adaptation within sets of niches or adaptive zones.

There are, finally, books that have become the classic texts of evolution, basing their evidence on broad ranges of experience by their various authors (e.g., Dobzhansky, 1955; Huxley, 1942; Mayr, 1963). These books could also offer an appropriate evolutionary background for a reader approaching the present work but uncomfortable about the sufficiency of his preparation on the topic of evolution.

To make this brief review of the popular and semipopular literature on evolution more complete, a few more texts should be cited. A recent book by Halstead (1969) contains helpful summaries of many issues. No analysis of relative size can overlook Thompson's (1942) classic "On Growth and Form" or fail to cite Huxley's (1932) seminal monograph "Problems of Relative Growth." A most important text is the series of essays on "Behavior and Evolution" edited by Roe and Simpson (1958). These essays range from good to superb and were a major inspiration for undertaking the present work. Although the emphasis of the essays was on behavior and the evolution of the brain, the cumulative effect of the book for me was to emphasize the fact that there was no useful definition of animal intelligence and no useful summary of the actual history of the evolution of brain size. For other reasons, cited in the previous chapter, I had come to recognize a relationship between animal intelligence and relative brain size, and immediate explorations of these evolutionary problems seemed to be called for. This book is one result of that exploration.

We can talk only so much about books without concern for the substance of their content. It will be unnecessary to consider any of the general issues of evolutionary theory, such as the meaning of selection pressures, mutation rates, fitness, or other variables that are familiar to the student of evolution. But texts on evolution generally slight the issue that is most important for us, namely, the specific selection pressures that produced unusual effects on the evolution of the brain, the sensorimotor apparatus, and intelligence.

In order to write on these themes one must have a necessary minimal geological vocabulary; a list of geological eras, periods, and epochs; and the best estimates of their duration and their chronology. Let us, therefore, begin this brief sketch of vertebrate history with a chronology.

GEOCHRONOLOGY AND HISTORICAL GEOMORPHOLOGY

The full history of the vertebrates probably began in the middle to late Ordovician period, about 500 m.y. ago. The history of the mammals dates from the late Triassic period, about 200 m.y. ago, according to the fossil record. The major expansion of the mammals (the "Age of Mammals") is the Tertiary period (65–3 m.y. ago) of the Cenozoic era (65 m.y. ago–present), and the earliest hominids may date from the Upper Pliocene epoch (about 5 m.y. ago) but are known mainly from the Quaternary period (the last 3 m.y.). I would want the reader to be able to read the concatenation of geostratigraphic names just presented with ease and familiarity, but doubt that many readers not already familiar with these terms would find such a presentation palatable. Let me, therefore, review my sources and references on geochronological issues.

Kulp's (1961) chronology and table were followed for dating before the Cenozoic era, and his basic table is reproduced here as Table 2.1. For Cenozoic dating Berggren (1969) and Evernden *et al.* (1964) were followed; Berggren's table includes data of other investigators and is reproduced as Table 2.2. The original papers and their lists of references should be referred to for information on how the absolute dating by radiocarbon methods was done.

When details of vertebrate history are presented in correlation with the data of Tables 2.1 and 2.2., it is important to keep in mind that the enormous span of years involved more than the mere passage of time. At the end of the Paleozoic era the world's map was strikingly different from what it is today, and, although details are still under debate, it is almost universally accepted that the great land masses were grouped in a radically different way from their present orientation into the familiar continental groups.

Perhaps less dramatic for our world view, but even more important from the point of view of the evolution of the brain, were the major climatic changes that took place during the history of the vertebrates. The Mesozoic era, for example, was a relatively mild period compared to the present, and this fact had clear implications for the kinds of life that could evolve at that time (Axelrod and Bailey, 1968). This relative mildness may have been correlated with yet other major instabilities in the history of the earth. These were the shift of the magnetic pole and the likelihood of shifts in the axis of rotation of the earth relative to the land masses, because of continental drift (Tarling, 1971). Such shifts probably did occur and had major implications for the diurnal cycles of different parts of the world. If one finds a fossil vertebrate embedded in a particular rock today, one can be reasonably certain that it died not too far from where the sediment that became that rock was at the time of its death. If the sediment was then near a polar region, however, the animal would have had to be adapted to climatic shifts of the sort characteristic of the poles of that time, and it also would have had to be adapted to light–dark cycles very different from the typical diurnal cycles of present everyday experience. Its life could have been somewhat like that of animals of the present northern areas with respect to winter-long nights, although the temperature may have been significantly higher even in the near-polar regions.

When specific evolutionary trends are considered, the issues of the type just raised are crucial for understanding why one development rather than another took place. We must know the physical environment, but we must

Table 2.1
Geological Time Scale[a]

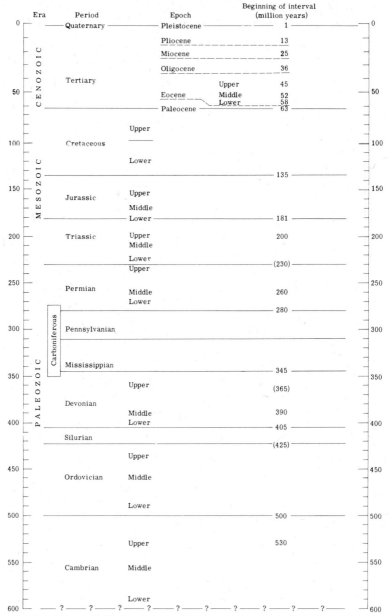

[a] Reprinted from J. Kulp, *Science* **133**, pp. 1105–1114. Copyright 1961 by the American Association for the Advancement of Science. In this book the Tertiary period is also divided into a Paleogene epoch, which includes the Paleocene, Eocene, and Oligocene, and a Neogene epoch, which includes the Miocene and Pliocene, and for some purposes (see p. 323) the Pleistocene.

Table 2.2
Cenozoic Time Scales[a]

Time in m.y.	Homes (1959)	Kulp (1961)	Evernden et al. (1961, 1965)	Funnell (1964)	Suggested radiometric boundaries and stage correlations (this paper)	
1	Pleistocene —(1)—	Pleistocene —(1)—	Pleistocene —(1)—	Pleistocene —(1.5)—	Pleistocene —(1.8)—	Calabrian
				—(3.5)—		Astian / Piacenzian
5	Pliocene			Pliocene	Pliocene	Zancian
		Pliocene	Pliocene	—(7)—	—(5.5)—	Messinian
				Upper	Upper	Tortonian
10	—(11)—		—(12)—	—(12)—	—(10)—	
		—(13±1)—			Middle —(14)—	Langhian
15				Middle		Burdigalian
	Miocene	Miocene	Miocene	—(18)—	Miocene	
20				—(19)—	Lower	Aquitanian
				Lower	—(22.5)—	
25	—(25)—	—(25±1)—	—(25)—	—(26)—		Chattian
			Oligocene		Upper	
30		Oligocene		Upper —(31)—		Bormidian
	Oligocene		—(33)—	—(32)—	—(32)—	
35				Lower	Lower	Rupelian
		—(36±2)—		—(36.5)—	—(36)—	Lattorfian
40	—(40)—	Upper		—(38)—		Bartonian
			Eocene	Upper	Upper	Priabonian
45		—(45±2)—		—(45)—	—(45)—	
		Middle		Middle —(49)—	Middle —(49)—	Lutetian
50	Eocene	—(52±2)—		Lower	Lower	Ypresian
				—(54)—	—(53.5)—	
55		Lower	—(55)—			
		—(58±2)—		Upper —(58.5)—	Upper	Thanetian
60	—(60)—	Paleocene	Paleocene		—(60)—	
		—(63±2)—		Lower	Lower	Montian S. S.
						Danian S. S.
65	Paleocene		—(67)—	—(65)—	—(65)—	
70	—(70)—					
	CRETACEOUS					

[a] As summarized by Berggren (1969). We will consider the Pleistocene as beginning 3 m.y. ago.

also know the biological environment. What plants were available for the herbivores, for example, and what were the predators and prey of each era?

INVASIONS OF MAJOR VERTEBRATE NICHES

Let us now consider the substantive events in vertebrate evolution. Certainly the first major event must have occurred when the first vertebrates appeared during the Ordovician period, perhaps 500 m.y. ago, to judge from the fossil record. This was the evolution of motility, a major change from chordate behavior in which the adults were typically sessile, anchored to rocks on the estuarine or river-bottom floor. The history is recounted and reconstructed in several of the texts cited at the beginning of this section.

There was an enormous difference implied by having to live out an entire life as a moving animal. Information about events at a distance would be at a premium in such species, and selection pressures to adapt to receive such information must certainly have been important. The coordination of movement with that information would also be important. And we have the basic ingredients (selection pressures) for the evolution of a rostrally dominant sensorimotor apparatus and a central nervous system in which the important parts were in the head.

We have no data about the brains of the first vertebrates, although there are some records of the sensory structures and the appearance of the brain in slightly later vertebrate species (frontispiece). These animals, known from the Silurian and Devonian periods, were ancestors of living lampreys and hagfish, jawless fish of the class Agnatha, that apparently lived as bottom fish either in fresh water or in estuaries of rivers emptying into the ocean.

An interesting feature of that early evolutionary time was the probably "explosive" evolution of vertebrates into fresh water and marine environments. An adaptive zone had opened and evolution could proceed rapidly into many niches. Although there has been continuing evolution and diversification in the morphology of fish since that time, a basic pattern of brain morphology and sensorimotor coordination seems to have been achieved relatively early, and a broad range of adaptations evolved rapidly to serve the several classes of fish until present times. The sensorimotor and brain adaptations that are most clearly different in the fish have to do with whether they lived as bottom fish, outside the range of photic stimuli, or whether they could use visual information. These differences may be seen in the oldest of the fossil fish endocasts that are available (Chapter 5).

The next major evolutionary change occurred when the vertebrates invaded terrestrial environments, and this occurred in the late Devonian period (350 m.y. ago) for the earliest amphibians. Some amphibians very quickly (in geological terms a period of perhaps 30 m.y.) evolved into the reptiles, and, as in the fish, the major adaptive radiation of land forms could also be characterized as explosive, since many and diverse forms appeared almost simultaneously.

The entry into the new terrestrial adaptive zone may be analyzed from two points of view. First, why was the terrestrial zone invaded only in late Devonian or Carboniferous times? Second, what special demands on

Vertebrate History 39

sensorimotor systems did life in such a zone entail? The first question is not answered simply, but apparently the invasion of the land depended on the presence of suitable environments on the land. The invasion of the land by vertebrates occurred only shortly after, or simultaneously with, the invasion by plants and by invertebrates. It was, in a sense, an expansion of all life to include terrestrial environments, although in the vertebrates we can recognize the special adaptations necessary for such a life as being somehow closer to our own experience with the exigencies of existence.

The second question is taken up in more detail in later chapters, although it is appropriate to foreshadow some of that discussion by considering a few examples of new demands on sensorimotor systems. The movements for swimming are not quite adequate for a life on land (although they are readily adapted to such a life). Similarly, the nature of photic and auditory stimuli is changed when the transmission medium is air rather than water, and, to the extent that a sensory apparatus had evolved to handle such stimuli, some modifications would be necessary if these sensory systems were to function on land in a way analogous to their functions in water. Other sense modalities, such as sensitivity to pressure or movement of water, as provided by the lateral-line system of fish, or information of the type provided by electric organs, may become quite useless for animals living on land.

With the invasion of terrestrial niches a new problem of temperature control occurs because the loss of heat from the surface of the body on land is quite different from that in water. Aquatic vertebrates would presumably have a body temperature close to that of the water in which they live, and their bodily functions would be adapted to that temperature. Terrestrial animals are much more at the mercy of temporary shifts in weather and seasonal shifts in climate. It is, therefore, not surprising that land animals evolved mechanisms to maintain body temperature. These were extrinsic behavioral mechanisms, such as moving in and out of the sun, in the case of poikilothermic reptiles, and intrinsic mechanisms in homoiothermic birds and mammals. Terrestrial animals in nocturnal niches would have a much more severe problem of temperature control than those in diurnal niches, since animals in diurnal niches can receive direct warmth from exposure to the sun (Heath, 1968).

The aerial niches are in many respects the most unusual and potentially most demanding with respect to sensorimotor controls. These niches were apparently invaded during the Mesozoic era, first by the flying reptiles and shortly thereafter by the earliest birds. The evidence is that flying reptiles and the earliest birds (*Archaeopteryx*) were contemporaries, and neither survived the Mesozoic era. *Archaeopteryx* was succeeded in the Cenozoic era by descendant species of birds, whereas the flying reptiles became entirely extinct, possibly to be replaced in their niches by soaring and gliding birds and perhaps also by bats. The latter is uncertain, however, since the pterosaurs, like most reptiles, were very likely diurnal animals and had well-developed eyes and optic lobes, whereas bats from their earliest fossils were certainly echolocators and were probably nocturnal as they are today. The inferior colliculi in bats, which correspond for audition to the optic tectum (superior colliculi) for vision, were remarkably developed in the earliest bats (Edinger, 1927, 1941) as they still are, suggesting that they used audition as a distance sense and did not rely as heavily on vision as do most reptiles and birds. (I do not refer to the fruit-eating bats, which are not usually echolocators and are not well known as fossils.)

Sketch of Vertebrate History

It is instructive to review when things happened and what was contemporaneous with what. This can be done rapidly by giving dates rather than geological stratigraphic terms.

The earliest vertebrates appeared perhaps 500 m.y. ago, and the earliest vertebrate brains are known from about 425 m.y. ago. The major radiation of the several classes of "fish" (Agnatha, Placodermi, Chondrichthyes, and Osteichithyes) occurred about 400 m.y. ago, followed within 50 m.y. by the amphibian and reptilian radiations. The "Age of Reptiles" (Mesozoic era) dates from about 220 to 65 m.y. ago, but the mammals and birds also appeared during that period. In fact, the major radiation of ruling reptiles such as the dinosaurs was only slightly in advance of the appearance of the first mammals about 200 m.y. ago. The flying reptiles or pterosaurs evolved somewhat, but not much, before the birds, which first appeared about 150 m.y. ago.

Birds and mammals are from entirely different reptilian stocks. Birds are derived from the major reptilian radiation of ruling reptiles and may be thought of as specialized relatives of the dinosaurs. They had found and exploited a particularly attractive adaptive zone. The mammals on the other hand are very distant poor relations of the ruling reptiles and may be thought of as isolated survivors of a group of reptiles (the mammallike reptiles) that lost the competition with the ruling reptiles for the preferred reptilian niches. I will repeat this assertion elsewhere in this book because it is one of my major hypotheses. Fully stated, this hypothesis is that the mammallike reptiles in losing the major reptilian adaptive zones to the ruling reptiles gave up the diurnal niches and settled into the, then relatively empty, nocturnal niches, for which reptiles were not normally well adapted. In their adaptation to these niches they became true mammals, developing night distance senses (hearing and smell) to replace the normal reptilian vision, evolving the ossicular chain of middle-ear bones, and having their brains enlarged to accommodate the information processing neural networks for their newly developed senses. (Visual information can be processed by the extensive neural networks peripheral to the brain, which are part of the retina of the eye.)

At the threshold of the Cenozoic era, about 65 m.y. ago, many major reptiles became extinct, and the mammals could radiate adaptively into the reptilian niches. Birds, likewise, found new aerial niches given up by flying reptiles. Among the reptiles themselves new niches were found and exploited, in particular by snakes. It is the mammalian radiation that will take up most of our attention because we will seek in that radiation the origins of intelligence and its later evolution. Near the end of the Cenozoic era, about 2 to 6 m.y. ago, the early hominids begin to appear as occasional fossils, and in the concluding chapters (Chapters 16 and 17) we will have to consider the adaptive zone that they invaded and how that zone led to the evolution of man and eventually language and culture.

MEASURING THE EVOLUTION OF THE BRAIN

The approach to measuring the evolution of the brain followed in this book is to emphasize general changes in total information-processing capacity as reflected in changes in relative brain size. We will not be contributing to systematics in which cranial structures related to the brain are used as evidence to establish phylogenetic relationships. The issue is, rather, to identify the effects of selection pressures on information-processing capacity within and among vertebrate species as reflected in gross structural changes in the brain. For this purpose there is much to be gained from the identification of a measure, the gross brain size, as associated with the total information-processing capacity of the brain. The identification seems to be appropriate for mammals: within a fairly broad representation of living mammalian orders the gross brain size is a kind of natural biological statistic that estimates the number of neurons and the degree of dendritic proliferation (Chapter 3), which are probably the biological bases for neural-information-processing capacity. Similar data are not available for brains of other living vertebrate classes and are necessarily lacking for fossil vertebrates. Yet, it is a permissible assumption (verifiable by studies yet to be performed for living birds, reptiles, and fish) that orderly relations obtain among the microscopic and gross characteristics of the brain and between gross brain and endocast size.

Like other organs, the brain is large or small in different species according to whether the body is large or small. Other factors potentially associated with brain size, such as dexterity, sensory prowess, or intelligence (however these are defined), must be understood to be differentiable from the "body-size factor," and the first requirement of the analysis is that a reasonable role for body size be identified. The problem is not simple. For example, there is some evidence that species with larger bodies are more "intelligent" than species with smaller bodies (Rensch, 1967), and body size and "intelligence" may, therefore, be confounded variables in the statistical sense. We would, nevertheless, generally consider body size, per se, as an isolable character with respect to adaptive efficiency. And we should look for measures of brain development and evolution that are either independent of body size or within which, at the very least, a role of body size is understood and explicitly recognized.

BRAIN:BODY "SPACE"

For this approach it is first necessary to map the generalized vertebrate condition, and this can be done for living animals by finding an appropriate set of brain and body weights. Data of this sort have been published; yet, in the many published compendia errors of the most elementary sort appear all too frequently. Specimens that are obviously juveniles or emaciated or desiccated by the method of preservation are regularly included in "normal adult" samples. Important differences in methods of collection and measurement are overlooked, and data from individuals within the same species and from among different species, as well as averages, are considered together and as equivalent. Trivial errors, such as the misplacement of a decimal point, are carried over from report to report as successive analyses attempt to evaluate brain:body data available in the literature. In the face of this confusion I chose to use, as baseline data for my general analysis, materials collected by a single team of investigators whose methods of reporting were reasonably consistent and who also published notes about their specimens, describing their condition and differentiating juveniles and infants from adults. These data were reported by Crile and Quiring (1940) and have been repeated in two other publications (Quiring, 1941, 1950); with a few exceptions they were based on the live weights of bodies and organs of 3581 vertebrate specimens.

In making use of their data, I have chosen one pair of brain and body weights for each species, generally data from individual specimens but in a few cases the average for a species. My minimum criterion for inclusion was that the body size reported appeared in the range of sizes as given in standard works such as Hall and Kelson (1959) or Walker (1964) for mammals and in similar published data on birds, fish, and reptiles (Hartman, 1961; Spector, 1956). In every instance the weights were for the heaviest animal if more than one specimen was reported. The rationale was that, since most of these animals were caught haphazardly during hunting

Measuring the Evolution of the Brain

and fishing expeditions, the typical (average) specimen would probably represent younger or weaker and more easily trapped individuals, and the heaviest specimen might be more representative of the full range of the phenotype than an average of the trapped specimens.

By following this criterion, I assembled brain:body data for a total of 198 vertebrate species and charted them on log–log coordinate graph paper, as shown in Fig. 2.3. These included 94 species of mammals (including 18 primates), 52 of birds, 20 of reptiles, and 32 of bony fish (class Osteichthyes). As a matter of general interest I have indicated the primate points separately.

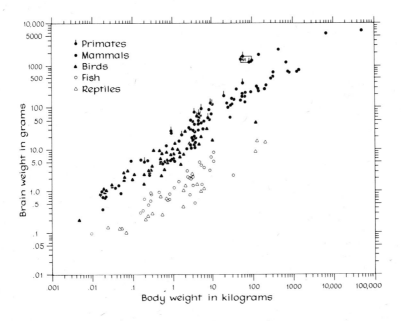

Fig. 2.3. Brain and body weights of the largest specimens of 198 vertebrate species collected by Crile and Quiring (1940), graphed on log–log coordinates. Rectangle M contains the full range of data on living men from their sample; four points on the borders of the rectangle are the heaviest and lightest individuals with respect to brain and body. From Jerison (1969).

It is apparent from Fig. 2.3 that one can hardly do more than separate the living vertebrate classes into two groups with respect to brain development. There are "lower" vertebrates, including fish and reptiles (one frog is also in that group, adding Amphibia), and "higher" vertebrates, including birds and mammals. Although in every other case a species is represented by only one point, I have made an exception for the data on man. These data, represented by a rectangle about the letter "M," are shown as four points on the edges of the rectangle. A total of 42 male human brain and body weights were presented by Crile and Quiring, and the four points are the extremes: the lightest and heaviest brains (1130 and 1570 g) and the lightest and heaviest bodies (36 and 95 kg). All the male human points would fall within that rectangle, and it is noteworthy how small is the region of the entire graph occupied by these apparently highly variable data. Graphing on log–log coordinates results in orderly arrays of data because it minimizes the effects of such variation.

In order to provide a better sense of the data in Fig. 2.3 and to identify some of the various animals some of the data points have been "named" in Fig. 2.4. The polygons, indicated by dotted lines, are minimum convex

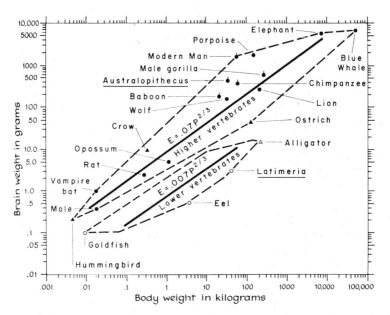

Fig. 2.4. Selected data from Fig. 2.3 to illustrate brain:body relations in familiar vertebrate species. Data on the fossil hominid *Australopithecus* and on the living coelacanth *Latimeria* are added. Minimum convex polygons are shown by dashed lines, enclosing visually fitted lines with slopes of two-thirds. From Jerison (1969).

polygons. They include all the data on "higher vertebrates" (birds and mammals) in one polygon and on "lower vertebrates" (fish, reptiles, and amphibians) in the other. Two points are added to Fig. 2.4 that were not shown in Fig. 2.3. These are of the brain and body of the recently discovered "living fossil" fish, the coelacanth, *Latimeria chalumnae*, as reported by Millot and Anthony (1956, plates 8–14), and of *Australopithecus africanus,* the Pleistocene hominid discussed most adequately by Tobias (1965). My point in adding these data was to indicate that "new" and unusual specimens could be placed on the maps; although *Latimeria* fell on one edge of a polygon, it was not necessary to make any changes in the maps to accommodate these specimens.

To supplement the picture in Figs. 2.3 and 2.4, we should consider additional data that have since appeared and that are known to be reliable. New data on living insectivores (Bauchot and Stephan, 1966) are included in Chapter 10, and these result in an extended and slightly broader mammalian polygon as shown in Fig. 10.5. With respect to fish, Ebbesson and Northcutt (in press) have recently reported that sharks and their relatives (class Chondrichthyes) are intermediate between and overlap both the fish and mammalian polygons and that the brains of these cartilaginous fish are complex and "advanced" in other ways as well. They also present the only data that I have seen on the lamprey (*Petromyzon,* class Agnatha), and the single point that they add to the map of brain:body polygons shows this jawless fish as clearly below the level of living bony fish. In other respects Ebbesson and Northcutt confirmed the boundaries of the lower tetrapod polygon with data on additional reptiles and several amphibians.

Analysis by Minimum Convex Polygons

In Figs. 2.3 and 2.4 we have drawn a general brain:body function for living vertebrates, which is a kind of map of "brain:body space." We created a functional mapping without writing equations giving least-squares fits for various brain:body data. Like others, beginning with Dubois (1897), I sometimes follow a curve-fitting allometric approach and will review the problem later (see White and Gould, 1965; Gould, 1966 for discussions of allometric analysis and of my approach in particular). But first I will describe the new and simpler procedure in which raw data provide a matrix of information against which new data may be evaluated. The new approach consists merely of presenting a set of points, the brain and body sizes of vertebrates, and looking at the set as a map of the present or past status of the evolution of the brain. Various subsets may be defined and enclosed in minimum convex polygons,[1] that is, polygons of minimum areas, with

---

[1] These polygons are discussed by mathematicians as two-dimensional convex sets, or convex hulls, and have been analyzed in connection with problems in linear programming (Kemeny *et al.,* 1959). A number of operations analogous to curve fitting can be performed on the polygons to define their shape, orientation, symmetry, and so on (Stenson, 1966; Knoll and Stenson, 1968). This includes placement of a principal axis in a polygon, which is equivalent to placing a regression line. Mathematically and biologically, a principal axis is a more proper line to characterize data of this type, which are not random deviates from the line, and the equation of the principal axis may be used as an allometric function if one seeks such a function.

vertexes on data points and with no internal angles greater than 180°. Inclusion of a new point within a polygon establishes it as a member of the subset, and exclusion of a point forces its rejection. Some biological insight may then be applied to help one in the interpretation of inclusion or exclusion, and one may also make judgments about errors.

Figure 2.4 illustrates both the analysis by "mapping" and the allometric method of analyzing brain:body data. In addition to the maps for "lower" and "higher" vertebrates, two straight lines within the polygons represent logarithmic transformations of the allometric functions of brain size as a function of body size for each group of vertebrates. Although it is common practice to use curve-fitting techniques to represent data of the type presented in Fig. 2.3 as deviations from a line, it is actually the case that maps of the sort suggested in Fig. 2.4 are more appropriate descriptive devices. In fitting a straight line to the data, the assumption is implicit that individual points are "random deviates" from that line. The true situation is that each of the points in Fig. 2.3 represents a most likely position of future samples from its population (species). It is, therefore, likely that the minimum convex polygon represents a region of possible brain:body relations, of "brain:body space," as it were, which should be used to characterize the groups of species within each region. The line is a simplifying device that may be useful to characterize the regions. Thus, the lines drawn in Fig. 2.4 were constrained to have equal slopes of two-thirds. In that way the vertical displacement involved in evolution from "lower" to "higher" vertebrates could be suggested to be approximately tenfold. That is, given the general form of the equation of these lines, $E = kP^\alpha$, with $\alpha = \frac{2}{3}$, it takes a tenfold increase in k to shift from the lower to the upper line.

The major drawback for the method of convex polygons is that it is difficult to apply to overlapping groups (although methods could be developed). In this book it is used whenever possible, because it is easy to justify both biologically and mathematically (Jerison, 1968, 1969, 1970a, 1971b).

The method of convex polygons cannot always be used because it does not always enable us to make fine enough distinctions among populations when we have only gross data. Since the direct evolutionary approach is limited to such data based on fossil materials, we will also use a method that is derived from allometric analysis, especially for the data on the evolution of the brain in mammals. Before discussing allometric analysis, its drawbacks will be illustrated by considering how it was applied by Quiring (1941) to the data on which Fig. 2.3 was based.

Quiring used the entire sample of several thousand individual specimens, described by Crile and Quiring (1940), to produce a set of best-fitting lines on log–log coordinates, probably by regression analysis. His results are shown in Fig. 2.5. Not only is information lost by his reliance on curve-

Measuring the Evolution of the Brain

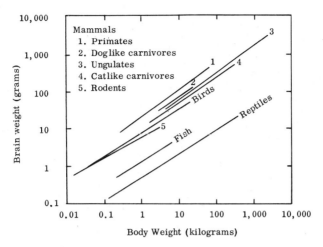

Fig. 2.5. Curves fitted by Quiring (1941) to data of the more than 3000 specimens from which the sample of Fig. 2.3 was drawn. If the several groups in this figure were treated as random samples from their respective populations, it would be possible to differentiate only two groups, statistically, the lower vertebrates (fish and reptiles) and the higher vertebrates (birds and mammals). The analysis may be contrasted with that of Fig. 2.6. From Jerison (1970a).

fitting procedures, but the curves are actually misleading. There is no indication of variability in Quiring's curves nor of the degree of overlap between mammals and birds on the one hand and between reptiles and fish on the other. It is apparent that much information from Fig. 2.3 is simply lost in Fig. 2.5 and that the true situation is represented much more adequately by Fig. 2.3.

The appropriate method of mapping those data by convex polygons is shown in Fig. 2.6, and those polygons will be the ones applied when we use the method. In Fig. 2.6 the sets of points for mammals, birds, fish, and reptiles are each enclosed in minimum convex polygons and illustrate the relationships among the four vertebrate groups. It is clear that the orientation of the polygons is approximately the same; that their shapes are similar; and, most importantly, that two, and only two, groups of vertebrates are completely distinct with respect to relative brain size. The mammals and birds form one natural grouping with overlapping polygons, and the fish and reptiles form a second grouping with overlapping polygons. But these two groupings do not overlap one another.

BRAIN INDICES AND ALLOMETRIC ANALYSIS

The brain indices to be discussed in the next chapter are developed as a result of an allometric analysis of relative brain size. The "allometric" approach, as discussed by Huxley (1932), was developed to study relative

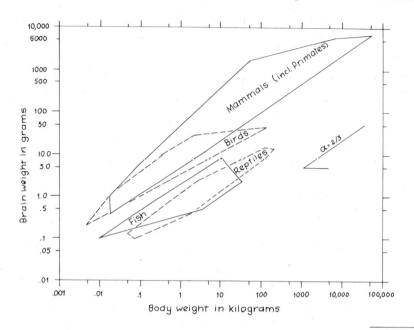

Fig. 2.6. Minimum convex polygons enclosing the vertebrate classes of Fig. 2.3. The polygons can be described by their principal axes (see text) to suggest best-fitting slopes, and these are 0.69 for mammals, 0.56 for birds, 0.50 for fish, and 0.62 for reptiles. It is also clear, however, that each polygon can be oriented reasonably well about a line with a slope of $\frac{2}{3}$. From Jerison (1970a).

growth of organs that developed at different rates. It was easy to apply to cross-species comparisons and to analyze organ size in species that differed in body size. It is obvious that larger animals will have larger body organs, and if one is to assign a number to an organ in one species that classifies it as larger or smaller than the corresponding organ of another species, the real concern is usually with the "relative" rather than "absolute" sizes of the organs. One often tries to estimate the importance of the organ in the economy of an animal by estimating its size relative to the animal's total size and comparing the result to that obtained in other species.

The method of allometric analysis, as indicated earlier in this chapter, involves curve fitting to data of the sort presented in Fig. 2.3, with all the hazards of curve fitting to such data. It is, nevertheless, possible to justify a simple set of assumptions for curve-fitting procedures. For example, we might justifiably think about a "typical" reptile or mammal without prejudicing our understanding of speciation or individuation. It may even be possible to meet the sampling assumptions for least squares by defining the

Measuring the Evolution of the Brain

population as the set of all species, for example, of mammals, and setting each measure as a randomly sampled species from that set.

The results of curve fitting are suggested in Fig. 2.4, in which axes were inserted into the brain:body polygons of "higher" and "lower" vertebrates. The advantage is immediately apparent—it can be suggested that the increase in relative brain size between "lower" and "higher" vertebrates was of the order of a factor of 10. This is the meaning of the shift in the values of k (Eq. 2.1, below) from 0.007 for lower vertebrates to 0.07 in higher vertebrates. The lines in Fig. 2.4 were not placed by the method of least squares. Instead they were chosen to have equal slopes of $\frac{2}{3}$ and were then placed visually through the approximate centroid of each array of data. This procedure is superior to conventional "objective" methods for our purpose, because it permits us to compare values of k with one another. In effect, we take advantage of our prior knowledge that the slope is likely to be $\frac{2}{3}$ and that only the intercept is unknown. We can compare intercepts for different sets of data as long as we assign the same slopes to them, and any reasonable value of the slope can be used.

For all the allometric analyses in this text, I have followed a dual strategy. I have always computed a best-fitting line by objective methods: the conventional method of least squares, assuming errors in both brain and body size, and by the less conventional, though really appropriate, method of determining the principal axis of a polygon (see footnote 1, p. 45). I have done this primarily to determine whether $\frac{2}{3}$ may be used as an approximation of the slope of the line actually used as the best-fitting line. And unless there is clear evidence (there almost never is in the data of this book) that a slope of $\frac{2}{3}$ is inappropriate, I will always consider a best-fitting line for brain:body data to be of the form:

$$E = k P^{2/3} \qquad (2.1)$$

E and P are brain and body weights in centimeter-gram-second (cgs) units (grams or milliliters), and k is a proportionality constant, the y-intercept of the equation when considered in its logarithmic form. The equation will appear in later chapters, especially the next chapter, and we need not discuss it extensively, especially in view of the excellent discussions by Gould (1966, 1971) in recent years. We should note that an exponent of $\frac{2}{3}$ implies a surface:volume relationship and may, therefore, be the basis for theorizing on the significance of the brain size.

ESTIMATING BRAIN SIZE

When either a natural or artificial endocast has been available for a particular specimen, its volume was determined directly by Archimedes' method of measuring its loss of weight in water. In the few instances for which such endocasts are available for reptiles, half of that volume was used as the estimate of brain volume. This ad hoc assumption is consistent with the literature on brain:endocast relations and our experience with a few larger lizards, as cited earlier in this chapter. It would be useful to have a more accurate set of data on such brain:endocast relations, but, as noted before and as we will see when the data on the evolution of the reptile brain are considered (Chapter 7), even errors of 50% or so in the estimation of brain size would not alter the basic conclusions to be derived from these data.

In the mammals and birds and the occasional species of other classes (pterosaurs, small paleoniscid fish) in which the endocast was obviously brainlike in appearance, the endocast was considered to give an exact estimate of the brain volume.

In many specimens the endocasts could not be measured directly, and I therefore developed the method of "double integration" in which the brain is modeled as an elliptical cylinder.[2] The length of the brain from the tip of the forebrain (excluding the olfactory bulbs and stalk) to the point of exit of N.XII (the hypoglossal nerve) is the altitude of the cylinder; the average width of the brain between those limits is the major axis of the ellipse, which is the cross section of the cylinder; and the average height of the brain between those limits is the minor axis of that ellipse. This particular model was used in a series of estimates in which published illustrations of endocasts giving dorsal and lateral views were available and in which the actual endocasts were also on hand. The difference between the volume measured by Archimedes' method and by this kind of double integration never exceeded 5%.

The average heights and widths of illustrations of endocasts can be determined planimetrically; if a planimeter is not available, the method shown in Fig. 2.7, in which an illustration of the endocast of *Tyrannosaurus* is the model, may be used. In this way the volume of an endocast may be measured from an illustration, and many endocranial volumes that I report were measured in this way.

MEASURING OR ESTIMATING BODY SIZE

The best way to determine body size is either to weigh an animal or to determine the volume of an accurate scale model. This has been done for living specimens and some fossil specimens to be described in this book. For most specimens, however, it would be unwieldy to attempt to reconstruct

---

[2] It has been pointed out to me by H. J. Bremermann that in developing this method I was rediscovering Cavalieri's Principle, known to geometers since the seventeenth century.

Measuring the Evolution of the Brain 51

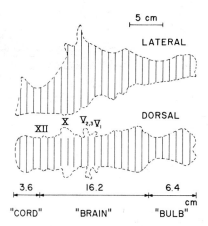

Fig. 2.7. Graphic double integration to estimate the volume of the endocast of *Tyrannosaurus* (cf. Figs. 7.3 and 11.3). Cranial nerves labeled in the dorsal projection are three trigeminal branches ($V_{1,2,3}$), the vagus (X), and the hypoglossal (XII). Mean lengths of solid lines drawn through the "brain" portion of the endocast are 4.8 cm and 6.6 cm for dorsal and lateral projections, respectively. Volume estimate = $(2.4)(3.3)(16.2)\pi = 404$ ml. Volume for the remainder of the endocast = 132 ml. Total volume = 536 ml, by graphic integration. Direct measurement by water displacement (Osborn, 1912) resulted in 530 ml. From Jerison (1969).

an accurate model, and for that purpose I have been satisfied to use weight:length equations, determined empirically with living species, as the basis for estimating the weight (or volume) of a fossil specimen. We will always assume a specific gravity of 1.0 for tissue. The possible error because of this assumption is certainly no more than 10%, and, since our scales are logarithmic, that error literally vanishes in graphic presentations used in this text (cf. data on man in Fig. 2.3).

With the exception of one set of data to illustrate our method, we will present the length:weight equations and their supporting data in the chapters in which they are used, but a few general remarks are in order here. Within each class of vertebrates there is a relatively small range of body shapes that we have to contend with, and one can literally map a mouse on an elephant, point-to-point, without extraordinary distortions. We may have been overly impressed with D'Arcy Thompson's famous and important diagrams of the distortions associated with evolution and growth, such as the changes in shape of the head of the horse, from its fossil ancestor, *Hyracotherium*, to its present genus, *Equus*. The distortions are important for detailed analysis, but actually make little difference when we estimate gross volume from body length.

I will illustrate the weight:length method by presenting an analysis to estimate body size from body length in carnivores and ungulates (Chapters 13 and 14) and show how this was applied to estimating body size of a fossil camelid, *Poëbrotherium labiatum* (Jerison, 1971b).

The fossil camelid in this example is known from a complete mounted skeleton, and measurement of body length was taken from the skeleton. The procedure was to measure head and body length, as illustrated in Fig. 2.8, because this gives a measure comparable to those published on living

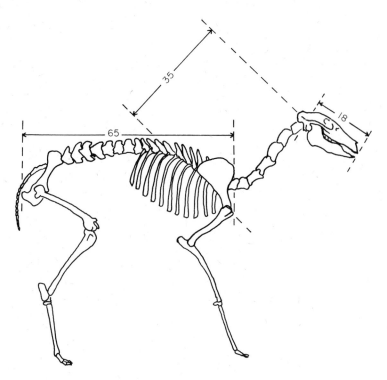

Fig. 2.8. Method of measuring body length illustrated on skeleton of *Poëbrotherium labiatum* from Scott and Jepsen (1940, Fig. 132). Numbers are lengths in centimeters, and their sum is the head and body length. From Jerison (1971b).

mammals. The measured total length was 118 cm. The body weight was then estimated from the body length to be 40 kg, by applying the formula

$$P = 0.021 \, L^{3.03} \tag{2.2}$$

where P is weight in grams and L is length in centimeters.

The formula was developed for this and related applications by a least-

Measuring the Evolution of the Brain

squares fit to all the usable carnivore and ungulate data in Walker (1964), which include data on living camelids. My procedure was actually to determine lines that connected the extremes of the ranges of the reported lengths and weights and then to fit an "average" line to those lines. It is illustrated graphically in Fig. 2.9, which is convincing evidence for the orderli-

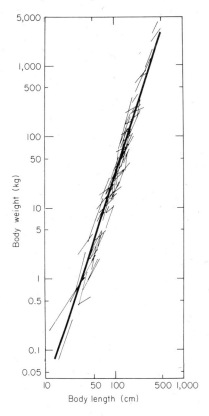

Fig. 2.9. Relationship of body weight to body length in all perissodactyls, artiodactyls, and carnivores described by Walker (1964) with data on the ranges of the measures; logarithmic coordinates. Each line connects minima and maxima of lengths and weights for a single species. The heavy line is Eq. (2.2) or (2.4). From Jerison (1971b).

ness of the data on which Eq. 2.2 is based. Note that Eq. (2.2) is approximated by $P = 0.025\ L^3$, and the exponent is dimensionally appropriate for a length–volume relationship (Thompson, 1942, pp. 33, 194, 1089, passim; Gould, 1966).

For empirical applications I have found it necessary to take into account the habitus of the specimen that was being evaluated, and for animals with heavy habitus, such as the rhinoceros, Eq. (2.2) was altered to

$$P = 0.043 \, L^{3.03} \tag{2.3}$$

I have also used the dimensionally exact versions of these equations in some chapters, since it is usually impossible to distinguish the effects. Thus,

$$P = 0.025 \, L^3 \tag{2.4}$$

is graphically indistinguishable from Eq. (2.3) in Fig. 2.9. Similarly, for heavily built animals, I have occasionally used

$$P = 0.050 \, L^3 \tag{2.5}$$

Chapter 3

Gross Brain Indices and the Meaning of Brain Size

Brain indices are numbers intended to indicate the relative size of the brain or its parts, taking body size into account. We will use them to make comparisons among species or higher categories. (They are not valid for comparisons among individuals of a single species unless there has been selective breeding of races or subspecies to cause them to differ greatly in body size.) All of the indices that use data on the weight or volume of the brain or its parts are related to Dubois' (1897) "index of cephalization," which was developed from Snell's (1891) analysis of brain:body relations. These relate brain size to two factors (a) body size and bodily functions, and (b) the "encephalization" of psychic functions. In this chapter I will discuss the relationships among gross indices (Jerison, 1970a) and the relationships between gross brain size and the microscopic characteristics of the brain (Jerison, 1963).

It is possible to use indices based on gross size of the brain (or its parts) because of the statistical orderliness of the geometry of the brain. Though seemingly almost random in small sections, measures of the average neuron density and fiber proliferation are a surprisingly orderly function of brain size, independent of species in mammals.

The plan of this chapter is to discuss, first, the rationale for gross indices involving the weight or volume of brains and bodies of various animals. I will then show why brain size may be used as a natural biological statistic that estimates (in the statistical sense) the value of more meaningful biological parameters, such as the numbers of neurons and glial cells and, perhaps also, the complexity of the neuronal networks of the brain. Most of the chapter is devoted to gross brain indices. These are the only indices that can be determined from fossil endocasts, in which the fine structure is irretrievable and the external surface and volume are represented only by compacted clay that replaced once living tissue.

GROSS BRAIN INDICES

Of the gross indices, I emphasize those based on direct measurement of the weight or volume of the brain and body. Indices that rely on other measures are sometimes suggested. These are always related secondarily to either brain size or body size. Anthony's (1938) suggested index, the ratio of the cross-sectional area of the corpus callosum to that of the medulla is an example of an index based on secondary measures because the former is related to brain size and the latter to body size. The biological problems posed and answered by the proponents of each variant measure may be important conceptually, but, because of the orderliness of the geometry of the brain and body, they may be of minor interest as the basis of an index. If Anthony's index is used, one automatically restricts oneself to placental mammals, in which the corpus callosum is present. Among placental mammals one can determine, however, that the relationship between the gross brain size and the cross-sectional area of the corpus callosum is so orderly (Bauchot and Stephan, 1961, Fig. 4) that it is appropriate in most instances to use one to estimate the other, depending on which is easier to measure. Since the gross brain size is usually easiest to measure, it is normally unnecessary to devote oneself to other measurements that give about the same amount of information.

The reason for concern with measures other than body size is the well-known variation in body weight associated with seasonal, ecological, and other factors. In my view, the importance of this variation is overemphasized. The variation would be significant if one were to make judgments about specific specimens. I reject the appropriateness of gross indices for that purpose; Sholl's (1948) criticisms seem particularly valid here. Indices should be used, I believe, only to buttress broad generalizations about relative brain size of the type that may occur in faunal groups isolated from one another either geographically or temporally. One might thus compare the South American fauna of the mid-Tertiary period with the Holarctic fauna of the same period (Chapters 13 and 14). Another comparison might be among successive faunas, such as a Paleocene–Eocene fauna with an Oligocene fauna (Jerison, 1961). For such comparisons errors in body size tend to counterbalance one another as one applies an approach like von Bonin's (1937) to the description of relative brain size for all the mammals in one's sample.

Indices based on the parts of the brain offer much promise for the future of comparative neuroanatomy in the quantification of many results. They are usually based on planimetric reconstructions of the type first used by Tilney (1928). The development of these indices has reached a completely new level of refinement in the work of Bauchot, Stephan, and their colleagues (Bauchot, 1963; Stephan, 1966; Stephan et al., 1970), who used allometric methods after Portmann's model for the analysis of the planimetric data. It is unfortunate that the analysis of the parts of the brain can be used only in a limited way on paleoneurological data from endocasts, and it is these important data that force us to continue the development of more gross indices (and related approaches such as mapping by convex polygons) for the analysis of fossil endocasts.

With the recognition of the problem of allometry—that organ size is related (nonlinearly) to body size—the development of brain indices returned to Dubois' approach while some of his errors were avoided. All workers in comparative anatomy who are concerned with the size of organs recognize a role of a general body size factor because of which large animal species tend to have large livers, hearts, and other organs, including brains. It is natural to want to compare the relative development of brains in animals of different body sizes by controlling the body size factor and then determining whether there is significant residual difference in relative brain development. This is, in effect, what Dubois attempted to do when he created his index of cephalization. Later advances on Dubois' work often consisted of mathematically minor changes in his index that seemed easier to rationalize biologically. The work of Portmann (1946, 1947) and his students and colleagues (Wirz, 1950; Stingelin, 1958; Portmann and Stingelin, 1961) is in this category, although they did achieve significant advances by beginning analysis of the relative development of the parts of the brain. Let us now examine the relationships among recent and older approaches that have used a measure of "cephalization."

INDICES OF CEPHALIZATION

All the methods can yield a gross brain size index, and all do it by relating brain size to body size, although the relation is sometimes hidden in the computations. Dubois (1897), who originally developed the index, began by accepting an allometric function for brain:body relations (Snell, 1891) of the form

$$E = kP^\alpha \tag{3.1}$$

in which E and P are brain and body weights or volumes measured in cgs units, and k and α are constants. By more or less ad hoc methods, Dubois estimated $\alpha = 0.56$ and could thus calculate k for any particular specimen (or species average) in which E and P were known. He called k the "index of cephalization" for the specimen.

To appreciate the meaning of the index one should note that in logarithmic transformation Eq. (3.1) becomes

$$\log E = \log k + \alpha \log P \tag{3.2}$$

This is a linear equation in log E and log P with a slope α. Log k is the intercept and has the value log E at log $P = 0$. The index of cephalization as defined by Dubois was, therefore, the value of k_i for a particular individual i with brain and body sizes E_i and P_i in Eq. (3.2) and with $\alpha = 0.56$.

If Dubois sought the index of cephalization for a cat with a 25-g brain and a 3000-g body, he could graph the point (25, 3000) on log–log paper, draw a line of slope 0.56 through the point, and read the intercept at $P = 1.0$ (i.e., at log $P = 0$) on the ordinate. He would find it to be 0.28. A lion with a 225-g brain and a 150-kg body would also have an index of cephalization of 0.28. (Dubois actually defined his value of α to produce such an equality because he believed cats and lions to be equally "cephalized.") In short, lions and cats lie on the same line.

Recognizing that Dubois' approach to evaluating α was ad hoc, von Bonin (1937) undertook the more objective approach of fitting a single straight line to the logarithms of brain and body weights of a relatively large, though haphazard, sample of mammals, following a regression analysis. A graphic presentation of data on over 100 mammals did look like a scatter plot and served to justify the procedure. With this procedure he found that $\alpha = \frac{2}{3}$, approximately, which agreed with Snell's (1891) theoretically determined value. (This results in an index of cephalization of 0.14 for the cat and 0.08 for the lion.) I was able to confirm the slope of $\frac{2}{3}$ (Jerison, 1955, 1961) on other data (using a least squares curve-fitting method somewhat more suitable to such data) by performing a functional analysis (rather than regression analysis) in which errors of measurement are assumed in both brain and body measures.

My work with the index began with the question of its reliability, and I asked first whether k_i was really independent of body weight. To my surprise (Jerison, 1955) I found that the von Bonin index, when applied to primates, was negatively correlated with body weight, and although the relationship was curvilinear, it was extremely orderly. I now realize that what I found could have been predicted from data then in the literature. Sholl (1948) had reported a much flatter slope for data among species of macaque monkeys ($\alpha = 0.18$) than either the Dubois value ($\alpha = 0.56$) or the von Bonin value ($\alpha = 0.66$). From other reports, such as Lapique (1907), it was well known that related species within a genus had flatter slopes than those resulting from fits to more disparate groups. My situation was effectively as diagrammed in Fig. 3.1, in which a series of brain:body "points," or "data," aligned along a line with a flat slope such as $\alpha = \frac{1}{3}$, are fitted by equations, such as Eq. (3.2), with $\alpha = \frac{2}{3}$. The result of such an erroneous choice of α is that points from lighter species will have higher values of k_i

Gross Brain Indices

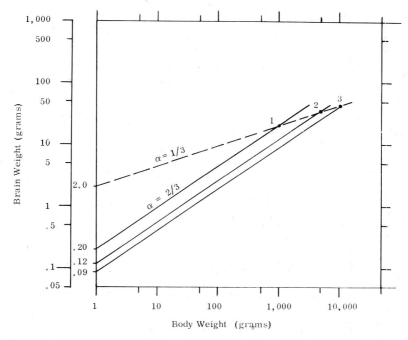

Fig. 3.1. Effect of choice of slope of the basic allometric brain:body function on the measure of the index of cephalization k. Three hypothetical animals related by a "true" $\alpha = \frac{1}{3}$ are placed as points 1, 2, and 3 on mammalian allometric functions with $\alpha = \frac{2}{3}$. The smaller animals have higher values of k (for animal 1, $k = 0.20$; for 2, $k = 0.12$; for 3, $k = 0.09$), which are seen to be the intercepts of the functions. An assumption that the index should be based on a "natural" relationship for these particular points would lead to a slope (α) of $\frac{1}{3}$ and the same index ($k = 2.0$) for all three species. The meaning of k as the intercept of the brain weight axis at $P = 1$ g ($\log P = 0$) is illustrated by the inner ordinate on which the several numerical values of k are indicated. From Jerison (1970a).

than points from heavier species, and the relationship between k_i and P will be orderly and negative.

These details are instructive because they indicate the effects of apparently minor differences of assumptions about the indices. Since brain:body data present arrays of scattered points, it is obviously an uncertain business to choose among possible functions that could fit subgroups of the points. But contrary to some critics, such as Sholl (1948), there is an obvious orderliness to brain:body data that cannot be ignored. Although the computation of indices for individual specimens may be questionable, the orderliness of the relationship in overall data may be useful in other ways. In my judgment indices may be used properly to discriminate among very broadly defined groups of animals as Wirz (1950) has done for living families of mammals and as I have done for an evolutionary succession (Jerison, 1961).

An obvious drawback in the use of an index of cephalization is that it has no numerical reference to anything of biological interest. It is actually a biologically impossible number. It refers to the brain weight of an impossibly small adult vertebrate that weighs 1 g. It is also an awkward number mathematically because it is not dimensionless. The latter consideration is beyond the scope of this text (see White and Gould, 1965; Gould, 1966, 1971), although, in brief, the problem is that the value of k in Eqs. (3.1) and (3.2) has a length in centimeters. If k were a pure number, the volume of the body would be raised to the $\frac{2}{3}$ power as the measure of the "volume" of the brain, but that measure would be, dimensionally, an area. It is only by having k as a length that the brain size is three-dimensional, as it must be. The point is not trivial, although it may sound odd to the nonmathematically inclined, and it has received detailed consideration, especially in the physical sciences (Bridgeman, 1931).

I have never felt awkward about using an index that happened to have the dimension of length. Nevertheless, in the process of developing methods for describing the results of my investigations into the evolution of relative brain size, I have begun using a measure that I have called the "encephalization quotient" or EQ. This measure has the agreeable quality of being dimensionless. I was aware of work by Bauchot and Stephan that had been going on for at least 10 years in which they used a similar measure, but did not immediately recognize the near identity of our two approaches. I believe, however, that the measure of EQ as I have used it in recent publications (e.g., Jerison, 1970b, and in this text) has some advantage in that it is related to "average" brain:body relations of the mammals as a whole rather than to the group of "basal insectivores" that Bauchot and Stephan have used as the source of their index. Let me, first, describe their approach.

Bauchot, Stephan, and their colleagues have used the data on total brain and body size of certain insectivores, characterized as "basal" because of their relatively small neocortical development, to derive the basic relationship between brain and body. This results in an equation such as Eq. (3.1), with $\alpha = 0.64$ or 0.63 (different values appear in different papers), and they relate the brain size of other mammals to the brain size of insectivores. They then convert any other brain:body data, for example, that on primates (Stephan and Andy, 1969), into percentages of the brain weight of a basal insectivore of the same body size.

The equations are straightforward. They assume the basal relationship

$$E_b = 0.033\, P^{0.64} \tag{3.3}$$

Gross Brain Indices

The constants are determined empirically, by curve fitting, with cgs units. (Their method is regression analysis, in which errors are assumed only for the brain weight. I consider this inappropriate, although it has trivial effects on the results.) They then consider the actual brain weight of any other species i, with body weight P_i and brain weight E_i, and compute their percentage measure as

$$\text{Index} = 100 \times \frac{E_i}{0.033\, P_i^{0.64}} \qquad (3.4)$$

The only differences between their index and my measure of the encephalization quotient, EQ, are that in the latter one does not multiply by 100; one uses $\tfrac{2}{3}$ instead of 0.64 as the exponent of the body weight; and one uses the value of 0.12 instead of 0.033 as k in order to have an "average" k.

ENCEPHALIZATION QUOTIENT

The encephalization quotient EQ is the ratio of actual brain size to expected brain size, with expected brain size defined by Eq. (2.1). The expected brain size is a kind of "average" for living mammals that takes body size into account. By using EQ, any living or fossil mammal may be compared with any other mammal with respect to relative brain size and with a minimum of confounding (in the statistical sense) by the body size factor.

In more formal terms let EQ_i for species i be the ratio of its brain size E_i to the expected brain size E_e in a mammal of the same body size P_i. Thus

$$EQ_i = E_i/E_e \qquad (3.5)$$

In a particular species of body size P_i we may use Eq. (2.1) to give expected brain size if we determine the average function, as it were, fitted to a broad sample of mammals. In such an average one finds $k = 0.12$. Thus, the expected brain size is given by

$$E_e = 0.12\, P_i^{2/3} \qquad (3.6)$$

Substituting Eq. (3.6) in Eq. (3.5), we have

$$EQ_i = \frac{E_i}{0.12\, P_i^{2/3}} \qquad (3.7)$$

This defines EQ for any species in which brain and body size are known. In terms of the index of cephalization $k_i = E_i/P_i^{2/3}$ and EQ_i is $k_i/0.12$.

As examples, suppose that we find the average body size of healthy adult male squirrel monkeys to be 1000 g and the brain size to be 24 g. (These are reasonable figures for these monkeys.) The denominator of

Eq. (3.7), the expected brain weight for a 1000 g mammal, is seen to be 12 g [$(10^3)^{2/3} = 10^2$ and $0.12 \times 10^2 = 12$]. Since the numerator is 24 g, the squirrel monkey as a species has $EQ = 2.0$. This means its brain is twice as large as that of an "average" living mammal. The value of EQ in *Homo sapiens*, if $E = 1350$ g and $P = 70,000$ g, would be $EQ = 6.3$. Our brain is, thus, about 6 times as large as we should expect it to be were we typical mammals, which puts us with the dolphins, at the head of the living vertebrates, with respect to relative brain size.

GROSS BRAIN INDICES AND HYPOTHESES OF BRAIN EVOLUTION

Inherent in the use of an index as a comparative device is some hypothesis about the way the brain changed in evolution. The index of cephalization and the encephalization quotient are based on a multiplicative hypothesis. This appears graphically as the assumption that steps in cephalization among mammals occur by displacements of a basic brain:body line (log–log coordinates) upward as a series of parallel lines. It is multiplicative because it signifies a multiplication of the brain weight by some constant. Referring back to Fig. 3.1, for example, one may note that indices of cephalization would have been computed as 0.09, 0.12, and 0.20 for the fictitious data. This means that for a given body weight a specimen with brain:body points lying on the line with $k = 0.20$ had a brain that was 2% times as large as that of a specimen of the same body weight with brain:body points lying on the line with $k = 0.09$.

The best known and most frequently cited version of this multiplicative hypothesis was presented by Brummelkamp (1940). He proposed that the brains of living mammals could be aligned to show stepwise increments of the Dubois version of Eq. (3.2), with $\alpha = 0.56$, by displacements of k in units of $\sqrt{2}$. This was analyzed statistically by Sholl (1948), who showed that the hypothesis was untenable. Despite Sholl's devastating criticisms Brummelkamp continues to be cited (e.g., Dodgson, 1962; Lenneberg, 1967), perhaps because there is a kernel of validity in his assertions.

Sholl also showed that Dubois' value of $\alpha = 0.56$ was not a demonstrable constant for related species of mammals, and he presented evidence that $\alpha = 0.26$ for individuals within a species with considerable variation in body size, as proposed by Lapique (1907), was not demonstrable in squirrels and rhesus monkeys. In fact, he found $\alpha = 0$ for those samples. Sholl extended his criticisms beyond the data that he had evaluated to make a sweeping condemnation of the whole process of deriving indices. (If he had used related species instead of individuals of the same species, he would have avoided his error.) This has had unfortunate effects in deflecting interest both from the obviously orderly brain:body relations that exist and are easily demonstrated and from the possibility of interpreting and using these relationships for comparative analysis. One of my purposes here is to correct that effect by showing where indices and related quantifications are clearly appropriate.

With respect to the stepwise increase in k or in some of the other gross indices, I have shown a number of years ago (Jerison, 1961) that empirical analysis of the fossil evidence suggested that there was an increase of that type. In the material of later chapters the nature of the increase is clarified and is shown to be more complex than implied by a simple multiplicative hypothesis because there are evolutionary effects on variances (diversities) as well as on means. Evidence from data on the anthropoids shows that significant additive factors must also be taken into account (Chapter 16).

From the point of view of an evolutionist the difficulty with Brummelkamp's (and Dubois') position is more philosophical than mathematical. They appeared to assert a kind of evolutionary principle—that brain size had to increase in certain ways as organisms evolved. The position is essentially orthogenetic, implying a directional force in evolution of the sort that fits better with theologically oriented analyses such as Teilhard's than with scientific approaches (Simpson, 1964, Chapter 11). The interpretation for the increase in relative brain size that I have offered in this text and elsewhere is straightforwardly evolutionary. I have assumed, with considerable support from the available evidence on the course of evolution, that the adaptive radiation of animals involved their invasion of many different kinds of niches, only some of which placed them under selection pressures toward enlargement of the brain. Progressive enlargement of the brain beyond the amount originally demanded for survival in particular niches did occur often, but to very different extents in different groups of species.

I have also considered the possibility of using an additive instead of a multiplicative hypothesis to explain the increase in relative brain size. This raises important technical issues about the meaning of brain size that are considered in more detail in the next section. My concern has been with an amount of tissue added to a basal brain size (Jerison, 1955) or, alternatively, a conversion of the measure of amount of tissue into a measure of the number of neural elements or "extra neurons" contained in that tissue (Jerison, 1963). My discussion of this topic is based on data from progressive mammals because little information is available on other groups. But, it seems reasonable to assume that the basic relations found in one group will be at least similar to those found in the others.

THE MEANING OF BRAIN SIZE

The weight (and, equivalently, the volume) of the brain is, of course, the sum of the weights of its constituent parts. If the cell is taken as the biologically significant unit, then the brain weight can be partitioned into contributions from each of the various types of cells in the brain and from the body fluids in the brain. Biologists may be reluctant to use a gross measure like brain weight because of the complexity of the brain's composition and the variety of its components, but it should be realized that, if the contribution of any subset of cells to the total weight is orderly, then the gross brain weight could be used to estimate the number of cells in the subset. It is shown in this section that the number of cortical neurons in the brains of mammals from a variety of orders can be estimated with fair accuracy from the brain weight alone. Stated another way, it is suggested that, even if one knows only that a particular number is a weight or volume of the brain of a mammal, it is possible to make a reasonable estimate about the number of cortical neurons in that brain.

Critics of quantitative analyses of the type that is about to be presented often raise the question of the complexity of the neuronal network in the brain. It is sometimes even asserted that the number of cells is unimportant and that only their interconnectedness should be of interest. We will see later in this analysis that there must be an intimate relationship between the number of neurons and the complexity of the neuronal network. We may appreciate why this must be so if we pause to realize that a neuron consists of a cell body and a network of processes. Only the cell body is usually observed in preparations used for cell counts, but the space between the cell bodies must be occupied by something. That something, of course, is the network of fibers and the glial cells packed around them (de Robertis *et al.,* 1960).

Cortical Volume, Neuron Density, and Brain Size

There have now been a number of determinations of the relationship between cortical volume and brain size. The most recent, by Elias and Schwartz (1969), is consistent with earlier determinations by Harman, by Shariff, and by Bok, which I have summarized (Jerison, 1963); a more complete summary of Harman's data is presented in Fig. 3.2, along with the new data of Elias and Schwartz. The results of Elias and Schwartz (1969) show the orderly relation between cortical surface area and total brain volume, and those of Harman show the simple relation of cortical volume to total brain volume.

The most important point about these results is that they show simple relationships among gross quantitative variables, which are independent of species. The mammalian condition may be described by a simple equation (cgs system), where c_1 is a constant and V is cortical volume:

$$V = c_1 E \tag{3.8}$$

Species with larger brains have more cerebral cortex, and the relation is

The Meaning of Brain Size

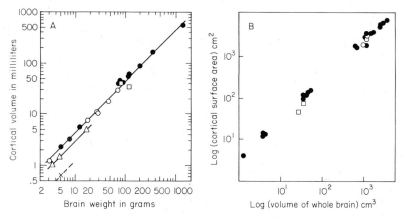

Fig. 3.2. A, Cortical volume as a function of brain volume. Lines are fitted by eye to have a slope of 1.0. Primate data include man. Data from Harman (1957) have been corrected following a personal communication. Primates (solid circles), carnivores (open circles), ungulates (squares), rodents (triangles), opossum (cross). B, Relation between volume and cerebrocortical surface area of mammalian brains. Upper group of points, whales and dolphins (solid circles) and man (open circles); middle group, carnivores (solid circles) and kangaroos (squares); lower group, opossums; lowest point, mouse opossum. Reprinted from H. Elias and D. Schwartz, *Science* **166**, pp. 111–113. Copyright 1969 by the American Association for the Advancement of Science.

linear. This equation is used later when we estimate the total number of cortical neurons as the product of the cortical volume and the neuron density.

The neuron density is the number of neurons per unit of cortical volume; if the total number of cortical neurons is N, neuron density is N/V. Estimates of neuron density in different species have all been based on sampling procedures in which an effort was made to work with samples of homologous cortex. Sections thick enough to contain whole neurons were cut, and actual counts were made from samples of such sections, which were stained to show cell bodies.

The relationship between neuron density and brain size is also dramatic, as shown in Fig. 3.3. It is clear that differences among species may be ignored and that neuron density in the cortex is simply a matter of gross brain size. Species with large brains have their neurons less tightly packed than small-brained species. The curves in Fig. 3.3 were derived from different laboratories and include a variety of species. Tower's data, for example, include mice, rats, carnivores, cattle, primates, an elephant, and a whale; yet a single function seems to fit all the data. The equation relating neuron density to brain size is

$$N/V = c_2 E^{-1/3} \tag{3.9}$$

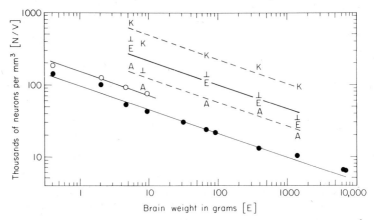

Fig. 3.3. Neuron-packing density as a function of brain weight. Shariff's (1953) data for eulaminate (E), agranular (A), and koniocortex (K) are given separately; dotted lines through K and A and heavy line through E and average ⊥ of his data for the entire cortex. Open points from Bok (1959) and filled points from Tower (1954). All lines fitted by eye for slope of $-\tfrac{1}{3}$. Species as identified by each author (read left to right along each line): Shariff—tarsier, marmoset, "cercopithecus" monkey, chimpanzee, and man; Bok—mouse, rat, guinea pig, and rabbit; Tower—mouse, rat, guinea pig, rabbit, cat, dog, macacca, "beef," man, elephant, and whale.

The constant c is not specified in either Eq. (3.8) or Eq. (3.9) because it was apparently affected by the different techniques used in the various laboratories, perhaps in determining which part of the cortex to consider, how to establish homologies, how thickly sections should be sliced (how accurate a microtome was used), and so forth. The uniformity of the slope, however, indicates that an underlying relationship between neuron density and brain size exists, just as an underlying relationship between cortical volume and brain size exists. The relationships can be combined to use brain size as the independent variable from which the total number of cortical neurons may be estimated as a dependent variable.

This is done by taking the product of Eqs. (3.8) and (3.9)

$$N = (V)(N/V)$$
$$= (c_1 E)(c_2 E^{-1/3})$$
$$N = c_1 c_2 E^{2/3} \tag{3.10}$$

The number of cortical neurons should, therefore, be proportional to the $\tfrac{2}{3}$ power of the brain size. This proposition can be tested with existing numerical estimates of the cortical neurons in various species. If these are graphed on log–log paper, one should expect the slope of the "best-fitting"

The Meaning of Brain Size

line to be about $\frac{2}{3}$. The test can be performed with Shariff's data and is illustrated in Fig. 3.4. A least-squares fit to the data gives the slope as 0.62 ± 0.09, which is sufficiently close to the expected slope of $\frac{2}{3}$.

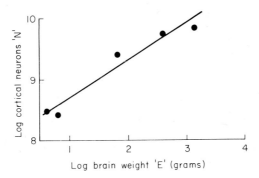

Fig. 3.4. Relation between number of cortical neurons and brain weight in five primates: tarsier, marmoset, mona monkey, chimpanzee, and man. Data from Shariff (1953). From Jerison (1963).

There are a number of reservations that should be indicated about this approach. Some have been mentioned in previous reports in which these data are presented (Jerison, 1963, 1970a). The most important reservation, from an evolutionary point of view, is that all the data were collected on "progressive" mammals. There are hints (Harman, 1957) that some of these relationships would be radically different in samples of insectivores or didelphids. Less important but still significant is the question of establishing homologies among parts of the brain. This is, basically, a hopeless task (but see Campbell and Hodos, 1970), and it would be much more appropriate in an analysis such as this, especially for the material of Fig. 3.3, to have a random sample from the entire brain and to state some packing constant for the brain. The evidence on specialized tissues in the brain that is available from the graph of Shariff's results (Fig. 3.3) suggests that such a packing factor exists and will affect the way cells are packed in nuclei throughout the brain of a particular species.

One peculiar misunderstanding has arisen from this approach. It has been asserted that brain size, or cranial capacity, was not a "suitable parameter" for the study of hominid evolution (Holloway, 1966, 1968). There is, of course, no disagreement about that. There is probably some misunderstanding of the meaning of "parameter," however, although the term is explicitly defined for statistical usage. I have repeatedly tried to describe brain size as a "statistic," that is, as a number used in the estimation of parameters. As such, brain size does not have many ideal qualities. Its sampling distribution has not been studied, and it has other technical faults. But it has one very good quality—it can predict values for interesting "parameters" such as neuron density and, as we will see later, other "parameters" such as the volumes of various cortical and subcortical structures. This does not exhaust the interesting "parameters" relevant for studies of human or vertebrate evolution, but it is an interesting lot. Furthermore, I would guess, simply because neural "information-processing" tissue has to be packed somewhere, that the facts of packing density will always be related to brain size, at least to some extent.

It is not my intention here to provide a definitive set of equations that enable one to estimate precise values for the parameters of the brain's constituents. Rather, I wish to indicate the tactics for the approach, the style of thinking that should be used, and the kind of regularities that should be sought. We must have research with better precision of cell counts, a broadened range of species on which counts are taken (in particular "primitive" as well as progressive species), and additional consideration of the so-called intercellular space, which is filled with glia and neuronal processes. A limited amount of data on this aspect of the problem can be assembled and is reviewed in the next section.

Neuronal Connectivity and Brain Size

Since neurons are less densely packed in larger brains than in smaller brains, there is more space for other structures between the perikarya. Wright (1934) pointed out that the diameter of the axons relative to the diameter of the cell body limits the number of neurons that can be packed in a vertical column of cortex to fewer than 20. This is because vertically oriented axons (diameter $= 6\ \mu$) from more than 20 neurons could not be packed into a column of the cross-sectional area of the cell body (diameter $= 40\ \mu$). Since dendrites and glia also have to be packed in, this further limits the number of neurons per unit volume. There is a clue here that explains the significance of the less dense packing of neurons in larger brains. Less densely packed perikarya could have more elaborate dendritic branching, and more glia could be packed around the dendrites. In short, a lower neuron density may imply increased connectivity among neurons.

If neurons are less densely packed, there will be fewer axons to squeeze into a volume and a greater possible dendritic extent per neuron. The neurons might, therefore, be expected to have longer processes, especially dendritic fields, in larger, less densely packed brains. The only data on this kind of question, which can be answered by comparing "homologous" neurons in large-brained and small-brained species, have been presented by Bok (1959, p. 208, Table XV). He has measured the lengths of the dendrite trees in homologous neurons of a mouse, rat, guinea pig, and

The Meaning of Brain Size

rabbit. Bok's measures of the lengths of the dendrite trees were taken from an elegant control system that he developed. This system recorded the motion of stage and eyepiece control knobs of a microscope while each of the dendritic branches of a Golgi–Cox stained neuron were tracked by a pointer mounted in the microscope eyepiece. I am uncertain, however, about Bok's criteria for designating neurons as homologous.

The relationships between Bok's reported lengths of the dendrite trees and brain size (left ordinate) and neuron density and brain size (right ordinate) are graphed in Fig. 3.5.

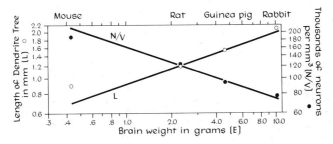

Fig. 3.5. Neuron-packing density and length of the dendrite tree, both as a function of brain weight. Lines fitted by eye to have slopes of $-\frac{1}{3}$ and $\frac{1}{3}$, respectively, illustrating the reciprocal relationship between these two "dependent" variables when brain weight is treated as the independent variable. Data from Bok (1959).

There is clearly a reciprocal relationship between the points (open circles) giving the length L of the dendrite trees in homologous neurons of different mammals and the points (filled circles) giving the neuron densities N/V for the same mammals, both as a function of brain size.

Implications of Neuronal Geometry

If it can be verified, the significance of the reciprocal relationship is obvious. Imagine a neuron that, except for its axon, is entirely contained within a cubic millimeter of cortex, receiving all its "information" from other neurons anywhere via boutons (or spines—the locus is of secondary importance) impinging on its dendritic branches. Ignoring boutons on the cell body and assuming that the boutons line up more or less uniformly along dendrites, the probability of synaptic transmission would be about equal[1] at all points along the dendrites. The probability of synaptic

---

[1] This assumption, necessary for preliminary theorizing, simplifies the true situation—that there is a gradient of excitability along a dendrite, with a maximum some distance from the cell body (P. D. Coleman, personal communication; Coleman and

activation (excitation or inhibition) of a neuron would, therefore, be proportional to the length of its dendritic tree. The total information processed in a unit volume of cortex would be proportional both to the number of neurons in that volume N/V and to the length L of the dendrite trees of those neurons. The reciprocal relationship in Fig. 3.5 suggests that as N/V decreases, with interspecies increase in brain size, the dendritic length L increases in a balancing way. The total activity or information processed by a unit of cortical volume may, therefore, be the same in any mammalian brain, regardless of the species from which the brain came.

A simple mathematical statement of the argument may make it easier to appreciate. An equation can be written relating the length L of the dendrite tree in homologous neurons to the brain size; this is the equation of the upwardly sloping line of Fig. 3.5

$$\log L = \log c_3 + \tfrac{1}{3} \log E$$

As a power function this becomes

$$L = c_3 E^{1/3} \tag{3.11}$$

This equation states the rule by which dendritic extent increases as a between-species effect as brain size increases.

The activity or information handling as discussed below would be defined as proportional to the product of the length of the dendrite tree L, as given in Eq. (3.11), and the neuron density N/V, as given earlier in Eq. (3.9). Thus,

$$\text{Activity} = (L)(N/V) = (c_2 E^{-1/3})(c_3 E^{1/3})$$
$$= c_2 c_3 E^0$$
$$\text{Activity} = c_2 c_3 \tag{3.12}$$

Brain size drops out as a variable in the equation. In Fig. 3.5 the numerical value of $c_2 c_3 = 1.5 \times 10^6$ per cubic millimeter of cortex. The implication of Eq. (3.12) is that the "information processing" activity in a unit of cortical volume V in different species is independent of brain size in the species. The result is surprising and seems to be a formal equivalent of Lashley's principle of "equipotentiality." One might describe it as a principle of "functional equivalence" of units of cortical volume, if these are

---

Riesen, 1968). Sholl (1956) and Berry et al. (1973) review many of the technical issues that arise in statistical approaches to functions of large aggregates of neurons, issues that could not be discussed in sufficient detail here. Like most anatomists, however, they minimize between-species differences and have not noted the potential simplifying role of brain size as an "intervening variable." That role is well illustrated in the derivation of Eq. (3.12).

The Meaning of Brain Size

sufficiently large. It should be noted, finally, that this analysis does not imply that contiguous cortex must be involved. The sampling of brain cells could be random with respect to location, provided that large numbers were sampled.

GLIA/NERVE CELL INDEX

The failure of the "glia/nerve cell index" can now be explained easily. This index, proposed by von Economo and others as a phylogenetically valid measure of relative brain efficiency independent of brain size (Haug, 1956), is the ratio of glial cells to neurons. Hawkins and Olszewski (1957) and Friede and van Houten (1962) have shown that, in fact, the index increases with brain size. The present view of brain geometry shows why this must be the case. It follows from three points. First, the number of neurons in a unit volume is known to decrease as a function of brain size. Second, essentially all the space in a unit cortical volume is considered to be occupied by neurons (perikarya) and glial cells. Third (a point not emphasized in this book), cell size increases only slightly with body size. Thus, as less space is taken up by neurons in larger brains, there must be more glial cells, and the ratio of glia to neurons must increase.

Tower and Young (1973) present neurochemical evidence which supports this analysis. They have shown that the glia/nerve cell index increases approximately as a function of the $\frac{1}{3}$-power of mammalian brain size, regardless of species. One of their illustrations is remarkably similar to our Fig. 3.5—showing a reciprocal relationship between neuron density and the glia/nerve cell index when both are correlated with brain size. The glia/nerve cell index is, thus, in the role of the "length of the dendrite tree" as a variable in Fig. 3.5. This makes sense, morphologically, because glial cells should be tightly packed about neural processes (axons and dendrites), and the number of glial cells per nerve cell should be a relatively simple (probably linear) function of the total length of these neural processes.

Other neurochemical evidence supporting this view has been in the literature for some time and is taken from studies of DNA in the brains of several species of mammals. The amount of DNA is determined by the number of chromosomes in cell nuclei, and when this is measured in the cerebral cortex it permits one to estimate the concentration of neurons plus glial cells. (Neuronal and glial DNA are technically difficult to separate.) The evidence is consistent with the hypothesis that the total number of neurons and glial cells per unit of cortical volume is the same in rats, guinea pigs, rabbits, cats, dogs, and humans (Heller and Elliott, 1954; Mandel et al., 1964). This is exactly the result to be expected from the analysis that led to Eq. (3.12). It reflects the reciprocal relations as shown in Fig. 3.5 and as reported by Tower and Young (1973).

Cortical Volume and Other Brain Measures

We are interested in the correlates of brain size, because we can determine little more than volume from a fossil endocast. It is, therefore, important to review some recent results on the multivariate analysis of the structure of the brain in mammals in which brain size was used as one of the "variates." I will later present an analysis I performed at about the same time as a more extensive analysis by Sacher (1970). He and I (independently) used similar data, of the type collected by Bauchot, Stephan, and their colleagues, and in my case the analysis was based on one of the tables published by Bauchot (1963) in his doctoral dissertation. The analyses are similar, and the main reason for presenting my own results is that they present a fraction of the analysis which is most relevant to the present discussion. Sacher's report can be referred to for details on method.

Let me first discuss Sacher's results; I will later emphasize points that he chose to understate. His purpose was to demonstrate the use of multivariate methods to analyze the complex array of tissues facing the neuroanatomist, and he, therefore, emphasized methodology and variables likely to interest the neuroanatomist today. He considered relative volumes of neocortex, cerebellum, diencephalon, and so forth, and only secondarily considered total brain size or total body size as variables. The analysis, fortunately, stands independently of the initial interest of the analyzer. My major concern is with the role of gross brain size, because that is the only measure available directly from the fossil material. I will reinterpret Sacher's results in terms of brain size and body size.

Using factor analysis as his method, Sacher determined 5 factors that accounted for 98.5% of the variance in a group of 31 measures of the brain. He used the volumes of the medulla, diencephalon, hippocampus, neocortex, gross brain size, and so on, as well as other measures, such as body size, a group of 15 ratios of the volumetric measures, and an arbitrary rating of phyletic standing of a group of insectivores, prosimians, and anthropoids (see Chapter 16; his were the same species as in my analyses, relying on the data of Stephan *et al.,* 1970). I would interpret the 5 factors somewhat differently from the way Sacher did. His Factor I was a clear brain size factor, which showed that the gross brain size could be used to estimate the size of almost all parts of the brain on which he had measurements. The olfactory bulbs were an exception, and the paleocortex volume was presumably not easily predictable from the gross brain size. The argument is simpler if one uses the results of his correlational analysis on which the factor analysis was based. The correlation between brain size and neocortex size was 0.989; between brain size and cerebellum it was 0.996; and between brain size and diencephalon it was 0.995. It fell to 0.945 in the correlation with body weight and to 0.941 in the correlation with paleocortex plus amygdala volume. But the only brain variable with which the correlation of the total brain weight was less than 0.941 was the size of the olfactory bulb and tract. That correlation was negligible (for real relationships) $r = 0.350$.

In essence, this means that if we know the gross brain size we can estimate many other interesting things about the constituents of the brain. That was the essential point of the previous discussion, and it was important to have it verified in the apparently sophisticated analysis offered by Sacher. His analysis is potentially much more important for the comparative neurologist than for the paleoneurologist. For the latter it merely provides reassurance that the gross brain size is, in fact, a "natural biological statistic" for estimating other, perhaps more interesting, "parameters."

The other factors identified by Sacher's analysis included only one additional factor that I would consider of neuroanatomical interest, a rhinencephalon factor that accounted for the sizes of structures such as the olfactory bulbs, paleocortex, and, to a lesser extent (loading = 0.61 and 0.62), of septum and schizocortex. In my analysis in Chapter 11, in which I examine the parts of the brain in fossil mammals for gross differences, I also found that olfactory bulbs behaved differently as a function of body size (or brain size) than did forebrain or hindbrain. It should be noted, incidentally, that body size did contribute to both the brain size factor and the rhinencephalic factor, more or less as suggested in Figs. 11.6 through 11.9, according to Sacher's analysis.

The other three factors discussed by Sacher are of little interest. One is a "brain/body" factor determined entirely by the simple ratio of brain size to body size. It demonstrates, if a demonstration is needed, that the simple ratio is essentially meaningless, as are statements about it such as "the brain/body ratio decreases as body size increases." The fact of Eq. (2.1), that is, that the basic metric for relative size is logarithmic, should somehow be learned by all who continue to be tempted to divide a brain weight by a body weight in order to remove body weight as a factor. The correct "quick method" to accomplish that makes use of Eq. (2.1) and takes the ratio $E/P^{\frac{2}{3}}$. Sacher included that ratio as a "variate" and found that it has a reasonable loading (0.77) on his Factor I. The brain weight or volume each have the loading 0.80. Among the direct measures neocortex has the highest loading (0.86), but the phyletic variate has a still higher loading (0.91). The interpretation of these loadings is not quite as simple as one would like; it is easier to interpret a correlation coefficient. This is due to the fact that the placement of a factor in "factor-space" involves a special procedure of uncertain biological significance. Although the particular procedure Sacher used is the one generally adopted, because it is easiest to rationalize mathematically (I also used it—the "varimax" method), there are arbitrary elements to it.

Sacher's remaining two factors have loadings only on several of the artificial ratios that he constructed (for example, medulla:brain volume and diencephalon:brain volume), and I interpret them as specific factors of little general interest at this time.

I will now summarize the results of my reanalysis of Bauchot's (1963) monograph on the volumes of diencephalic nuclei of insectivores and galago. (The insectivores included the tree shrew, *Tupaia glis,* which is intermediate between insectivores and primates in some respects; see Chapter 16.) The main result of this reanalysis is to show that the measure of total brain size accounts for almost all the variance in the measures of the volumes of 29 diencephalic nuclei as determined by Bauchot.

In the analysis I will use the concept of a factor score (see Sacher, 1970, and references cited there). The concept is derived from factor analysis, which was originally developed to analyze intelligence tests and led to the designation of special factors of intelligence, such as spatial ability, numerical ability, memory, and so forth. The concept of a factor score in that context is essentially the concept of how much of a particular ability a particular individual has. It is useful for intelligence or personality measurement if a series of tests are given that represent a set of factors differentially, because one can then estimate a "profile" for an individual.

A very different application occurred in my factor analysis of Bauchot's data in which he gave the volumes of 29 different diencephalic nuclei in 17 species of animals. The first step was to determine how many factors were represented. It turned out that about 95% of the common variance of the 17×29 matrix of data could be accounted for by the action of a single variable, the gross brain size. (This was clearly the same as Sacher's Factor I.) Thus, a single factor could be extracted as a common factor score, and it could be related to the brain size. I considered the remaining variance trivial. The common factor was differentially represented in the 17 species, of course, and in Fig. 3.6 we can see how the common factor scores were related to the brain sizes of the various species. It is clear that brain size could predict the volume of the diencephalic nuclei and that the prosimian (*Galago*) and the advanced insectivores (the tree shrew and the elephant shrew) were not differentiated from other species. Brain size, again, proved itself to be a good statistic, a natural biological statistic that gave information about finer properties of the brain.

GROSS BRAIN SIZE AND BRAIN FUNCTIONS

If the gross brain size is related to the number of elements in the information-processing system of the brain and the degree to which the elements

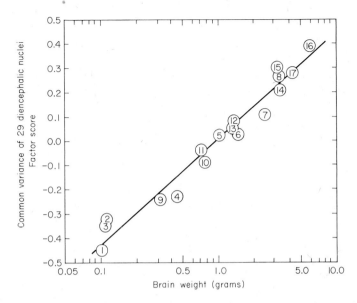

Fig. 3.6. Factor analysis of data from Bauchot (1963), and the factor scores of 17 species (insectivores and a prosimian) as a function of brain weight: 1, *Sorex minutus*, 2, *Sorex araneus*, 3, *Crocidura russula*, 4, *Crocidura occidentalis*, 5, *Talpa europaea*, 6, *Setifer setosus*, 7, *Tenrec ecaudatus*, 8, *Erinaceus europaeus*, 9, *Neomys fodiens*, 10, *Chlorotalpa stuhlmanni*, 11, *Chrysochloris asiatica*, 12, *Galemys pyranaicus*, 13, *Elephantulus fuscipes*, 14, *Galago demidovi*, 15, *Tupaia glis*, 16, *Rhynchocyon stuhlmanni*, and 17, *Potamogale velox*.

can affect one another, the amount of information processed (which must also be determined by these elements in the formal information-theory sense) may be a relatively simple function of the size of the brain. This was implied in our earlier result (p. 70) and should be developed now. In considering this view, nothing is suggested about the kind of information that is processed or the locus of the processing. There may be auditory information stored or processed at a brain stem, collicular, geniculate, or neocortical level and olfactory information processed via a system that is essentially paleocortical. There may be information associated with short-term memory processed by neocortical sensory systems and hippocampal paleocortical systems, and there may be attention systems that are largely subcortical and paleocortical in their more accessible (for study) portions. But the total information-processing capacity may be the sum of the capacities of these and other part systems and their interactions.

One's model for brain–behavior relations may affect the kind of brain index considered relevant. If the model is punctiform, with independent physiological functions associated rather specifically with specific structures, indices suggesting an overall information-processing capacity may be of little interest. On the other hand, recent evidence points to the waking brain as being a complex interactive system in which truly isolated functional systems probably never occur (Magoun, 1963). The isolation of systems is, in a sense, an artifact of experimental methods that artificially isolate parts of the system in order to study them more easily. The historically interesting view that was skeptical of localization of function (Flourens, 1842) was made more modern and precise by Lashley (1950) who could recognize some localization while emphasizing the importance of action of the mass of neural tissue in the brain. We are, perhaps, ready for further refinement in which the interacting subsystems, localized anatomically or physiologically, can be recognized as elements of a total system for which total capacity for information handling may be defined. And that total capacity may be associated with gross brain size.

In many instances a definition of total capacity would be of great interest, even where the basic concern is, properly, with the mechanism of a single system or part of a system in the processing of a particular kind of information (e.g., localized attention functions in the visual system as discussed by Killackey and Diamond, 1971). By understanding the total system, one can better evaluate the role of the part system as contributing to the whole. For evolutionary analysis it is particularly important to consider broad advances in grade as well as specific advances or adaptations for niches. Orderly changes in total information-processing capacity, measurable by changes in average brain size of faunal groups from different geological strata (Jerison, 1961), raise issues for this kind of "progressive" evolution that can resolve conflicting interpretations of selection pressures and the brain.

The problem of deriving and using indices to measure broad faunal changes is different from their use to suggest a precise ordering of closely related living species with respect to brain development. It is apparent from graphical analyses of the type presented earlier in Fig. 3.1 that the numerical values of computed indices are very much influenced by, and sensitive to, mathematically minor differences in certain constants. Since these constants (e.g., α) are of uncertain biological significance and are really empirical constants derived from possibly biased data and uncertain curve-fitting procedures, the indices that are derived are inevitably ad hoc to some extent. Only when a rationale can be presented for specific values for the constants and for the computations, can one judge the utility of indices as ways to make sense of data.

The difficulty with indices arises from the fact that they are numbers, and we are accustomed to think of numbers as precise. It is only in recent years that students of the life sciences have become aware of the possibility that numbers vary in their precision. Most of us are now sensitive to the fact that some numbers are little more than labels for categories and imply no ordering with respect to an ordered dimension, whereas other numbers can be used in the usual sense of physical weights and measures (Stevens, 1946). Indices should be, and are usually expected to be, numbers of the latter type in which an index of 4 is twice as great as 2, and 2 units away from it, both numbers being referable to a zero point on the index scale or dimension. The ad hoc nature of actual indices limits the inferences possible, however, and in many cases the scale of measurement is not an appropriate one. For example, if there is a doubling of brain size relative to a basal line, we may properly hesitate about considering a doubling from a basal size of 3 g to an enlarged size of 6 g as being equal to a doubling from a basal size of 50 g to an enlarged size of 100 g. Many more elements are added in an extra 50 g than are added in an extra 3 g, and presumably there would be a shift in information-processing capacity proportional to the absolute, rather than the relative, number of elements added. It was this consideration that led me to try an additive rather than a multiplicative index (Jerison, 1955, 1963) for analyzing relative brain development in the order Primates.

This argument leads to the conclusion that the numerical methods associated with the development of indices always need careful analysis. It would be useful, for example, if one could work from a theoretical model of the growth and development of the brain and its nuclei, within and between species. This model should be in the form of precise equations relating gross brain size to body size and the size of the parts of the brain to both brain and body size. One would then have a rational basal equation against which deviation in particular species could be projected and analyzed. It is likely that several such equations would have to be developed and that the basic equation would contain parameters associated with orders of mammals.

Until such procedures are developed (and the work of Bauchot, Stephan, and their associates seems to be bringing the quantitative data to a point at which this kind of development may become a relatively easy exercise in quantitative morphology, as Sacher has shown), particular indices can be quite misleading as an array of numbers. It is much more useful to present, along with the indices, the data on which they are based. For example, an array of the brain and body weights presented graphically, along with the basal equation to which the points are to be related, is an excellent device to eliminate undue dependence on indices. In the case of Fig. 3.1 it would be immediately apparent that a natural relationship existing among the points is overlooked in the computation of the indices.

APPROACHES TO A THEORY

The following theory is presented elsewhere in essentially the same form (Jerison, 1963) and is included in this book because it appears to be a useful way to examine the evolution of the primate brain (Fig. 16.10). It is a "model" theory because it suggests the levels of precision that should be sought. It is overly simple in that no account is taken of the different contributions of the parts of the brain, and thus, like other presentations in this book on the evolution of brain size, its main relevance is for the analysis of fossil endocasts. It is used for that purpose in Chapter 16, when the evolution of the size of the primate brain is considered. I believe, however, that the evolution of brain size in all taxa can, in principle, be analyzed in the same way as in this theory. It will be necessary only to add a few dimensions to the analysis, which, in its present form, is essentially a three-dimensional analysis. Its dimensions are gross brain size, basal number of "cortical neurons," and number of "extra" cortical neurons. The additional dimensions that are needed should refer to sex (because of dimorphism) and to taxonomic groups, with a division at least between "primitive" and "progressive" groups.

In my first analysis of the problem (Jerison, 1955) the major primate groups were differentiated on the basis of a brain weight factor that was assumed to be independent of body weight and dependent on relative brain development (Dubois' "psychic factor," more or less). The total brain weight was assumed to be the result of the addition of this factor (E_c) and a second brain weight factor (E_v) that was dependent on the body weight. The analysis was made quantifiable by assuming that the second brain weight factor was exactly equal to the total brain weight of a primitive mammal, specifically, that the second factor could be estimated from the body weight by Eq. (3.1) with $k_i = 0.03$.

The implication for the evolution of the hominid brain was that the variable factor E_v could be estimated for fossil hominids from estimates of their body sizes, and the level of brain development achieved by these hominids could then be stated quantitatively by calculating the constant factor for the endocranial volume. The relationship, described in Eq. (3.10), between brain weight and number of cortical neurons modifies the analysis and makes it somewhat more elegant as an exercise in theoretical biology.

A Theory of Brain Size

The mammalian brain size (weight in grams or volume in milliliters) can be analyzed into two independent components, E_v, determined by body size, and E_c, associated with improved adaptive capacities. The total brain size E is the sum of these two components. Thus, the first hypothesis is

Hypothesis 1. $\quad E = E_v + E_c \quad$ (3.13)

Analogous measures are available for the total number of cortical neu-

Gross Brain Size and Brain Functions

rons N and the neurons in E_v and E_c, which will be designated N_v and N_c; a corollary to the first hypothesis is

Hypothesis 1a. $\qquad N = N_v + N_c \qquad$ (3.14)

A specific relationship between N and E is stated as the second hypothesis

Hypothesis 2. $\qquad N = 8 \times 10^7 E^{2/3} \qquad$ (3.15)

The empirical version of Eq. (3.10) is adopted here as a hypothesis, after determining by a visual fit that $c_1 c_2 = 8 \times 10^7$. The exponent $\frac{2}{3}$ in Eq. (3.15) means that the number of cortical neurons is proportional to the cortical surface. This is reasonable because the cortex is an outer, or surface, layer of cells, no more than a few millimeters thick, and the cortical neurons are all in that "outer layer," or cortex.

If Eq. (3.15) holds for the total brain weight and neuron number, it cannot hold for both N_v and N_c. It is, therefore, necessary to state that Eq. (3.12) also holds for N_v as follows

Hypothesis 2a. $\qquad N_v = 8 \times 10^7 E_v^{2/3} \qquad$ (3.16)

The reason for Hypothesis 2a is apparent when one considers that primitive mammals have a brain size $E = E_v$ sufficient to maintain vegetative, sensorimotor, and related behavior, and Eqs. (3.15) and (3.16) state the number of neurons in such a brain. When one assigns a portion of the brain size of a progressive mammal to activities that it has in common with primitive mammals, it is natural to assume that the number of neurons associated with these activities should be the same in the progressive as in the primitive animal.

It may not be obvious that E_v in the progressive mammal's brain will have to be more massive than in the primitive mammal. This "theorem" is derived as follows. From Hypothesis 1 additional neurons in the E_c component would increase the total number of neurons in the progressive mammal's brain, and by Eq. (3.15) the total brain weight would be greater by a corresponding amount. From Eq. (3.9) we note that neuron density is lower in larger brains, and, therefore, that the number of neurons N_v in the progressive mammal would be less tightly packed and, hence, would have to be fitted into a larger mass of brain tissue than E_v, the total amount of brain tissue in the primitive mammal of similar body size.

In all mammals a brain weight factor E_v can be estimated, conservatively, from the condition of primitive mammals (Chapters 10 and 11); thus

Hypothesis 3. $\qquad E_v = 0.03 \, P^{2/3} \qquad$ (3.17)

(The value 0.03 rather than 0.02 was chosen to be consistent with previous work.) In contemporary mammals this brain weight factor can be used to estimate the number of cortical neurons associated with primitive behavioral functions, by following the argument of Hypothesis 2a and applying Eq. (3.16).

These three hypotheses provide the basis for precise statements about relative brain development because N_c, as developed here, is a numerical measure of progressiveness in brain development beyond the level required by increasing body size. N_c can be estimated when information on gross brain and body weight is available.

Of the three hypotheses, only the first has no direct empirical correlate. The second hypothesis is based on the empirical result presented earlier as Eq. (3.10). The third hypothesis, as stated in Eq. (3.17), is related to an empirical result obtained on a sample of archaic Eocene mammals and the opossum (Jerison, 1961) and was retained in spite of the possibility that the multiplier might be slightly too high.

The first hypothesis, despite its failure to be associated with a direct empirical result, is a very common one in discussions of brain development. It may first have been stated by Manouvrier (1885) and also in evolutionary, but nonquantitative, terms by Edinger (1885). Dubois (1920) argued against the additive aspect, preferring a multiplicative jump by a factor of $\sqrt{2}$, as discussed and rejected earlier in this chapter. Bok (1959) stated it as follows: ". . . our measurements clearly point to the conclusion that the total number of cortical nerve cells in the various animal species is defined by two influences: the size of the body and the degree of cephalization" (p. 241). We have now stated it as a simple quantifiable additive hypothesis. Numerical computations based on this hypothesis have been presented by Tobias (1965, 1967, 1971) and are reviewed in Chapter 16.

It must be the case that mammals with highly developed brains differ from their less cephalized relatives in the number of cortical neurons, but it seems unlikely that they would also differ significantly in the manner of functioning of large aggregates of cortical neurons. The more advanced forms might be more competent in information storage, in decision-making behavior, in sensorimotor coordination, and so forth, but the neural mechanisms for such activities should be similar in progressive and primitive mammals. Thus, the efficiency of the brain should be reflected in the number of neural elements, and additional components (neurons) in the progressive brain can be considered merely as additional elements of the same type that occur in the primitive brain. This point is related to one made by von Neumann (1951), which is that the capacity of a computer can, in a general way, be stated by the total number of elements that it contains.

A curious issue should be raised in this connection because of the unusually large number of neurons in the cerebellum. Eccles (1973) reviews recent cerebellar neuron counts in the domestic cat and reports totals of more than 2 billion (2×10^9). If Eq. (3.15) is true for the cat, we would expect something less than 1 billion (10^9) neurons in the cerebral cortex of this species. The limited evidence from quantitative comparative neuroanatomy (Blinkov and Glezer, 1968) suggests similarities between the statistics of the cerebellum and cerebrum in living mammals, including a correlation between the gross sizes of these structures and the inverse relationship between gross size and neuron density in both structures. Evidence for a correlation between the increase in size in the evolution of forebrain and hindbrain is presented in Chapter 11. As mentioned in Chapter 1, all this suggests some "structural" cerebellarization as well as corticalization. The quantitative correlations imply that overall cerebellar functions are closely related to overall cerebral functions. Eccles' view of the cerebellum as a "neuronal machine" that controls and smoothes fine movements is attractive from my point of view on the evolution of intelligence, in that it makes it unnecessary to introduce additional complications into the numerical analysis of the meaning of brain size. There is no reason for the "supporting" system not to have more elements than the basic "integrative" system of the brain. We may, nevertheless, want to change some of our equations in order to estimate the total number of brain neurons. I would predict that these are related in a fairly simple way to gross brain size and to the total number of cortical neurons.

In recent years there has been some emphasis on the reorganization of the brain in "higher forms" or in man. That emphasis is misplaced. It is likely that as we learn more about the wiring diagrams of various brains we will recognize more and more differences. These differences should impress us no more than the behavioral differences that are much more easily observed. They are impressive, of course, but we should expect them and recognize that they will not enable us to derive ordered measures of differences among species. A notion such as total information-processing capacity is, in my judgment, more likely to be productive of an understanding of a dimension of evolution and differentiation. Reorganization, on the other hand, corresponds to a notion of change associated with speciation. We should expect adaptations to various niches to be made possible by the evolution of appropriate structures and functions, and reorganizations of the brain would be no more than the neural equivalents of species-specific behavior patterns.

Chapter 4

Beginnings: Habits and Brains

The recorded history of the brain began about 425 m.y. ago when jawless armored fish living in fresh water estuaries left their heads to posterity. The fossils of these ostracoderms are occasionally whole, but usually consist only of the fossilized bony armor of their heads. We will begin with their story, which can be told briefly, and leave the description of the major radiation of the other classes of fish for the next chapter. We will consider at the conclusion of this chapter just what it is one sees in a "fossil brain," how this can be related to a living brain, and, of course, how it is related to behavior. This chapter is, in short, a transition from method to substance. We will begin with the description of the fossil material and then undertake a description of how that material and more advanced vertebrate fossil brains have been analyzed. We will, later, review the general anatomy of the brain as it is inferred from surface features.

THE FIRST VERTEBRATES

The armored ostracoderms were jawless fish (class Agnatha), relatives of modern lampreys and hagfish. Their niche was probably either like that of small catfish or that of the ammocoete larva of the lamprey, although their appearance, in life, would have been quite different (Fig. 4.1). We must imagine one or two possibilities—a motile form or a semisessile form. In the first instance they would have lived as bottom fish, shoveling microorganisms, such as bacteria and small crustaceans, from the mud as they scraped the estuarine floor, which was their most likely habitat. Less likely, as sessile forms they may have siphoned the bottom mud, sifting detritus to retain foodstuffs, as does the ammocoete. The evidence for these guesses is in their generally flattened shapes, their mouths, which were directed downward rather than forward, and the structure of their gills (Watson, 1951). They may have occupied yet another possible habitat for a motile form, that is, near the surface of the water where they could skim algae and other microorganisms from whatever floated at the surface. This is

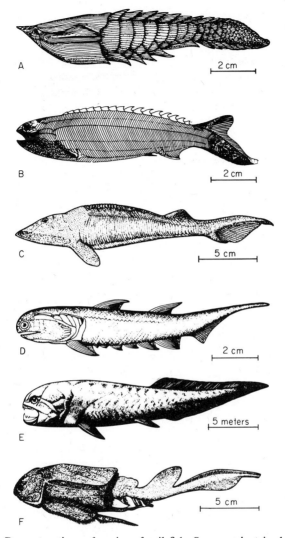

Fig. 4.1. Reconstructions of various fossil fish. Some ancient jawless fish (agnathans), *Poraspis* (A), *Pterolepis* (B), and *Hemicyclaspis* (C); an early bony fish, *Climatius* (D); and placoderms, *Dinichthys* (E) and *Bothriolepis* (F), illustrate the probable size and appearance in life of these early forms. The reconstructed brain of Fig. 4.2 is for a form similar to *Poraspis*. Romer (1966, 1968) indicates the technical nomenclature for B as *Pterygolepis* and for E as *Duncleosteus*. From Romer (1969).

suggested by their reversed heterocercal tails, in which the spinal column entered the ventral, instead of the dorsal, part of the tail fin, an adaptation that would tend to produce lift during normal swimming motions.

These speculations about habitats and feeding habits are examples of how "fossil" behavior is usually reconstructed. The evidence is from the body—the adaptations of the head, teeth, and skeleton. Although there are instances in which the morphology of the brain is a critical source of information (we will see some in later chapters), it is the rare behavioral adaptation that can be read from brain morphology. In general, we can obtain more "behavioral" information from the individual fossil skeletal remains, and we may get additional information about the environment within which fossil forms lived from the variety of associated fossils that have been retrieved with them.

On the assumption that the ostracoderms were bottom fish, we can reconstruct their differences as well as their similarities to living species. The ostracoderm adaptive zone was different from that of Recent bottom fish. Different predators had to be avoided, and the material on which to feed was undoubtedly also different. The late Silurian and early Devonian waters (400 m.y. ago) in which they lived were populated with a diversity of predacious arthropod and molluscan life. These invertebrates, some as large as men in bulk, were often equipped with sturdy probes and massive pincers. They could be resisted either by speedy evasive action or by sturdy armor. We can guess that both were tactics of the ostracoderms. Certainly with their carapaces, which covered as much as half their bodies, they would have been able to resist many of these predators.

Earliest Endocasts

Two groups of armored fish from this early period in vertebrate evolution provide information on the origins of the brain. They are the relatively generalized Heterostracids and the more specialized Cephalaspids. The Cephalaspids had bony skeletons as well as bony head armor, and both were preserved in fossil form. Stensiö (1963) has laboriously reconstructed the Cephalaspid cranial cavity from serial sections of the head. The results of this effort, which are considered first, have to be compared with reconstructions from serial sectioning of modern representative Agnatha. A special difficulty occurs in the reconstruction of the cranial cavity of fish (and to a lesser degree of amphibians and reptiles) in that the brain does not occupy the entire cranial cavity; the reconstruction is really of the space in which the brain lies. In presenting the cavity as an indicator of the appearance of the brain, one must imagine the space as having contained a significantly smaller brain. This act of imagination is a necessary part of the description of the most primitive vertebrate brains. In the specialization of the mammalian brain this is a minor problem because almost all mammal brains fit tightly in their braincases, and an endocast usually looks very much like a brain. The volume of a mammal's endocast is essen-

The First Vertebrates

tially the same as the volume of its brain, permitting many important extrapolations.

To return to the earliest brains, and accepting the fact that the reconstructions are sometimes more imaginative than one would like, there remain lessons to be learned both about the brains and about the nature and limits of all inferences about brain evolution derived from fossil endocasts, including those of mammals. Stensiö's reconstructions include a cephalapsid endocast that is sufficiently brainlike to permit the major features of the brain to be named and identified as homologous with brain structures in modern vertebrates. There is no question at all about most of the cranial nerves; there is little question about the major divisions of the brain into forebrain, midbrain, and hindbrain or about the identification of the rostral end of the spinal cord. Further detailed analysis, as presented in Stensiö's reconstructions of the brain, is shown for the cephalaspid, *Kiaeraspis,* in the frontispiece and is also discussed in Chapter 5.

The picture that can be developed from the fossil evidence is of the sort indicated by Whiting and Halstead Tarlo (1965), who have presented a reconstruction of the brain of the more generalized heterostracid

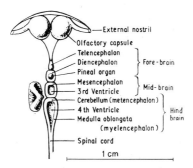

Fig. 4.2. A restoration of the central nervous system of Heterostraci, modeled after an ammocoete brain (cf. Tables 4.1 and 4.2). This can serve as model of generalized vertebrate brain. From Halstead Tarlo (1965).

fish and compared it with a sketch of the brain of the pride, or ammocoete larva, of the lamprey. Their reconstruction is less detailed than Stensiö's, but there is virtue in this vagueness because it is consistent with the fundamentally vague evidence. The basic information for this earliest of vertebrate brains, shown in Fig. 4.2, was derived from a few depressions, or pits, in the underside of the fossil head armor. The method of reconstruction is, essentially, to ask the following: "What kind of brain can reasonably be imagined to have lain under and extended into those pits?" Of course, Stensiö's endocast made from the fossilized bony cranium of the cephalaspid was helpful in answering this question. However, it was the good identification of the Heterostracids and Cephalaspids as members of the class Agnatha, or jawless fish, that was really critical. This identification suggested that evidence about the internal structures of the fossil forms should be compared with similar evidence obtained from contemporary Agnatha. Both groups of surviving agnathans, the lampreys and hagfish, are very specialized as adults. They are parasites that attach themselves to other fish and feed off their body juices. The ammocoete larval form of the lamprey, on the other hand, approximates much more nearly a hypothetical, primitive free-swimming vertebrate condition, and it is natural to compare the remains of the fossil forms with data on the contemporary ammocoete as have Halstead Tarlo (1965) and Whiting and Halstead Tarlo (1965).

BRAINS OF LIVING AGNATHANS

The fossil forms illustrated earlier are so clearly modeled (by the paleontologists) after living agnathans, such as *Petromyzon* or its larval ammocoete, that one can almost refer to those living species to describe the brain. In this text morphological descriptions of brains or endocasts are not usually presented since the emphasis is on problems of relative size. But this chapter is an exception and is designed to provide the reader with a few benchmarks for his own analysis of the illustrations of specimens discussed later.

The comparative gross neuroanatomy of the agnathans was a subject of considerable interest at the turn of the century, but relatively little significant material on the topic has accumulated in recent years, and none with which I am familiar on brain:behavior relationships. I can, therefore, discuss the morphology of the brain of living agnathans to provide some comparisons with those of the fossils but can contribute only by inference to a consideration of the significance of that morphology for behavior. My main sources are Papez (1929), which is a general textbook useful for this kind of material, and a paper by Herrick and Obenchain (1913), which will provide a useful illustration. More complete material is available in the encyclopedic volumes by Ariëns Kappers *et al.* (1936).

In the illustration of the brain of the lake lamprey, *Ichthyomyzon concolor* (Fig. 4.3), Herrick and Obenchain have provided an unusual reconstruction of the brain of a living animal. This is actually a wax model reconstruction, made in a way similar to the reconstructions by Stensiö illustrated earlier. Sections were made corresponding to cuts on a slide,

The First Vertebrates

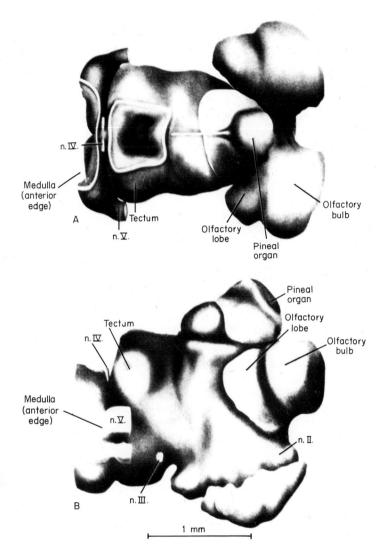

Fig. 4.3. Reconstruction of brain of lamprey, dorsal aspect (A) and lateral aspect (B). After Herrick and Obenchain (1913).

and then the wax sections (at a magnification of about 75) were mounted one atop the other.

There are no remarkable qualities about this brain or endocast. Herrick chose to label most of the forebrain "olfactory bulbs," but in other respects this is a relatively unspecialized brain, with no outstandingly enlarged structures of the sort that are shown later in this book. In terms of relative size the lateral view is most useful, showing the relatively large medulla, the mesencephalon, including tectum or optic lobes, also relatively enlarged (although the lamprey is not a significantly visual animal), a thalamic area in the central portion, and the small telencephalon (in lateral view), which consists entirely of the olfactory lobe and is connected directly to the olfactory bulb. There is one peculiar structure—the double pineal eye (epiphysis), which is not clearly visible in the views of the wax models constructed by Herrick and Obenchain (Edinger, 1956b).

Among the more notable features of the brain, which almost always appear in an endocast, are the points of exit of the cranial nerves. These may sometimes be displaced in the endocast if the nerves run alongside the brain in the supporting tissues before reaching their characteristic points of exit from the cranium.

With respect to functional specialization, there are few comments to be made on the illustrated brains of living agnathans. Papez noted the elongated medulla as related to the importance of the fifth to tenth cranial nerves. But we may, in general, think of the brain as illustrated in Fig. 4.3 as representing, perhaps, the minimal brain, with a few systems secondarily smaller than they would be in a generalized vertebrate, because these brains are from parasitic forms. Once attached to their hosts, lampreys require little information for further guidance of their motion.

THE GENERALIZED VERTEBRATE BRAIN: A PRIMER FOR ENDOCASTS

Two illustrations in this text should be referred to as guides to brain:endocast relations. These are the illustration of the brain and endocast taken from the same cat (Fig. 2.2) and the brain and two endocasts of the iguana (Fig. 2.1). The latter is especially revealing in that it shows the difference between an endocast of a reptile taken with the cartilagenous portion of the skull in place and that taken when only bony cranium is retained.

For the primer I will use sketches of the endocast of a primitive mammal in which the elements that are sometimes seen on endocasts are displayed. The sketches show views of the olfactory bulbs and tract, the forebrain, the superficially visible cranial nerves, the superior and inferior colliculi, and the cerebellum, pons, and medulla.

I will also review the main functions associated with the visible structures. If one of the structures appears enlarged in a particular species, then it is presumptive evidence for the importance of that function in the species. As an example, let us consider the approximate point of entry of

The Generalized Vertebrate Brain: A Primer

the tenth cranial nerve, which is the vagus nerve. This complex nerve, involved in autonomic functions and with both sensory and motor aspects, has, in some species, a greatly expanded point of entry in the medulla, so much so that it is called a vagal lobe. In the buffalo fish (Fig. 5.6) the vagal lobe is the largest lobe of the brain and indicates the important role of chemical sensitivity in this fish. This is an example of the principle of proper mass (Chapter 1).

To illustrate the structures visible in endocasts and to indicate their functions, it is appropriate to use fossil data, and this may be done with sketches of the archaic ungulate of 55 to 60 m.y. ago, *Phenacodus primaevus,* by Simpson (1933a). These illustrations (Fig. 4.4) were chosen because most of the structures visible and identifiable in any vertebrate endocasts are visible in this mammalian cast. The major defects of the

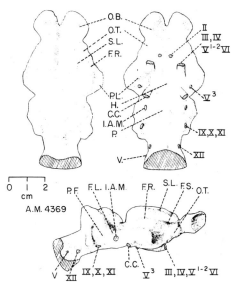

Fig. 4.4. Endocast of the archaic ungulate *Phenacodus primaevus* (AMNH 4369), indicating detail visible in well-preserved fossils. Structures are shown from dorsal, ventral, and lateral views. II, Optic nerves (filling of optic canal); [II], III, IV, V_{1-2}, VI, the common canals of these cranial nerves, and filling of the anterior lacerate foramen; V_3, mandibular nerve (filling of foramen ovale); IX, X, XI, points of exit of these nerves (filling of foramen lacerum posterius); XII, point of exit of this nerve (filling of hypoglossal canal or condylar foramen); C.C., carotid canal; F.L., "flocculus," or cerebellar lobule in petrosal anterior to internal auditory meatus; F.R., rhinal fissure; F.S., fossa sylvii; H., filling of fossa hypophyseos; I.A.M., internal auditory meatus (nerves VII and VIII); O.B., olfactory bulb; O.T., olfactory tubercle; P., pons; P.F., petrosal fossa (of cast, not a fossa in the petrosal); P.L., pyriform lobe; S.L., lateral sulcus; V., minor vascular foramina. From Simpson (1933a).

illustrations are that they fail to show the richness of the representation of the convolutional pattern in the endocasts of some progressive vertebrates, and they indicate too fine a representation of the brain in the endocast compared to its usual status in lower vertebrates. One should also refer to Fig. 11.4 for the view of *Hyopsodus,* another archaic ungulate (Gazin, 1968), in which the inferior and superior colliculi are exposed in much the same way as in the endocasts (and brains) of bats. *Phenacodus* serves, simply, as a good generalized mammal. The sketches are presented in Fig. 4.4, and the labeling follows that originally used by Simpson (1933a).

We can note, first, that many of the cranial nerves are identifiable in the endocast. The nerves, their names, and general functions are summarized in Table 4.1. Our interest in the cranial nerves is secondary; only

Table 4.1
Summary of Cranial Nerves[a]

No.	Name	Brain entry	Functions
I	Olfactory	Anterior forebrain	(s) Smell, chemical sense
II	Optic	Thalamus (forebrain)	(s) Vision
III	Oculomotor	Midbrain	(m) Eye movement, eyelid and pupil movement, lens and pupil control; (s) sensory feedback from lens and iris muscles
IV	Trochlear	Midbrain	(m) Eye movement
V	Trigeminal	Hindbrain (medulla and pons), and Gasserian ganglion in braincase but outside of brain. A very complex nerve in structure and function.	(s) Branches to head, face, and tongue (anterior portion) through *opthalmic, maxillary,* and *mandibular* nerves; (m) *maxillary* and *mandibular* in jaw movements
VI	Abducens	Hindbrain (medulla)	(m) Eye movements
VII	Facial	Hindbrain (medulla)	(m) Facial muscle control, gland innervation, middle ear muscle innervation; (s) deep facial sensation
VIII	Auditory and vestibular	Hindbrain (medulla)	(s) Hearing, balance
IX	Glossopharyngeal	Hindbrain (medulla)	(s) Gills and lateral line sensitivity in fish, posterior taste buds of tongue in mammals; (m) parotid gland and pharynx

The Generalized Vertebrate Brain: A Primer

Table 4.1—(cont.)

No.	Name	Brain entry	Functions
X	Vagus	Hindbrain (medulla). Another complex nerve, generally visceral but with some somatic components.	(s) General viscerosensory for thorax and digestive tract, somatosensory for skin of ear; (m) visceromotor for heart, lungs, digestive tracts usually via peripheral ganglia
XI	Spinal accessory	Hindbrain (medulla)	(m) Neck and upper body
XII	Hypoglossal	Hindbrain (medulla)	(m) Tongue movement

[a] This table is intended as a quick reference for use in viewing endocasts illustrated in the text. Statements are summaries of what are, for most of the cranial nerves, much more complex relations with peripheral structures and structures in the brain, with the letter (s) for sensory and (m) for motor function. From the point of view of visibility of cranial nerves in endocasts, the following points should be kept in mind: (1) The illustrations of the reconstructed "brains" of primitive fish (Figs. 4.2, 5.1, 5.2) and of various mammalian endocasts (Fig. 4.4 and many others) are often labeled with respect to visible points of entry or exit of cranial nerves. It is usually possible to see N.I. (at the tip of the olfactory bulbs), N.II, N.V (with several branches), N.VII and N.VIII (at the internal auditory meatus), N.X (along with N.IX and N.XI), and N.XII in a good endocast of a mammal. (2) N.XII is generally used to demark the posterior end of the medulla as represented in a mammalian or reptilian endocast. (3) In the usually distorted endocasts of larger fish the cranial nerves conclude with N.X. The more posterior cranial nerves (N.XI and N.XII) are not distinguished in fish.

in the account of the whale endocast (Chapter 15) is there any special concern with the specific placement of these nerves, although we are concerned with the roles of parts of the brain, including the olfactory bulb, and we will consider the roles of cranial nerves and their representation in the medulla as aspects of the adaptive radiation of fish (Chapter 5).

Our main interest here is in the parts of the brain visible especially well in *Phenacodus*—the forebrain and the hindbrain. Another concern is the midbrain visible in *Hyopsodus*. The particular markings of the endocast that are of interest include the placement of the rhinal fissure, which would separate neocortex, dorsally, from paleocortex, ventrally. The lobe beneath the rhinal fissure is usually referred to as the pyriform lobe. The functional significance of many structures has been summarized in Table 4.2.

It is particularly difficult, and perhaps even inappropriate, to present the data in the form of Table 4.2 because of the implication of specialized functions for each structure. Despite the doctrine of localization it is, nevertheless, a gross simplification to assign functions such as olfaction either to the forebrain, or even to the pyriform lobe. The entire brain may be

Table 4.2
Functions of Brain Structures Visible in Endocasts[a]

Structure	Function
Major structures (anterior to posterior)	
Olfactory bulbs	Sensory terminus for olfactory nerves (N.I); generally enlarged in animals that rely on smell.
Forebrain	Cerebral hemispheres usually only visible portion. Integration of information from sensory and motor systems. Primarily olfactory and chemical information in "lower" vertebrates, audition and vision as well in "higher" vertebrates.
Midbrain	Reflex control integrating motor behavior with auditory and visual information, postural reflexes. Much expanded as optic lobes of reptiles and birds.
Hindbrain	Cerebellum is associated with the maintenance of complex motor integration associated with movement, postural control.
	Medulla is associated with reflex control of visceral activity and with the primary relaying of information to and from "higher" centers for hearing, taste, head motion and position, limb and body position, and touch. Continues posteriorly, outside the brain, as the spinal cord.
Other structures (in alphabetical order)	
Amygdala	Lobe of forebrain visible in lateral view in some mammals but best seen in ventral view as medial portion of pyriform lobe (see below).
Cerebellum	See hindbrain, above.
Cerebrum	See forebrain, above.
Colliculi	Dorsal portion of midbrain, revealed in some primitive mammals. Superior colliculi (s.c.) control elaborate visual reflexes including those involving binocular effects; inferior colliculi (i.c.) control comparable auditory reflexes, including sound localization.
Diencephalon	Forebrain structure delivering information to cerebral cortex in mammals and birds. Not visible in endocast.
Epiphysis	See pineal body.
Hypophysis	See pituitary.
Medulla	See hindbrain, above.
Mesencephalon	See midbrain, above.
Olfactory tract	Posterior continuation of olfactory bulb, merging posteriorly with pyriform lobe.

The Generalized Vertebrate Brain: A Primer

Table 4.2—(cont.)

Structure	Function
Optic lobes	Much expanded midbrain, associated with complex visual behavior, visible in endocasts of birds (Chapter 9) and occasionally suggested in appearance of reptilian endocast (Fig. 2.1).
Pineal body	Visible in endocasts of many fossil fish and some fossil reptiles. Photosensitive structure related to photoperiodicity and probably other (not fully understood, biochemical and hormonal) functions.
Pituitary	Represented by distinct depression visible in ventral views of endocasts of large land vetebrates. The "master gland" of the body, controlling many hormonal functions of other glands, as well as growth. Size of "pituitary fossa" (depression in base of skull in which pituitary lies) is related to body size of a species.
Pons	Fiber complex visible in ventral view of hindbrain of mammalian endocasts. Carries pyramidal (motor cortex efferent) fibers, fibers of N.V and others.
Pyriform lobe	Lobe of forebrain (part of cerebral cortex) visible beneath rhinal fissure. Considered a more primitive part of the cortex; part of paleocortex or rhinencephalon.
Rhinal fissure	A fissure in the cerebral cortex, visible in most endocasts, and dividing neocortex, dorsally, from paleocortex, ventrally. The relative amount of cortex visible ventral to rhinal fissure is generally considered as an index to the primitiveness of a mammalian species. (e.g., Fig. 4.4 and Fig. 13.1. Note that this fissure is almost invisible in endocasts shown in Fig. 2.2; it may be seen as a continuation of olfactory tract in cat brain and as a corresponding region of *Hoplophoneus* endocasts, at lower margin of each picture.)
Tectum	Dorsal wall of midbrain, visible in many endocasts.
Telencephalon	"Higher" forebrain—olfactory integration in "lower" vertebrates; cerebral cortex and corpus striatum in "higher" vertebrates, involved in most advanced brain: behavior relations.
Thalamus	See diencephalon.

[a] Like Table 4.1, this is a much simplified outline of brain structures and functions, intended as a quick guide for viewing many of the endocasts illustrated in this book.

thought of as a complex, interacting, integrative center that controls the body and handles most of the information used by the body. It will be helpful, nevertheless, to have some reminders of the major specializations of the various parts of the brain visible in an endocast.

The picture of the mammalian brain is, of course, not adequate for other classes of vertebrates. It is possible, nevertheless, to use the data of Table 4.2 to understand the general relations among the parts of the brain of other fossils illustrated elsewhere in this book. The main peculiarity of the bird brain (Fig. 9.9), for example, is in the exposed optic lobe. This lobe is homologous to the superior colliculus of mammals. In reptilian endocasts one sometimes sees a structure reminiscent of an optic lobe (and this was one reason that Edinger was misled into considering the endocast of *Archaeopteryx* as reptilian). Comparison of a brain and an endocast shows, however, that the true optic lobe in reptiles is generally represented as part of the "forebrain" in an endocast. The main structures visible in a reptilian endocast are usually the olfactory tract, anteriorly, and the forebrain, if the cranium is ossified in that region. The rear portion of the forebrain and the midbrain and hindbrain (including the cerebellum and the medulla) are usually present in very vague outline, but a good representation of the pituitary is often available, especially in the giant reptiles (Fig. 7.3).

In the next chapters we cover the story of the evolving brain from the agnathan level just discussed, through the mammalian level shown in Fig. 4.4 (covered in Chapter 11), and then to the progressive mammals, including man and other primates (Chapter 16). A reasonable number of endocasts are illustrated in these chapters. The absolute size of the endocast should never be ignored if one seeks insights into the function of the brain that it represents. In few of the illustrations are enough details shown to indicate the various parts of the brain described in Fig. 4.4. For those it may be helpful, though not much more useful, to examine the original. A photograph is almost never adequate for the purpose, and a skilled artist who can copy an endocast accurately is rarely available for any particular specimen. It might be kept in mind by scholars that most museums are willing to lend specimens if they are not unique, and it is also frequently possible to make a copy of an endocast that is almost as accurate as the original. My work with the endocast of *Archaeopteryx lithographica* was with a copy provided by the British Museum (Natural History), and in that case it was actually better to have the copy than the original because more of the endocast could be viewed (Chapter 9; also Jerison, 1968).

It is also becoming increasingly possible to have endocasts especially prepared from some skulls if the scientific purpose to be met is sufficiently important. Gazin's endocast of the Middle Eocene primate *Smilodectes gracilis* (Gazin, 1965b) involved just such a preparation (Chapter 16). More such preparations will probably be made in the future. For example, the results in Chapter 14 indicate that it would be very important to determine the size (and appearance) of the endocasts of larger predatory marsupials of the Tertiary period of South America. The skulls are available, and it is necessary only to find the student sufficiently devoted to prepare an endocast. The complete excavation of the endocast of *Archaeopteryx* is another example because one should be able to retrieve a complete medulla and cerebellum and also obtain a full picture of the ventral surface of the brain.

It is conventional and appropriate to discuss methodological issues conservatively, and we should maintain our perspective as we read later chapters of this book by recognizing the inherent limitations of this approach. The evidence of relative size is, essentially, statistical, and the approach of this book can result in no more than a general map of a territory that is being explored with many of the tools of modern neuroanatomy and neurophysiology. Our evidence, limited to the details of the surface of the endocast, is on its surest ground when a simple measure of size is used. This information is part of the necessary background of a story, but to the extent that details will be forthcoming, they will have to come from the comparative neuroanatomy and neurophysiology of the brains of living animals, and the study of their behavior.

Part II

The Basic Vertebrate Radiation

In the following four chapters we will be concerned with the first several hundred million years of the history of the vertebrates. We will consider, in particular, the history of the brain in the major vertebrate groups that were known about 300 m.y. ago, going back in time an additional 100 m.y. or so, and forward in time almost to the present, to determine the fate of these early vertebrate groups. There are four classes of fish to be discussed; three are still represented by living species, and the fourth (class Placodermi) has been extinct for about 350 m.y. The history of the amphibians and reptiles, including the flying reptiles of the Mesozoic era, is discussed in separate chapters.

The history of the brain in these groups is somewhat difficult to chart because their brains did not fill their braincases except in a few exceptional groups or in very small specimens. We will, nevertheless, attempt to review that material from a quantitative point of view. I will try to be explicit about the assumptions that have to be made to permit such an analysis.

Chapter 5

The Lower Vertebrates: Fish

Despite the variety of brains in the living classes of lower vertebrates, including fish, amphibians, and reptiles, the direct evidence of their evolution, derived from fossil material, is limited to about 100 endocasts that are not easy to interpret. Unlike birds and mammals, most living and many extinct "lower" vertebrates have endocasts that are only vaguely like their brains either in size or shape. This is probably a side effect of differential relative growth of skull versus brain in lower vertebrates. According to Aronson (1963, p. 223; see also Brown, 1957), both structures continue to grow throughout life in fish, but in maturity the rate of growth slows down more for the brain than for the skull. In mature individuals of species that attain moderate sizes (over 30 cm), the brain becomes mirrored rather poorly by the endocast, and there may be major differences in the volumes of these two structures.

Latimeria, the coelacanth that was discovered in 1938 off the coast of Madagascar, provides an example of the difficulty. *Latimeria* is the only surviving species of an order of bony fish (Crossopterygii) that had previously been thought to have become extinct in the late Cretaceous period, about 75 m.y. ago. In most respects a study of the anatomy of the soft parts of this "living fossil" verified expectations of a uniformitarian hypothesis in which facts about living animals are used to interpret the fossil record. This was true for the brain, as well as other characters, because in so large a fish of any species (*Latimeria* weighed about 40 kg) the endocranial cavity would normally be much larger than the brain. The volume of the endocranial cavity of *Latimeria* was in fact about 300 ml, and its brain was only about 3 ml (Fig. 5.9). As Stensiö (1963) remarks, most of the endocranial volume of Latimeria "is filled out with a rigid, strongly adipose connective tissue" [p. 84 (see also Mlllot and Anthony, 1956, 1965)].

The situation in fossil forms may not be quite so acute because fossil lower vertebrates are believed to have been more completely ossified than many of their descendants. Sharks, for example, are only secondarily cartilaginous; their ancestors were probably extensively ossified (Romer, 1966). To quote Jarvik (1955):

> The ancient vertebrates of the Silurian and Devonian have a quality in common of great importance for the paleozoologist: their skeleton is generally well ossified and on the whole more complete than in later vertebrates where it often may be represented by tissues devoid of lime impregnation. Besides—a matter frequently overlooked by students familiar only with material of recent vertebrates—the high degree of ossification in the early vertebrates means that the cavities (for the brain, the ear, and other organs) are comparatively narrow, reflecting fairly well the external shape of the organs they housed . . . . When applying the modern paleozoological technique it is therefore often possible to obtain an intimate and safe knowledge of the skeleton with its cavities . . . .[1]

Some of that knowledge is neither as "intimate" nor "safe" as Jarvik suggests, and this will become clear from our quantitative analysis of Stensiö's reconstructions of fish brains. In general, the methods used by Jarvik and Stensiö will be most useful for suggesting the shapes of the brains in the fossil fish. They may be completely useless as a source of information about size relationships.

The lower vertebrates with respect to brain evolution include three[1a] classes of fish, the class Amphibia, and the class Reptilia. The evolution of the brain of jawless fish, class Agnatha, which includes the extinct Heterostraci and Cephalaspida, was described in Chapter 4. Now the evolution of the brain in all three classes of gnathostome, or jawed, fish will be reviewed and analyzed. Amphibia and reptiles are treated in Chapters 6–8.

BRAINS AND BODIES OF FISH

The extinct class of fish, Placodermi, was composed of armored fish that had an extensive adaptive radiation before their extinction in early Carboniferous times (Mississipian period) about 325 m.y. ago. Their largest recovered fossil was *Dunkleosteus* (*"Dinichthys"*), which may have reached a length of 10 m, and they also included species no more than 5-cm long (Fig. 4.1E). All had bony armored head regions. The living classes of fish are Chondrichthyes, or fish with cartilaginous skeletons, such as the sharks, and Osteichthyes, or bony fish, which include most contemporary fish.

---

[1] Reprinted from E. Jarvik, *Scientific Monthly* **80**, pp. 142–143. Copyright March 1955, by the American Association for the Advancement of Science.

[1a] New information (Ebbesson and Northcutt, in press) was received as this book went to press that some sharks and rays of the remaining class of fish, Chondrichthyes (cartilaginous fish), fall within the range of relative brain size of birds and mammals. In our terms, the class Chondrichthyes would be intermediate between "lower" and "higher" vertebrates. It was possible to take this information into account in the section on Chondrichthyes in this chapter and on p. 45, Chapter 2. When unqualified generalizations about fish appear elsewhere in the text these should be understood to refer to the classes Agnatha, Placodermi, and, especially, Osteichthyes.

Brains and Bodies of Fish

Skeletal and stratigraphic analyses show that jawed fish, and thus all jawed vertebrates (gnathostomes), evolved from the basal jawless class Agnatha in Silurian times, about 425 m.y. ago. The radiation of the placoderms, in particular, was very rapid, as evidenced by the diversity of types in Lower Devonian strata of 400 m.y. ago. The rapid radiation of the placoderms probably included forms in which bone became reduced and probably also some with skeletons that were entirely cartilaginous. These could have fathered the Chondrichthyes. Other bony, finned placoderms could have fathered the Osteichthyes. The evidence reviewed by Romer (1966) is that all three classes, Placodermi, Osteichthyes, and Chondrichthyes, appeared at nearly the same time. Fossils of all are recovered in significant numbers in Lower Devonian levels, the beginning of the "Age of Fish," 400 m.y. ago. Only one group of jawed fish is represented in the fossil record prior to Devonian times, the acanthodians, which are now considered by Romer as members of the class Osteichthyes. The direct evidence alone would, therefore, identify the bony fish as the first jawed vertebrates. However, considerations of the complexity of the skeletons of the Devonian types, as well as the relative frequency of fossils in Devonian beds, led Romer and others to the conclusion that the placoderms were the earliest of the jawed fish and the probable base from which the other living gnathostomes evolved.

The Placoderm Brain

We first consider the brain of a placoderm (order Arthrodira) *Kujdanowiaspis,* as reconstructed by Stensiö (1963) from its endocranial cast. His procedure was similar to that used with the cephalaspids (Chapter 4). To create the endocast (Fig. 5.1A), Stensiö made 715 sections by serially grinding the original specimen and tracing the cross sections revealed in

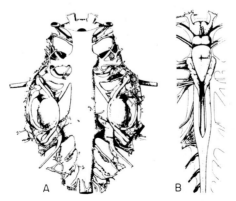

Fig. 5.1. A, Endocast; B, reconstructed brain of the arthrodire *Kujdanowiaspis*. After Stensiö (1963), as modified by Schaeffer (1969).

each of 715 shaved edges. He built up the figured wax model from these serial sections, and, although this endocast is clearly not a picture of a brain, it shows certain endocranial systems beautifully. These include the cranial nerves and the labyrinthine system.

Stensiö indicated how the brain could fit into the endocast by using the relative positions of the cranial nerves and analogies of these positions to those of the cranial nerves of modern sharks. His brain picture (Fig. 5.1B) suggests that the brain of this early vertebrate was relatively unspecialized, with no massive lobules on the medulla. The cerebellum is shown only slightly expanded, as in small sharks such as the dogfish. This model of an early brain is consistent with our expectations about the shape of early unspecialized vertebrate brains. The important limitations on the interpretation are on matters of relative size.

Size of Placoderm Brains: *Kujdanowiaspis*

It is possible to estimate the size of the brain of any fish if one can estimate either its body length or body weight. It is necessary only to assume that modern and fossil lower vertebrates followed approximately the same morphological patterns (Chapter 2). To the extent that this assumption leads to a consistent picture of the evolution of the brain, it may even be possible to test the uniformitarian hypothesis in a limited way. We will see, in a later section, that natural endocasts of fossil bony fish (paleoniscids) are about the right size and shape to be treated as brains, and the right size is defined as the size of a brain of a modern fish that would be appropriate for a modern fish's body.

The argument is quite simple; in the next paragraphs I will develop it from the equations of brain and body size. Graphs such as Fig. 5.2 on weight:length relations in fish (cf. Fig. 2.9) are the basis for estimates of body size. I will present the computations in what may be embarrassing detail for some, but I have learned to hide my embarrassment when reading similar presentations; it is sometimes nice to be led by the hand through a numerical maze.

Using the method of "least squares," we can determine Eq. (5.1) to relate the weight P of a fish to its length L. We had previously (Fig. 2.4) related a fish's brain weight E to its body weight P, and we can write Eq. (5.2). These equations (cgs system) are

$$P = 0.12\, L^{2.53} \tag{5.1}$$

and

$$E = 0.007\, P^{2/3} \tag{5.2}$$

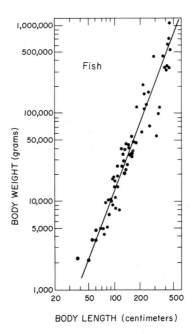

Fig. 5.2. Weight:length relations in fish. The line is a least-squares fit, assuming errors in both weight P and length L, $P = 0.12L^{2.53}$. The fact that the exponent is less than 3.0 signifies that shape changes as body size increases; as length increases the shape changes to provide less volume per unit length. This effect would occur if the cross section became less symmetrical, for example, if the animal was less rounded and more elliptical (with greater difference between major and minor axes) in cross section. Data from Spector (1956) on world-record fish.

We now ask: If *Kujdanowiaspis* were a living fish, how much would its brain be expected to weigh? The upper limiting value of the estimate tests the adequacy of Stensiö's reconstruction of the brain by asking how large a brain a particular animal could possibly have, and how would such a brain fit into the endocranial cavity.

The head armor of *Kujdanowiaspis* was about 5- or 6-cm long; Stensiö had, in fact, chosen this species because in his words it was a "comparatively small form" and could be studied by the very exact grinding method. From various reconstructions (e.g., Romer, 1966, p. 30) we may assume that arthrodires had between a third and a half of their length covered by their head armor. This would limit *Kujdanowiaspis* to a maximum of 18 cm in length, and we will use that as an upper limiting value for estimating the body weight of the animal. Let us work it out, using Eq. (5.1) in logarithmic form with $L = 18$.

$$\log P = \log 0.12 + 2.53 \log L$$
$$= \overline{1}.0792 + [(2.53)(1.2553)]$$
$$= 2.2575$$
$$P = 181 \text{ g}$$

To state an upper limit on body size in round numbers (which implies an upper limit on expected brain size), let us say that *Kujdanowiaspis* weighed a maximum of 200 g. The lower limit, incidentally, assuming the head armor to cover half of a 10-cm body, leads to a minimum estimated body weight of about 40 g. These computations now enable us to apply Eq. (5.2) to estimate the expected brain size in a normal, modern lower vertebrate weighing either 200 g or 40 g. The dimensions of the model brain in Fig. 5.1B can then be used to see whether they would be appropriate for the expected brain size in a normal, modern lower vertebrate weighing either 200 g or 40 g, that is, for *Kujdanowiaspis* were it a modern bony fish.

The result of solving Eq. (5.2) for a body weight of 200 g is an "expected brain weight" of about 0.25 g. The computation uses the logarithmic version of Eq. (5.2) as follows

$$\log E = \log 0.007 + \tfrac{2}{3} \log P$$

We have decided to take $P = 200$, which leads to

$$\log E = \overline{3}.8451 + \tfrac{2}{3} (2.3010)$$
$$= \overline{1}.3791$$
$$E = 0.239 \text{ g or about } 0.25 \text{ g} \qquad (5.3)$$

Assuming a specific gravity of 1.0 for tissue, this leads us to expect a maximum volume of brain in *Kujdanowiaspis* of 0.25 ml.

Let us now consider the volume of Stensiö's conjectural *Kujdanowiaspis* brain by imagining a cylinder that would be approximately the same size as the pictured brain. Such a cylinder would be about 4-cm long, with a diameter of about 0.6 cm. The volume of this cylinder is $4\pi(0.3^2) = 1.13$ ml. Our best guess is, therefore, that the brain placed by Stensiö in the endocranial cavity of *Kujdanowiaspis* was more than 1.0 ml in volume, which is at least 4 times too large. Had we used the minimum body weight estimate of 40 g in the equation for expected brain size, this size would be only about 0.08 ml, and Stensiö's overestimation would be fourteenfold.

This application of principles of the analysis of relative size, which follows naturally from a biological orientation like Thompson's (1942), is clearly of some importance in understanding the evolution of the fish brain. The method has virtues of both simplicity and elegance.

I have pictured the same analysis geometrically instead of algebraically in Fig. 5.3. We may begin with Stensiö's marvelous reconstruction of the endocranial cavity of this specimen (Fig. 5.1). There can be no quarrel with the endocast made this way; it is a result of direct reconstruction from serial sections of the original fossil. Our procedure is to place various "model brains" inside this construction. In Fig. 5.3A, the brain of Fig. 5.1B is replaced by a cylinder 4-cm long and 0.6 cm in diameter. This was the cylindrical "model of a brain" used in the calculations to estimate the volume of Stensiö's reconstruction of the fish's brain, and one now gets a sense of its adequacy for understanding size relationships. It is, of course, possible to produce other simple volumes as models of a brain. In one (Fig. 5.3B) the cylinder is as long as the first cylinder, and I have simply reduced the diameter to produce a 0.25-ml volume as required by Eq. (5.3). In another (Fig. 5.3C) I have produced a 0.25-ml volume by using somewhat more typical brain dimensions, with a ratio of 3:1 of length to diameter instead of the rather elongated brain implied by the 20:3 ratio of the first

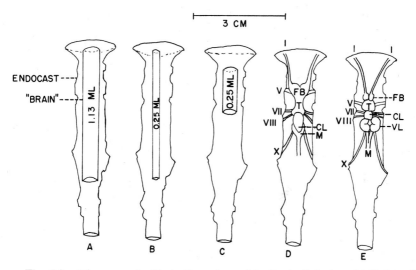

Fig. 5.3. Diagrammatic illustration of possible "brains" that could fit into the endocast of *Kujdanowiaspis*. The endocast is indicated by the outlines of each sketch. A, A 1.13-ml cylinder, with dimensions comparable to the "brain" fitted by Stensiö into the cranial cavity; B, smaller cylinder, equal in volume to the largest expected brain size, which is 0.25 ml for a living fish of the body size of *Kujdanowiaspis*; C, cylinder of appropriate length and width of a brain that is also within the possible brain size range appropriate to a living fish the size of *Kujdanowiaspis*; D, elasmobranch brain (the skate, *Raja* sp.) scaled down to about 0.25 ml in volume; E, specialized buffalo fish brain of *Carpiodes velifer* (Fig. 5.6A) scaled down to about 0.25 ml volume to indicate that very unusually shaped brains could have fit into the endocast. Cranial nerves; FB, forebrain; T, midbrain or tectum; CL, cerebellum; VL, vagal lobe; and M, medulla are labeled in D and E.

cylinder. That 3:1 ratio is approximately the one in fish brains generally, as can be seen by the final constructions (Figs. 5.3D and 5.3E) in which scaled down brains of a ray (*Raja* sp.) and of a specialized teleost (the buffalo fish, *Carpiodes velifer*) are placed inside the basic cavity.

This pictorial analysis demonstrates that the endocranial cavity in the placoderm *Kujdanowiaspis* was unlikely to have provided a particularly good mold of the brain, and one could have put brains of appropriate size in that cavity, even if the brains were notably specialized by having enlarged vagal lobes, cerebellum, or forebrain. We must reject Jarvik's claim, at least in the instance of this placoderm, about the adequacy of the molding of the brain by the endocranium.

Macropetalichthys AND ITS BRAIN

The second placoderm genus in which the endocast and brain were reconstructed by Stensiö was *Macropetalichthys,* in which the shield, or head armor, was about 23-cm long. As another arthrodire, this fish, too, would have had its armor covering from a half to a third of its body, and it must have been between 46- and 69-cm long. Applying Eq. (5.1) to a fish of this size, we estimate a minimum body weight of 2 kg and a maximum of 5 kg. We may calculate the maximum expected brain size as before from Eq. (5.2), and this leads to an expected maximum brain volume of 2.1 ml.

Stensiö's reconstruction of the endocast and brain of *Macropetalichthys* is shown in lateral view (Fig. 5.4) to provide a different perspective from

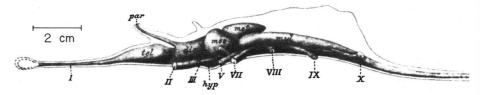

Fig. 5.4. Reconstruction of brain of *Macropetalichthys* placed in outline sketch of endocast; *par,* parietal opening for pineal; *hyp,* hypophysis; other labels as in Figs. 4.1 and 5.3. From Stensiö (1925).

the earlier reconstructions. As before, his reconstruction of the brain was guided by the entries of the cranial nerves, which are accurately pictured in a good endocast. The critical question is how accurately does the reconstructed brain reflect the volume of the brain of the animal when it was alive. We might note that this species was several times as large as *Kujdanowiaspis,* and, if a uniformitarian hypothesis were to hold, we would expect the brain to fill an even smaller portion of the endocranial cavity. We would expect Stensiö's reconstruction, which again fits the brain rather

Brains and Bodies of Fish

tightly into the endocast, to be even less adequate than that for *Kujdanowiaspis* with respect to size and to have an endocranial cavity relatively larger than the brain in *Macropetalichthys*.

The quantitative estimates follow the same modeling procedure used before. The endocranial cavity of *Macropetalichthys* is represented by a cylinder with a radius of 1 cm and a length of 8.1 cm. Stensiö's reconstruction of its brain could be represented by a cylinder of 0.9-cm radius and 7.9-cm length, almost the same sizes as the endocasts shown in Fig. 5.4. These data lead to 25.5 ml for the endocranial volume and 21 ml for the brain volume. Stensiö's reconstruction of the brain is apparently at least 10 times too voluminous if *Macropetalichthys* was no more developed with respect to its brain than a modern fish, because a typical modern fish of its size would have had a 2.1-ml brain. The lower limit of its expected brain size, that is, the expected brain size of a modern fish weighing 2 kg rather than 5 kg, is only 1 ml; the upper and lower limits are pictured in Fig. 5.5. The overestimation in Stensiö's reconstruction is twentyfold with respect to the lower limit of expected brain size.

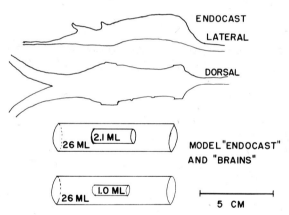

Fig. 5.5. Outline of *Macropetalichthys* endocast in lateral and dorsal view, indicating limits of expected brain size by inscribed cylinders (see Fig. 5.4).

The differences between the "errors" in the large *Macropetalichthys* and the small *Kujdanowiaspis* are consistent with our expectations. The "error" of a ten- to twentyfold overestimation in brain volume in the large genus is greater than the "error" of a five- to fourteenfold overestimation in the small fish, assuming the validity of the method used to get correct brain size estimates in lower vertebrates. *Macropetalichthys*, as a large fish, should have had proportionately less of its endocranial cavity filled than did *Kujdanowiaspis*, a small fish.

Shape of the Placoderm Brain

In view of the gross difference in volume between brain and endocast in these primitive jawed fish, the justifiable inferences about the shapes of the brains of these animals are limited. As indicated earlier, the positions of their cranial nerves are normal and comparable to those of living fish and, in fact, to all living vertebrates, including man. Differences between these ancient creatures and modern animals with respect to cranial nerves cannot be understood as associated with a primitiveness–progressiveness dimension. They are no more marked than differences among existing progressive species, and the placoderms are witness to the general proposition that the cranial and spinal nerves of animals changed only with respect to emphasis or deemphasis of information channels appropriate to the adaptive niche occupied by an animal. Their basic pattern was fixed with the evolution of jaws and the neural apparatus of the gnathostomes.

In one respect the placoderm brain is specialized, and perhaps primitive, and this may be seen most adequately in the lateral view of the brain of *Macropetalichthys*. There was a clear channel for a parietal, or pineal, eye, a condition that remains today in the cyclostomes, in *Sphenodon* (tuatara), the relict reptile of New Zealand, and in lizards. There are significant remnants of this organ, in a probably nonfunctional state, in many other living lower vertebrates (Edinger, 1955b). The size of the parietal opening suggests a functioning organ in at least some placoderms.

Any special suggestions about the shape of the major divisions of the brain in placoderms (Figs. 5.1, 5.4), in which forebrain, midbrain, and hindbrain were modeled, seem to me an uncertain exercise. I have tried to show that very unusually shaped brains could be squeezed into the available space if the brains were of the expected size (Fig. 5.3E). But since there is absolutely no hint in the endocasts that the brains they contained were other than normal, the most likely shapes are certainly those suggested by Stensiö in his reconstructions if they are scaled down considerably. We would do best to accept a null hypothesis, as it were, in which we assume that the placoderm brain was not different from the least-differentiated brain of a jawed vertebrate of today. Since the placoderms lived in a fish's niche, the best choice is the elasmobranch-type brain depicted by Stensiö but reduced as in Fig. 5.3D to conform to the expected brain size.

BRAIN EVOLUTION IN LIVING CLASSES OF FISH

There is essentially no fossil evidence on the evolution of the brain in cartilaginous fish that has been analyzed in a way analogous to that of the

Brain Evolution in Living Classes of Fish

placoderms or, as will be discussed presently, of the bony fish. Stensiö considers the placoderms as "Elasmobranchomorphs" and would apply all the preceding analysis to the evolution of the brain in sharks, rays, and chimeras. This is, of course, reasonable in view of the consistency of the data with the assumption that the fossil placoderms had brains similar in shape to the brains of modern elasmobranches of the same body size. The only fossil evidence relevant to this issue comes from the fossilized crania of a few Carboniferous sharks that are incompletely described (Romer, 1964; Gross, 1937). Obvious difficulties lie in the way of analysis of this class of fish because of the rarity of petrification of cartilage. From new evidence on living sharks and rays (Ebbesson and Northcutt, in press), one should expect the fossil forms to have relatively larger brains and endocasts than those of other classes of fish.

EVIDENCE ON CHONDRICHTHYES

The work of Ebbesson and Northcutt (in press) has extraordinary implications for the analysis of the evolution of the vertebrates and the role of brain size in evolution. Almost all authorities have considered the brains of cartilaginous fish as generalized and close to a primitive condition, and this was reflected in the use of such brains as models for the brain in extinct placoderms (Figs. 5.1 and 5.3D). Although there is no evidence of unusual specialization in elasmobranch brains, the fact that these are presently larger than those of (other) lower vertebrates is important for our analysis. We must recognize, however, that we know nothing about the history of the enlargement of the brain in cartilaginous fish, and it is possible that the enlargement occurred as part of later evolution of sharks and their relatives rather than in the early adaptive radiation of the group.

There must have been a significant selective advantage for the enlarged brain in cartilaginous fish, since evolution is hardly so erratic as to permit the appearance and persistence of so major a deviation from the typical pattern of "lower" vertebrates presented in Chapter 2. The cartilaginous fish have persisted relatively unchanged since their earliest known Devonian fossils, although advances in grade have been identified, and their greatest adaptive radiation was probably in the later Mesozoic and the Cenozoic eras (Schaeffer, 1967). Their braincases, to the extent that these cartilaginous "skulls" have been recovered and analyzed in fossils, are among the more conservative features of the group. In fossil and Recent species, major identifying structures such as the entry points of cranial nerves are similar. One would, therefore, guess that the brain became enlarged early in the evolution of this class of vertebrates and that the niches found by the earliest species remained viable for animals organized as they are.

Another implication of the new evidence is for the correctness of

Stensiö's reconstructions of placoderms. Although I believe that my criticisms of these reconstructions are valid, and that the best models for their brains with respect to size are from the bony fish, formulas other than Eq. (5.2) would have to be used to estimate their brain size if Stensiö is correct in associating placoderms with elasmobranches.

There is a more general issue that should be considered. How should we think about the dichotomy between "lower" and "higher" vertebrates, and what does this imply for the analysis of the cartilaginous fish as a major vertebrate class? The dichotomy is heuristic in enabling us to classify and remember most of the information we now have about brain size. It suggests, also, a fundamental aspect of brain evolution, namely, the conservatism of brain size as an evolutionary trait. Thus, the facts presented in Chapters 5–8, that the extensive adaptive radiation of bony fish, reptiles, and amphibians could be accomplished without major changes in the relative size of the brain, indicate that brain size did not typically change much even when many different niches were invaded and occupied successfully. In these vertebrate classes there were no adaptations of the brain for "intelligence," that is, for plasticity in the face of a changing environment.

In this context I am most comfortable with the idea that the adaptive zones of elasmobranches did involve some advantage for plasticity and concomitant enlargement of the brain. The correct analogy may be with the history of the enlargement of the brain in predator species of large land mammals (Chapters 13 and 14).

It is, perhaps, unfortunate that the sharks and their relatives are commonly thought of as merely another group of "fish." We should emphasize that they are a separate class of vertebrates, different from jawless and bony fish in the same sense that they are different from birds or mammals. Considering them as intermediate between "lower" and "higher" vertebrates with respect to relative brain size is then easily accepted as a fact of natural history. We can dispense with the myth of the "primitiveness" of this very successful class of vertebrates, whose success is evident from the persistence of basic adaptations for some 400 m.y. They are primitive only in the sense that they apparently found their niches early in the history of the life of vertebrates.

Brains of Osteichthyes

Although the bony fish are not at the same level as the sharks and rays with respect to brain evolution, there is more diversity of specialization in the brains of this group. Modern bony fish exist in a great variety of adaptive niches, and, following the general vertebrate rule, their adaptive radiation with respect to habit patterns is reflected in the morphology of their brains (Aronson, 1963; Evans, 1940; Marshall, 1966). Nevertheless, the gross size of the bony fish's brain as a function of body size is consistent with the interpretation that they, and all lower vertebrates, were at the same level of brain evolution, although the directions of specialization within that level may have differed (Fig. 2.4).

Modern bony fish that are bottom feeders generally have swellings on their brains in those regions (facial and vagal nerve nuclei in the medulla) to which their taste cells project (Fig. 5.6A). This reflects the role of taste receptors and analyzers in discriminating the edible from the inedible as they sift through river, estuary, or ocean bottoms. Fish that use olfactory cues extensively are characterized by moderate forebrain development because their forebrain is an olfactory integrative organ (Fig. 5.6B). Visually oriented fish such as trout or pike have enlarged mesencephalic optic lobes (Fig. 5.6C). A particularly unusual specialization in bony fish occurs in mormyrids (Fig. 5.6D), in which a massive growth of the valvula of the cerebellum completely overshadows other features of the brain, with this region appearing almost like cerebral cortex to the uncritical eye. This development is probably associated with the mormyrid specialization of electric organs as senders of social information and the specialization of some lateral-line system cells (mormyromasts) as receiving organs (Lissmann, 1958; Marshall, 1966, pp. 155–163). There is a massive input to the cerebellum in these fish from the seventh and eighth cranial nerves, which are involved with the lateral line and labyrinthine systems.

Modern bony fish are, thus, highly differentiated and specialized with respect to brain development, even though it is not necessary to think of them as operating at a higher level of integration than other lower vertebrates. The anatomy of the forebrain of the bony fish is different in some respects from that of other vertebrates in that it is embryologically "everted" rather than "inverted" (Nieuwenhuys, 1962, 1966). Although this results in smaller incomplete ventricles, there are no quantitative studies that would indicate a difference in information-processing capacity relative to other vertebrate brains. It is likely that the relationship between the amount of neuronal versus nonneuronal tissue in the brain is different in animals with everted brains, and it may affect the kind of inferences that can be made from brain size about level of brain evolution in bony fish. I will assume, however, that the effect would be small because the largest fraction of neuronal tissue is probably in mid- and hindbrain structures.

Fossil Brains in Osteichthyes: Paleonisciformes

Endocasts (Fig. 5.7) that are clearly "fossil brains" have been retrieved from bony fish. They mirror the external surfaces of a brain, and there is no need for fanciful constructions to guess at the appearance of the brain of the original fossil. These endocasts mirror brains that look modern. From their

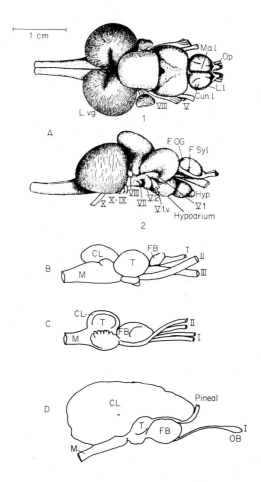

Fig. 5.6. Sketches of brains of living fish to indicate a variety of specializations reflected in the configuration of the brain. A, Buffalo fish *Carpiodes velifer,* dorsal (1) and lateral (2) views, specialized for chemical sensitivity by expanded vagal lobe of medulla, from Herrick (1905). Lateral views of other species: B, bullhead catfish *Ameiurus nebulosus,* a bottom fish with olfactory specialization reflected in expanded forebrain, after Polyak (1957). C, Northern pike *Esox lucius,* a visually specialized surface fish, the visual habits of which are reflected in expanded optic lobes (midbrain), after Polyak (1957). D, *Mormyrus kanume* with expanded "cerebellum" or electric lobe (CL) so large that it appears comparable to the mammalian cerebrum—specialized for electrical sensitivity, after Franz (1911). Scale uncertain for *Mormyrus.*

morphology they could have come from existing ganoids or teleosts rather than from paleoniscids of 300 m.y. ago. The outstanding characteristics of the known endocasts are the apparent expansion of the optic lobes, some

Brain Evolution in Living Classes of Fish 113

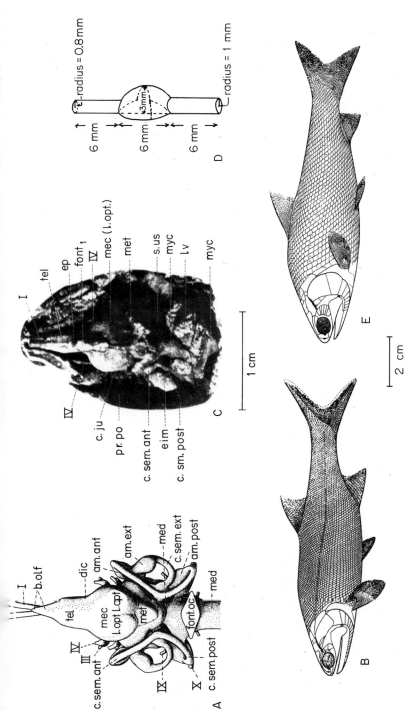

Fig. 5.7. Endocasts of small bony fish of 200–300 m.y. ago, with analysis of endocast volume and illustrations of their bodies. Endocasts and bodies to different scales, as indicated. A, Endocast of *Pteronisculus stensiöi*, from Stensiö (1963); B, body of *P. cicatrosus*, from Piveteau (1966); C, *Boreosomus gillioti* endocast, from Beltan (1957); D, model of endocast of *B. gillioti* used to estimate its volume; E, body of *B. gillioti*, from Piveteau (1966).

increase in the size of the cerebellum, and the appearance of medullary structures that may be either vagal or acoustic tubercles. The major characteristics of the contemporary fish brain appear to have been established early in the phylogeny of the fish.

A quantitative analysis of the size of the fossil endocasts (or brains) of the paleoniscids relative to their bodies follows the same procedures used with placoderms. The question is somewhat different now because the fossil is a natural endocranial cast that is clearly brainlike in shape. If it can be shown that it is also brainlike in size, the endocast can be analyzed with respect to its morphology as if it represented a brain.

All the data developed in the analysis are shown in Fig. 5.7, and the analysis is demonstrated with the one specimen for which good data on body size are available. The recovered specimen of *Boreosomus gillioti* was about 12 cm long, which from Eq. (5.1) suggests an animal weighing about 65 g. Because of expansion of the optic lobes in the endocast relative to the forebrain and hindbrain, a cylindrical model is not quite appropriate to estimate the volume of this structure from a two-dimensional figure. Instead it was modeled with cylinders for forebrain and hindbrain and a hemisphere for the midbrain. (In later chapters graphic double integration is used; it gives similar results.) The dimensions of these segments of the model of a brain were chosen to approximate those of the pictured endocast, and they were, of course, modeled independently of the determination of "expected" brain size for a 65-g body in lower vertebrates. The volume of the geometrical "model" brain is 0.09 ml. The expected volume of a brain of a modern 65-g fish, according to Eq. (5.2), is 0.11 ml. Thus, the volume of the endocast of *B. gillioti* is consistent with the volume for a modern fish of the same body size.

This gratifying result suggests that at least one fossil fish endocast can give information about the external configuration of the brain at an early stage of the evolution of the bony fish. *Boreosomus gillioti* was a paleoniscid, a bony fish of Devonian period vintage, considered to be ancestral to living bony fish, with the exception of the lungfish and *Latimeria*. The evidence that the configuration of its brain was similar to that of modern fish and that its size was consistent with a modern "lower vertebrate" level of brain evolution implies an important generalization for the evolution of fish and their brains. The adaptations of the fish brain that can be studied by the methods of comparative anatomy may be assumed to be no different from those suggested by the evolutionary fossil record. This reinforces the proposition that early in their evolution the lower vertebrates had invaded the basic niches open to them with a set of behavioral adaptations controlled by neural systems in the same general way they are controlled today. From the evidence of that early period in their evolution, some 300 m.y. ago, one could assume that few new approaches were taken by fish to solve the problem of neural control of behavioral adaptation to niches.

In the paleoniscids whose fossil brains have been recovered the typical adaptation appears to have been to use visual information; this was associated with the remarkable expansion of the optic lobes of the midbrain, as in a modern "visual" fish such as the pike (Fig. 5.6C). It takes little imagination to see comparable adaptations in yet undiscovered paleoniscids with respect to myelencephalic and cerebellar structures and even with the forebrain as an olfactory control system. Thus, the types of behavioral adaptations seen in modern fish (Baerends, 1957; Marshall, 1966, Chapter 13) could have appeared early in the evolution of these animals because some of the necessary associated brain specializations (e.g., development of an enlarged optic lobe) are known to have occurred. Beyond this, little of evolutionary significance about behavior can be added from the fossil record of the brain, and one must turn to the analysis of affinities and relationships among existing fish with respect to behavior patterns to trace more detailed phylogenies.

Another important implication of this analysis is that selection pressures among bony fish since the time of *B. gillioti* have not been toward the further expansion of their brains. This agrees with the result of the comparative quantitative analysis of relative brain size presented in Chapter 2. If fish, amphibians, and reptiles can all be subsumed under a single function relating brain to body size, the first guess would be that no special evolution of the brain was favored within or between any of these groups. Put more conservatively, this points to the proposition that the niches invaded by these lower vertebrates did not demand sufficiently new or different behavioral adaptations to be reflected in expansion of enough of the brain to appear in the gross brain:body analysis. In addition this implies that most present adaptations are probably of great antiquity, since a 300-m.y.-old fossil presents as specialized a picture of the brain as does a modern fish.

POTENTIAL FOR LIFE ON LAND

The rapid vertebrate radiation of 400 m.y. ago in the Devonian period included the evolution of forms that were "preadapted" for a life on land. The bony fish had lungs, as did at least some placoderms (Carter, 1967, p. 80). In their present descendants these lungs have generally disappeared; they are believed to have evolved into the swim bladder, which is normally a hydrostatic organ (Marshall, 1966). One major group of Devonian bony fish, the lobe-finned fish of the subclass Sarcopterygii (Choanichthyes), were prepared skeletally, as well as by their lungs, to move about on land. Their fins were reasonably sturdy as limbs, and it was from the order Crossopterygii in this group that the class Amphibia evolved.

The Devonian period was marked by seasonal droughts in which lakes and streams became partially dry, and forms that had functional lungs could survive during the short move over land from one pool of water to another. There was some advantage, also, in strong limbs that could aid movement and not force dependence on a slithering crawl that was probably not to be evolved efficiently for land locomotion until the appearance of snakes. The preadapted crossopterygians included the rhipidistians, from which evolved the closely related ancestors of the amphibians and the coelacanths.

Romer (1937) has described the endocranial cavity of *Ectosteorhachis* (*"Megalichthys"*) *nitidus,* a late Carboniferous rhipidistian of about 300 m.y. ago. If the brains of these groups had evolved during the 100 m.y. between their fathering of the earliest amphibians and the appearance of the particular *E. nitidus* that Romer studied, then *E. nitidus* would not be quite an appropriate model of the ancestral preamphibian preadaptation of the fish. However, the general conclusion from this survey of the brains of fossil animals is that the lower vertebrates as a group do not show major advances with respect to level of brain evolution; the occasional differences among them are best expressed as directional differences in adaptation within a "lower vertebrate" level. The only problem in the use of this specimen as a key to an evolutionary progression would arise if there were specialized adaptations of the brain reflected in the endocast of *E. nitidus*. We will see that this was not the case.

In addition to this close relative of the amphibians, the persistence of one coelacanth, *Latimeria,* to present times offers an additional and unusual opportunity to study the anatomy of the brain of a living member of this group. *Latimeria* is specialized in several ways compared to Devonian coelacanths, however, probably with regard to habit as well as structure. Perhaps most importantly, it is a very large (40 kg or more) deep-sea fish rather than a freshwater form. It is a predator, apparently, living on other fish. The ancestral forms were freshwater fish and considerably smaller, in most instances, although giant marine forms are known from post-Devonian, especially Cretaceous, levels.

Among the lobe-finned fish the Dipnoi, or lungfish, are most distant from the amphibian line, but they, too, may be considered with profit. They not only add to the picture of the background of tetrapods but also provide information about an unusual radiation of the fish, its relationship to the external appearance and gross size of the brain, and its significance for the evolution of the brain.

Brain of *Ectosteorhachis nitidus*, a Relative of Ancestral Amphibians

The endocast of *E. nitidus* was prepared by Romer (1937) using a serial sectioning method of the same general type used so successfully by Stensiö. Romer pointed out that the endocranial cavity of *E. nitidus* was grooved, and the grooves probably molded a venous or arterial vascular system. This would suggest that the brain itself was also molded by the cavity and that the endocast of *E. nitidus* could be studied as a "fossil brain."

A volumetric analysis of the endocast (Fig. 5.8), following the same procedures used before, supports that conclusion. The endocranial volume was about 3 ml for *E. nitidus*. Romer's view of its brain, shown inside the endocast in lateral view, has a volume of about 1.9 ml when modeled by a set of cylinders. I could find no precise data on the length of *E. nitidus*, but Romer (1966) indicates the lengths of fish in that group and from that (Carboniferous) stratum as between 1 and 2 ft. These lengths (30 and 61 cm) lead to body weight estimates of 660 and 4000 g, respectively, according to Eq. (5.1). The expected brain sizes for fish of these body weights, according to Eq. (5.2), are 0.5 ml and 1.8 ml. The expected ranges of the maximum brain weight, estimated by using the known ranges of modern fish weighing 660 and 4000 g, as indicated in Fig. 5.2, are 0.25—1.4 ml for the smaller size and 0.52—4.2 ml for the larger size. The smallest modern fish that would be expected to have a 1.9 ml brain would weigh 1 kg and be about 37-cm (14.5-inches) long. The 1.9 ml brain constructed by Romer for *E. nitidus* is, therefore, a reasonable reconstruction with respect to size.

The outstanding feature about the brain of *E. nitidus* in Romer's reconstruction is the absence of specialization. Unlike the paleoniscids described earlier, with their expanded optic lobes, *E. nitidus* may suitably be outfitted with a very generalized lower vertebrate brain of the type used to illustrate comparative anatomy texts. No structure is unusually expanded, and the ridge of tissue dorsal to the medulla is identified as chorioid plexus, analogous to that in contemporary lungfish (Romer, 1937).

There are two unusual features in Romer's reconstruction. First, and validly, it is clear that *E. nitidus* had a pineal–parietal eye which may have been functional. In this respect, a really primitive vertebrate brain may be fitted into the cavity because it is only in this regard that an unspecialized brain may be regarded as primitive. The second feature in Romer's reconstruction seems somewhat uncertain. This is in his sketch of divided "olfactory lobes" that separate and enter separate anterior cavities of the endocast. If one keeps in mind that olfactory lobes are, in fact, the telen-

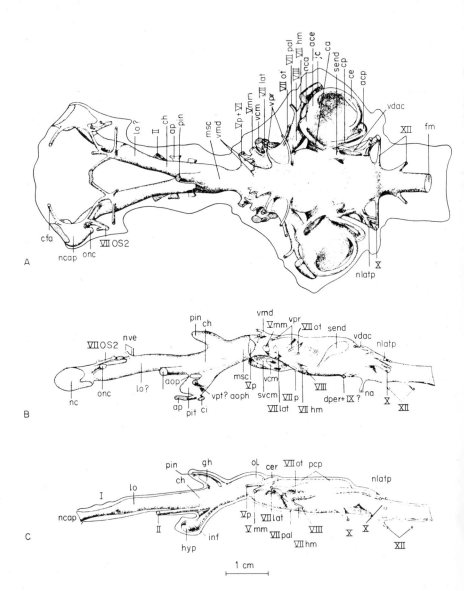

Fig. 5.8. A, Dorsal view; B, lateral view of endocast; C, lateral view of reconstructed brain, all of *Ectosteorachis nitidus;* forebrain at left. From Romer (1937).

cephalon of fish (Ariëns Kappers *et al.,* 1936, p. 1245), then this construction immediately strikes one as unusual, though not necessarily incorrect (cf. Fig. 4.3 of the lamprey brain). It seems more appropriate, however, to consider these evaginated structures as associated with the olfactory tracts and to terminate the brain at the point of their division.

One may conclude from the analysis of *E. nitidus* that the brain of bony fish ancestral to the land vertebrates did not tend toward specialized adaptations for particular niches as did those of other bony fish. The nonspecialized brain, with respect to the enlargement of special structures, may be thought of as a persistent "conservative" character in the lobe-finned fish. It persisted to present times in their descendants, including the lungfish and *Latimeria,* as shall be seen shortly, and it was characteristic of their other descendants, the amphibians and reptiles. Not until the emergence of birds and mammals from appropriate reptilian forebears did the tendency for the brain to show a species' specializations in the enlargement of superficially visible structures reappear as a common phenomenon.

The Brain of *Latimeria*

Full accounts of the superficial and internal structures of the nervous system of *Latimeria* have recently become available (Millot and Anthony, 1965); the present discussion is based on earlier statements (Millot and Anthony, 1956) and on Stensiö (1963). The first outstanding feature about the brain of *Latimeria* is its small size. It weighed about 3 g in a 40 kg specimen. According to Fig. 2.4, the expected brain size in such a fish is about 8 g, and the expected range is between 2.9 and 11.5 g. The brain of *Latimeria* is not so strikingly small when viewed in this perspective. Crile and Quiring (1940) report *Promicrops itaiara,* caught off the coast of Florida, as weighing 35 kg with a 2.3-g brain.

A second outstanding feature of the brain of *Latimeria,* which is especially important in this context in which so much reliance must be placed on endocranial casts, is the disparity already mentioned between the size of the brain and the endocranial cavity. Carter (1967) mentions the figure of 1% as the fraction of the cavity occupied by the brain, and published illustrations (see Fig. 5.9) confirm the fact that the fraction is small. Stensiö believes that this is a regressive feature in *Latimeria.* He bases his argument on the fact that the endocast of *Latimeria* is not at all brainlike and that the positions of certain channels, such as for the pituitary, are more normal in ancestral coelacanths such as *Nesides schmidti* of the Devonian period. He illustrates the endocast of *Nesides* (Fig. 5.9), which does, indeed, look brainlike, but one can neither support nor argue with his conclusions, based on this specimen, since he gives no data on absolute size of either endocast or whole fish. From other sources (Piveteau,

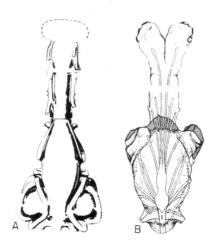

Fig. 5.9. A, Dorsal view of endocast of Devonian coelacanth *Nesides schmidti*, from Schaeffer (1969), after Stensiö (1963); B, dorsal view of endocast (brain and cranial nerves shown as dashed lines) of living coelacanth *Latimeria chalumnae*, from Schaeffer (1969), after Millot and Anthony (1965). Note the small size of the brain in the endocast, and compare with Figs. 5.3 and 5.5.

1966, p. 306) it seems clear that *Nesides* was a very small coelacanth, perhaps no more than 10- or 15-cm long. From general experience (Aronson, 1963) one expects small fish to have brains and endocasts that are more comparable in size and appearance than is the case in large fish. Stensiö may have been misled by not taking into account absolute size in interpreting morphological data.

The greatest interest in the *Latimeria* brain is, of course, in the information available from its internal anatomy and the correlation between internal and external anatomy—phylogenetically important under a uniformitarian hypothesis. Interest in *Latimeria* as a fish is exactly analogous to interest in *Sphenodon* as a relict reptile and in insectivores and opossums as relict mammals. The main difference is that *Latimeria* is a relict largely in its membership in the order Crossopterygii, which had been thought extinct. But it is clearly a rather specialized descendant of the order. Information about its brain must, therefore, be used with care. Any unusual structures or unusual arrangement of structures should be considered, first, as possible results of specialization, and, if that idea can be discarded, we can accept them as the primitive condition.

Present reports are that the brain of *Latimeria*, like that of *E. nitidus*, is a generalized "early vertebrate" brain. There are no remarkably expanded structures, and the only unusual feature appears to be the elongation of the olfactory bulbs and the infundibulum. Both effects in this very big fish are readily understandable in terms of the distance between the main body of the brain and the pituitary fossa at the base of the endocranium on one hand and the olfactory epithelium in the snout on the other. In short, these adaptations are associated with the increase in body size in *Latimeria* and the requirement that the brain "service" the body's organs. The pineal "eye" of *Latimeria* does not penetrate the skull although it does enter the fatty tissue that acts in place of skull as the immediate envelope of the brain.

THE DIPNOAN BRAIN

I will conclude with a brief review of the brain of lungfish. Some lungfish of present times attain sizes comparable to that of *Latimeria*. We can first consider the Australian lungfish *Neoceratodus forsteri* (Fig. 5.10),

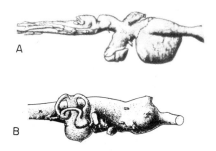

Fig. 5.10. A, Lateral view of endocast of living lungfish *Neoceratodus forsteri;* B, lateral view of fossil of 350 m.y. ago or more, *Chirodipterus wildungensis*. From Schaeffer (1969), after Stensiö (1963); forebrain at right, not to scale.

as described and figured by Bing and Burckhardt (1905), Säve-Söderbergh (1952), and Stensiö (1963). The special feature of the brain is its non-specialized appearance. The outstanding specialization is an enlarged forebrain. Also important is the disparity in size of the brain and endocranial cavity. The brain is only about 10% as big as the endocast, and although this does not seem as extreme as the case of *Latimeria* with its hundredfold difference, it is still large enough to make it difficult, though not impossible, to understand the brain's anatomy from an endocranial cast. The saving feature in the dipnoan endocast is the fact that the forebrain is rather tightly fitted into the braincase, and that structure may, therefore, be interpretable from the endocast.

Fossil Dipnoi are represented by a carefully studied specimen, *Chirodipterus wildungensis*. Stensiö (1963) presents a reconstruction of its endocast within an outline of its head, with other versions of the endocast of *Neoceratodus,* the living Australian lungfish (Fig. 5.10). There is a remarkable similarity in the shapes of the endocranial cavities of these fish, despite the great difference in their sizes. The major difference in shape of the brain, a difference emphasizing the primitiveness of *Chirodipterus,* is the presence of an opening for the pineal eye in the Devonian fossil. If the brain of *Chirodipterus* fits tightly into its braincase, there is some suggestion of enlargement of its forebrain.

CONCLUSIONS

The ray-finned bony fish, Actinopterygii, which are the dominant group of the present time, are the only fish that show obvious special adaptations in their brains. These adaptations are in accordance with observed habit patterns and are displayed, as pointed out earlier, in superficially visible enlargements of parts of the brain. Such enlargements may include, singly or in combination, the facial, acoustic, and vagal lobes of the medulla, the optic lobe of the midbrain, the valvula cerebelli, and the forebrain structures associated with olfaction. The only fossil actinopterygians that have been studied with respect to endocranial anatomy are members of the most primitive order of paleoniscids, which is now entirely extinct. It is, therefore, notable that this group was uniquely characterized by an enlargement of an appropriate portion of the brain, in this case, the optic lobes of the mesencephalon. The actinopterygians as a group responded to selection pressures by selective enlargement of parts of the brain that enabled a species to occupy an adaptive niche with special success. Among the remaining forms there appears to be no similar special adaptation. To an extent, dipnoans may be thought of as having somewhat enlarged forebrains. Thus, there was some tendency for development of visible signs in the brains in this case, presumably associated with the use of olfactory information in a special way. However, it does not strike one as forcefully as the ray-finned fish adaptation of the brain to behavior.

The crossopterygians, from which the land vertebrates evolved, are almost peculiar, both among primitive forms and in the unique "living fossil," *Latimeria,* in the degree to which their brains may be described as following a generalized and primitive vertebrate pattern. That pattern persisted in the lower vertebrates that descended from these groups, the amphibians and the reptiles, as is seen in the next chapters.

The possibility that there was some specialization of the closest contemporary relatives of the crossopterygian ancestors of the amphibians, that is, the dipnoans, toward forebrain development may be a matter of some interest and a basis for speculation. One may properly imagine the vertebrate brain as responding to selection pressures for adaptive behavior

Conclusions

in unusual niches. The adaptation chosen might differ in different groups depending on the characteristic capacities for evolution in the groups. The presently most successful fish, the Actinopterygii (and especially the teleosts among these), are characterized by an adaptive radiation visible in structures of the brain, in which various midbrain and hindbrain structures develop to unusual extents.

Unlike the bony fish, some sharks and rays presently have brains significantly larger than those of (other) lower vertebrates. The history of the brain in cartilaginous fish is unknown, making it difficult to judge the evolutionary significance of this fact. One should, perhaps, consider the cartilaginous fish, class Chondrichthyes, as unique vertebrates with respect to brain evolution, differentiating them from "true" fish as the taxonomist differentiates Chondrichthyes from Mammalia on the one hand and Osteichthyes on the other. From the point of view of the analysis of relative brain size the cartilaginous fish form a class of vertebrates intermediate between "lower" and "higher" vertebrates.

Chapter 6

Invasion of the Land: The First Tetrapods

As a dramatic event in evolutionary history, the vertebrate invasion of terrestrial niches is surely unsurpassed. It is documented in the successive geological strata of Devonian and Carboniferous times. The earliest amphibian remains that have been recovered are about 365 m.y. old, and complete fossil skeletons have been retrieved from 350 m.y. old strata (Fig. 6.1). Skeletal similarities between the late Devonian *Ichthyostega* and its crossopterygian relatives are striking, and it is clear that the lobe-finned fish were elaborately preadapted for the amphibian role.

ADAPTIVE RADIATION

Our perspective in time makes us dramatize the invasion of the land as a previously unoccupied zone, because we know how extensive the radiation of the tetrapods was. It is easy, however, to exaggerate the importance of radical systemic changes in some of the products of that radiation, such as birds and mammals, as if such changes were always necessary for the successful invasion of the terrestrial niche. Such thinking leads to orthogenetic traps, in which one might assume that brains, for example, must become bigger and better to enable animals to survive in radically new niches. Actually, very little really had to happen to their brains to enable crossopterygian fish to evolve toward, and eventually cross, the fish–amphibian boundary.

Consider the analogy of artificial adaptations that have been developed in present explorations of space. The "evolution" of these man-made adaptive systems, including artificial intelligence, enables men to survive in hostile and strange extraterrestrial environments. An astronaut in orbit is encapsulated in a spaceship (and suit) that keeps him in an essentially earth-surface condition and shields him from the harsh environment that he has invaded. Major constructions are required to maintain the barrier between the astronaut and the vacuum of space. These are analogous to the skins of vertebrates and add up to a kind of second skin; their development corresponds to the evolution of scales, feathers, hair, and skin and fatty, muscular, and skeletal supports. Of course, a man in a space capsule

Adaptive Radiation 125

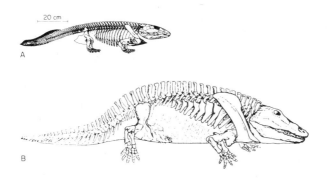

Fig. 6.1. Reconstructions of fossil amphibians, approximately to scale. A, *Ichthyostega,* the oldest known complete amphibian skeleton (late Devonian, from Greenland). B, *Eryops* (late Permo-Carboniferous, Texas Red Beds). These large amphibians were between 1 m and 2 m in length. Reprinted from "Vertebrate Paleontology," 3rd ed., by A. S. Romer, The University of Chicago Press, Chicago, Illinois, copyright 1966.

is not quite as self-contained as a man on earth. In space he is part of a larger system maintained by an elaborate communications and control network which can control and guide him and his capsule through prescribed paths in their environment. This is a "brain:body" system, as it were, extended through space to include the capsule's control system, the astronaut's hands and brains, computers at the surface of the earth, and teams of technologists whose brains also act as parts of the system. But, this elaborate "brain:body" system is necessary only because of the major energy expenditures and unusual control problems that are involved in placing a capsule into varying orbits about the earth or the moon or into trajectories for ascent and descent. If the paths were simpler, elaborate control systems would not be necessary for the simple task of invading and existing in the new environment.

The analogy is especially apt for one unique feature of the exploration of space, human adaptation to zero-gravity existence. Astronauts can maneuver themselves for significant periods of time in free space as "almost-free" vertebrates in that unusual environment. There have also been numerous experiments on brief periods of exposure to zero gravity by many men less encumbered by second skins, during airplane flights in a parabolic trajectory, which produces a zero-gravity field with respect to the airplane cabin. Films of the motions of men on zero-gravity flights and reports by the astronauts of their space walks show that men adapt quickly and easily to the problems of motion through space with zero gravity. They do it by "swimming" through space, moving their arms, legs, and torsos in a way somewhat like normal swimming, but actually taking advantage of an intuition about mechanical systems in which Newton's third law is exploited. They also move through space by using mechanical control devices such as small reaction motors, and they quickly learn to use these devices without special computers (Henry, 1966). This demonstrates that there is very adequate preadaptation of the human neuromuscular control system for maneuvering the human body through free space. No special computers have to be used, that is, no new "artificial intelligence" has to be evolved, to enable men to cope with this environment.

This digression into outer space can be brought back to earth if we imagine a member of a species that ichthyologists living in the Devonian period might have classified as a crossopterygian fish and that hindsight permits one to assign to the class Amphibia as opposed to the class Osteichthyes. Consider this animal's task in moving over land, and compare it to its motion as a fish in water (Fig. 6.2). The movements are controlled at a spinal level but are, nevertheless, a useful basis for thinking about the evolution of nervous control in the face of new environmental challenges. Sir James Gray has described the gross motion of fish as follows:

> As observed by the human eye, the motions of various types of fish appear to vary considerably from one species to another. At one extreme is the eel, which, during motion, is characterized by distinct waves of curvature which pass alternately down each side of the body from head to tail. At the other extreme is the mackerel or trout which appears to progress by means of transverse strokes of the expanded caudal fin. An examination of successive instantaneous photographs shows, however, that the nature of these two types is essentially the same, for in all cases, waves of curvature pass along the body with increasing amplitude as the hind end of the fish is approached.[1]

The movement of modern amphibians is really not radically different from the above description (Coghill, 1929); Watson (1951) believes that the tracks of fossil amphibians preserved in coal deposits of Czechoslovakia suggest very similar locomotion in the ancestral forms. It should be kept in mind that many early amphibians were large animals, 2-m long or more and probably weighing 70 kg or more. The problems of moving such a weight over the surface of the earth are clearly greater than simple swimming, in which larger bodies can be advantageous—big marine animals can swim much faster than their small relatives (Brainbridge, 1958).

The distinguished British paleontologist D. M. S. Watson described the probable method of locomotion of the 5-m long Carboniferous amphibian *Pteroplax* (*"Eogyrinus"*), using the tracks of its belly and feet preserved in the Czech coal measures (Fig. 6.2).

---

[1] Reprinted from "The Life of Fishes" by N. B. Marshall, Universe Books, New York, 1972, and Weidenfeld & Nicholson, Inc., 1966, p. 13.

Adaptive Radiation

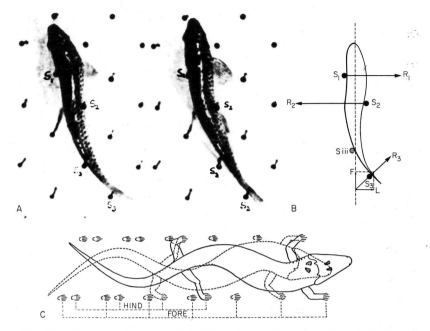

Fig. 6.2. Patterns of movement in fish and amphibians. A, The undulations in fish normally push against somewhat resistant water, with the resultant movement in a direction opposite to the force of thrust of the body. Pegs (S_1, S_2, S_3) serve the same function as water resistance, letting fish "swim" on land. B, The propulsive force (F) is the forward component of the pressure (R_3), which the peg S_3 exerts against the caudal fin; the lateral component (L) is equal but opposite to the resultant of the forces (R_1, R_2), which the pegs S_1 and S_2 exert against the front end of the body. C, Legs in an early amphibian served as pegs to replace the resistance of water, permitting body undulations to result in forward motion. The amphibian is *Pteroplax,* and its movement pattern was reconstructed from fossil tracks. Fish from Gray (1953) and reprinted from "Animal Locomotion" by Gray, Universe Books, New York, and Weidenfeld & Nicholson, Inc., London, 1968; *Pteroplax* from Watson (1951).

> The limbs are designed to support the body but it is evident that so long and flexible a trunk cannot have been carried free from the ground; the limbs themselves are too short and weak, and the pelvis is too feebly attached to the backbone, to have enabled the creature to move on land in any other way than by wriggling with the belly resting on the ground, as do many elongated living amphibia of similar shape though of much smaller size. In such a form of locomotion the hands and feet merely act as fixed turning points about which the animal moves. They have separate toes to give a more secure grip, and the limb bones and muscles need only be strong enough to act as struts, connecting the moving body to the fixed track [Watson, 1951, p. 85].

The control of movement in amphibians could, therefore, have involved about the same neural apparatus that controlled movement in fish. Effectively, the amphibians really "swam" on land. Analogously to the absence of a "selection pressure" for artificial intelligence in free-space maneuvers, no peculiar evolution of the nervous system was necessary to accommodate the adaptations that actually occurred in the invasion of the land niche by these early vertebrates. When systems did evolve for innervating the limbs to permit walking, they were directly related to swimming movements (Coghill, 1929, p. 29). With this background no one should be surprised at the descriptions of the fossil brains of two Carboniferous amphibians, *Edops* and *Eryops,* to be presented shortly. The extraordinary thing about the endocasts is their ordinariness.

Before describing the brains and bodies of these two forms I will present a brief summary of vertebrate evolution across the land barrier.

From Lobe-Fins to Amphibians

There are radical differences in the soft anatomy of lungfish and salamanders, differences that must have evolved at some time in their radiations (Carter, 1967). But according to the fossil record, much less extensive changes had occurred at the time of the transformation of Devonian crossopterygians into the ancestors of modern amphibians. The differences among the modern forms are, therefore, attributable mainly to subsequent evolution within each group rather than to the initial conditions associated with the invasion of partially or fully terrestrial niches.

In the amphibians, as in the jawless fish, the number of modern species is much reduced compared to their earlier extensive radiation, but unlike the agnathans, there are no modern amphibians that are good representatives of ancestral skeletal adaptations. The anurans (frogs and toads) are particularly unsuitable representatives because they are highly specialized in their niches and relatively late in their branching within the amphibian line. The urodeles (salamanders), though more like the ancestral forms, are also specialized in being degenerate with respect to both niche and size. Most urodeles are small, shy animals and hardly evoke the image of their ancestors, which were the dominant land vertebrates and major predators on the freshwater life of their times. The original invasion of the land was by forms that looked very much like their crossopterygian ancestors, even retaining a tail fin. *Ichthyostega* (Fig. 6.1), which was preserved as an entire skeleton, lived at the Devonian–Carboniferous border, about 350 m.y. ago. It is a good example of an early stage in the amphibian radiation; a moderately large animal (about 1-m long), it had short legs that were probably only marginally effective for moving about. Like the later *Pteroplax,* described by Watson (1951), it can be thought of as a fish that swam on land, as do living fish, occasionally (Gray, 1968).

The evolution of the brain in amphibians was probably even more conservative than in the fish. There is no suggestion, either in the fossil record or in comparative studies of the modern amphibian brain, of any

Adaptive Radiation

major departures from the primitive vertebrate condition. Nothing analogous to the specializations of some bony fish (e.g., teleosts) has been found. This fact was instrumental in Herrick's (1948) choice of the tiger salamander as the living species in which to analyze the "fundamental pattern" of the vertebrate brain. Herrick explained:

> It is probable that none of the existing Amphibia are primitive in the sense of survival of the original transitional forms and that the urodeles are not only aberrant but in some cases retrograde (Noble, 1931; Evans, 1944); yet the organization of their nervous systems is generalized along very primitive lines, and these brains seem to me to be more instructive as types ancestral to mammals than any others that might be chosen. They lack the highly divergent specializations seen in most of the fishes; and, in both external form and internal architecture, comparison with the mammalian pattern can be made with more ease and security.[2]

Herrick's comments on the fish brain refer to the Actinopterygii, the teleosts in particular, and, as shown in the previous chapter, the specialization of brain structures in this group of vertebrates is very ancient indeed.

The large body size of many of the fossil forms is a critical structural fact for an analysis of the evolution of the brain in amphibians. It is also important that in many early amphibians, especially the labyrinthodonts ancestral to reptiles, the cranium was extensively ossified (cf. Jarvik, 1955), and complete endocranial casts can be produced. These have been studied in two specimens, *Edops,* which had a body about 2-m long and *Eryops,* which was shorter and perhaps somewhat stouter. There is no complete skeleton of *Edops,* but both a skeleton and reconstruction of *Eryops* are illustrated (Fig. 6.1). These amphibians were primitive, from the Permo-Carboniferous border, and are, thus, about 280 m.y. old. *Edops* is in a structural stage ancestral to *Eryops,* although they overlap somewhat; the oldest *Eryops* specimens are found in the same Texan fossil beds as some *Edops.*

The Amphibia, as a class, first appeared in the Devonian period and divided early into three main groups, or superorders—the labyrinthodonts, from which the reptiles eventually evolved; the lepospondyls, which have no modern descendants and are not considered; and the lissamphibians, from which modern amphibians are descended. There is information only about the fossil brains of labyrinthodonts, but, as Romer and Edinger (1942) indicate, it is unlikely that significant additions to knowledge of the evolution of the amphibian brain would be forthcoming from the other orders. The story, in summary, is that of a class of vertebrates in which all species tended to have generalized brains.

---

[2] Reprinted from "The Brain of the Tiger Salamander *Ambystoma tigrinum*," by C. J. Herrick, The University of Chicago Press, Chicago, Illinois, copyright 1948, p. 16.

BRAINS AND BODIES OF AMPHIBIANS

A very complete discussion of the brain and endocast of *Edops* is presented by Romer and Edinger (1942). Except for the lack of appropriate emphasis on size relationships between brains and bodies, their report is a model of completeness and is a careful exercise in the kinds of deductions that can be made about a brain when it is represented by an endocast in which it fits somewhat loosely. Although I will not review it in detail I will present a volumetric extension of their analysis. Sawin (1941) presented a less detailed but also excellent analysis of the endocast of *Eryops* as part of his study of the skull of this fossil. His work was included in Romer and Edinger's also, and these are the basis for my discussion.

To prepare their argument, Romer and Edinger studied endocasts of three living "large" amphibians—the familiar mudpuppy of the comparative anatomy class, *Necturus maculosus*; the bullfrog, *Rana catesbeiana*; and the hellbender, *Cryptobranchus alleghaniensis*, which is the largest American amphibian. They decided from a comparison of endocasts and brains of these living forms that, although little could be learned about the mudpuppy's brain from its endocast, there was sufficient correspondence of endocast to brain in the other living amphibians to permit significant inferences about the general shape of the brain. For example, they considered the expansion of the mesencephalon in the bullfrog to be noticeably reflected in the endocast. This is not as obvious to me in viewing their figures (cf. Fig. 6.3), but it is clear that the tapering endocasts of both the bullfrog and the hellbender are appropriately formfitting with respect to the brains. The mudpuppy, on the other hand, tapers in the wrong direction, expanding rostrally instead of becoming smaller as the olfactory bulbs are approached.

My only negative criticism is that they neglected to report quantitative data with precision. Absolute size may not be ignored in the analysis of organ relationships and in the study of the evolution of body and organ systems. In his instructive concluding discussion of "Paleontology and Modern Biology," D. M. S. Watson (1951, p. 200ff.) revealed his own "discovery" of the importance of size relationships in the analysis of evolutionary processes; Bonner (1965) has written a book on this topic. My own work in this area has been mainly on the importance of size relationships in understanding the evolution of the brain.

In the case of Romer and Edinger, two of the most distinguished evolutionists of our time, their uncertain attitude toward scale factors in their figures, reflected in their use of illustrations of endocasts and brains drawn to different and inexact scales, made the present work harder. A minor example was their statement of "$2\frac{1}{2}$" as the magnification of the brain of *Necturus* in one of their figures not reproduced here when a comparison with the original of their figure (McKibben, 1913) suggests a scale factor of $2\frac{1}{4}$. This minor fault in a basically qualitative analysis is less an error in their presentation than an invitation to err for those who want to use their illustrations as data for quantitative analysis.

Let us now consider the data on living amphibians exemplified in Fig. 6.3. Our analysis is similar to that used in the previous chapter in which

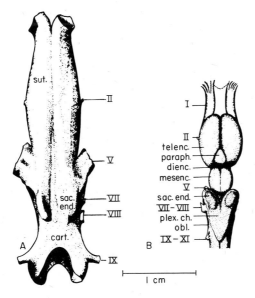

Fig. 6.3. Endocast (A) and brain (B) of a large living amphibian, *Cryptobranchus*. From Romer and Edinger (1942).

there are two basic questions; does the brain fit into the endocast, and is it the right size for the body? I summarize the analysis in Table 6.1, which includes body lengths given by Romer and Edinger for their specimens and the best estimates of brain and endocast volumes that I could make from their figured data. I did not use the methods of previous chapters to estimate body weight because the proportions of these amphibians are different from those of fish. Instead, I considered the mudpuppy and hellbender to be most accurately modeled by a cylinder equal to their body length with a diameter one-tenth of their body lengths, relations suggested by photographs in Bishop's (1943) handbook. For the bullfrog I used measurements of actual weights reported by Donaldson (1898) and Crile and Quiring (1940). Why these two sets of measures on frogs differ so greatly is unknown, but they suggest a great range of variation in adult specimens.

Table 6.1
Quantitative Brain and Body Relations in Three Large Modern Amphibians

	Species		
	Bullfrog (Rana)	Mudpuppy (Necturus)	Hellbender (Cryptobranchus)
Body length[a] (cm)	13	23	49
Body weight[b] (g) measured or estimated	500 313	320	960
Expected brain size [g or ml; Eq. (5.2)]	0.44 0.28	0.35	0.70
Estimated endocast volume (ml); see text	0.71	0.42	1.3
Estimated brain size (g); see text	0.24	0.22	0.37
Measured brain weight[b] (g)	0.46 0.22	—	—

[a] Romer and Edinger (1942).
[b] Data on bullfrog from Crile and Quiring (1940), upper value, and Donaldson (1898), lower value.

For the sake of this exposition I have used the term "estimate" to signify a calculation of a measure for a particular specimen, based on a model tailored to the dimensions of that specimen. I used the phrase "expected size" when I estimated a size from a regression equation such as Eq. (5.2). I used the phrase "actual size" when direct measures on the specimen are involved.

The expected brain sizes for amphibians of the body sizes listed in Table 6.1 are of the order of half the volumes of the braincases. This is not really a loose fit. The brain would be expected to be $0.5^{\frac{1}{3}}$ times the length of the endocast in any linear dimension, i.e., about four-fifths as long, wide, and high as the endocast.

I will describe, briefly, the method of estimating some of the brain sizes in Tables 6.1 and 6.2 with the example of the bullfrog. Applying Eq. (5.2) ($E = 0.007 \, P^{\frac{2}{3}}$) for a body weight $P = 500$ g, results in an expected brain size $E = 0.44$ ml. At $P = 313$ g, $E = 0.29$ ml. These values are entered in Table 6.1. To estimate the brain size of the bullfrog, I modeled the forebrain illustrated by Romer and Edinger (1942) as a $4 \times 4 \times 6$ mm ellipsoid, the optic lobes as two spheres, 3.6 mm in diameter, and the "brainstem" as a 4-mm diameter cylinder, 13-mm long. This led to an estimate of a brain size of 0.24 ml. That estimate is within the range of actual weights, which were 0.46 g and 0.22 g for the specimens of Crile and Quiring (1940) and of Donaldson (1898), respectively, and validates

Brains and Bodies of Amphibians

the use of simplifying geometric models to estimate volumes from planar drawings. The estimate of 0.24 ml is also quite close to the expected brain size of 0.44 or 0.28 g, which is based on Eq. (5.2), derived from the brain:body data in lower vertebrates (Fig. 2.4).

Romer and Edinger concluded from their comparison of modern amphibian brains and endocasts that:

> ... in modern amphibians the brain, contrary to the general assumption, is a factor in moulding the skull. Except for such a degenerate form as Necturus, the braincase, despite its non-neural contents, is demonstrably a capsule built for and modelled by the brain. We are therefore justified in concluding that the endocranial contours of the braincases of Paleozoic amphibians, more highly ossified and less degenerate than the modern forms, will yield significant data regarding the structures which they enclosed [1942, p. 369].

My quantitative supplement to Romer and Edinger's data on modern amphibians can be extended to their fossil amphibian endocasts and their projections of the fossil brains. We do this by establishing a relation between brain and body size in the fossil forms and evaluating these forms with respect to a brain:body coordinate system of the kind used in Fig. 2.3, established for other vertebrates. It is instructive to determine where they lie on a graph like Fig. 2.4. It was for the analysis of these data that it occurred to me to use the method of minimum convex polygons (Chapter 2). The method of this chapter and Chapter 5 for modeling brain size by various geometrical forms eventually evolved into the method of graphic double integration. It should be appreciated that the latter methods are, essentially, equivalent to one another.

ENDOCASTS, BRAINS, AND BODIES OF FOSSIL AMPHIBIANS

The endocasts of *Edops* (Romer and Edinger, 1942) and of *Eryops* (Sawin, 1941) are similar both in size and shape, despite the body size differences of these animals (Fig. 6.1). In modeling a brain to fit the endocast of *Edops,* Romer and Edinger chose an anuran brain, similar to that of the bullfrog, because they detected enlargements of the mesencephalic region of the endocast, which they interpreted as due to enlarged optic lobes. I cannot quarrel with their reproduction, and add only that the brain they modeled (Fig. 6.4) is relatively undifferentiated compared to the brains of most modern vertebrates. It is similar in shape to the brains of sharks, fish other than the bony fish, and modern amphibians.

The only more or less primitive feature in the brains of the two large fossil amphibians, *Edops* and *Eryops,* are the openings for the pineal, or parietal, organs in the two braincases and skulls. The endolymphatic sacs, which were presumably associated with the labyrinthine system then as

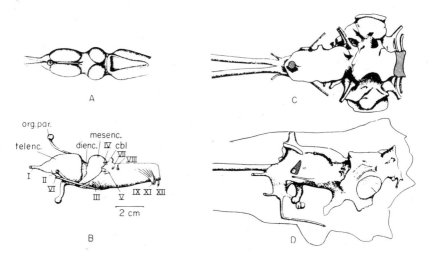

Fig. 6.4. Reconstruction of brain and endocast of the fossil amphibian *Edops*. A, Dorsal view of brain; B, lateral view of brain; C, dorsal view of endocast; D, lateral view of endocast. From Romer and Edinger (1942).

now, are large in both *Edops* and *Eryops,* although they follow the modern amphibian pattern in being contained within the braincase rather than being in a separate chamber that communicates with the endocranial cavity. This is a notable amphibian feature and not a primitive feature.

Romer and Edinger present considerably more detail about the endocast, with respect to the morphology of the cranial nerves and the vascular system, as these would be handled by the cranial foramina. Their discussion suggests no major variation from the contemporary vertebrate pattern other than the enlarged parietal opening just mentioned.

I have summarized the size relationships of endocasts and bodies in these two Paleozoic amphibians in Table 6.2. I assume that the bodies of both *Edops* and *Eryops* were basically cylindrical and of the same diameter, with *Edops* a somewhat longer animal. This is consistent with available data on their skulls and on the body of *Eryops* (Fig. 6.1).

The quantitative analysis is notable for its consistency with expected data as determined on modern animals. In viewing the tabled data one should keep in mind that the two species were almost certainly fat and sluggish. This is taken into account in the body size estimates. The expected brain sizes for the body sizes, according to Eq. (5.2), are 15 and 13 ml, about half the endocranial volume. This agrees with the situation in modern Amphibia (Table 6.1) and supports a uniformitarian view of the relation between brain size and endocranial capacity in amphibians. Romer and Edinger's illustrated reconstruction of the brain of *Edops* appears to be

Table 6.2
Quantitative Brain and Body Relations in Two Paleozoic Amphibians[a]

	Edops	Eryops
Body length[b] (cm)	200	150
Skull length (cm)	63	44
Body size[c] (liters or kg)	100	75
Expected brain size (ml or g) [Eq. (5.2)]	15	13
Endocast volume (ml) (Fig. 6.4)[d]	39	24
"Brain" volume (ml) (Fig. 6.4)[e]	19	12

[a] Data from Sawin (1941) or Romer and Edinger (1942) on endocasts. Based on Fig. 6.4, for *Edops*.

[b] Body length in *Edops* estimated from skull length, after *Eryops*.

[c] Body size estimated by cylinder, 32-cm diameter and three-fifths of the body length (cf. reconstruction of *Eryops*, Fig. 6.1).

[d] Endocasts modeled by cylinders, 24-mm diameter and 75-mm long for *Eryops*.

[e] Modeled by cylinder, 18-mm diameter and 75-mm long for *Edops*; one-half of volume of endocast for *Eryops*.

slightly larger than one would expect the brain to be if the computed volume from their planar figure is correct, but their construction is of the right order of magnitude. It is clear, in any event, that the brain did fill a significant fraction of the braincase, and Romer and Edinger's conclusion that the endocast should mirror the brain is supported by this quantitative analysis. On the linear measures there would be no more than about 20% unaccounted for in length, width, or breadth in which the brain could float. The brain must have had, approximately, the form of the endocast.

To indicate the approximate position of the Paleozoic and modern amphibians relative to those of contemporary vertebrates on the brain:body axis, I have placed a set of points in Fig. 6.5, giving the brain and body volumes estimated in Tables 6.1 and 6.2 as overlays on the vertebrate polygons presented earlier in Fig. 2.4. These estimated brain sizes of the two Paleozoic amphibians would be the "average" expected brain sizes for modern "lower" vertebrates of their body size.

CONCLUSIONS

A somewhat circular set of conclusions is suggested. First, amphibians, like ancient fish (or, perhaps, as ancient fish!), evolved early with what we now would recognize as the generalized vertebrate brain pattern. Second, they have remained at the generalized level until the present time,

6. Invasion of the Land: The First Tetrapods

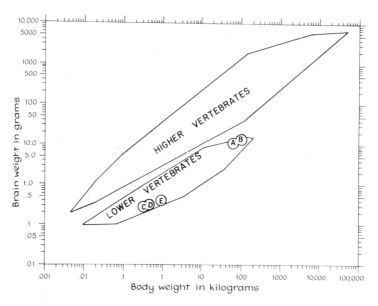

Fig. 6.5. Brain size and body size in modern and fossil amphibians, superimposed on the graph of brain:body relationships in modern vertebrates (cf. Fig. 2.3). A, *Eyrops*; B, *Edops*; C, bullfrog (*Rana*); D, mudpuppy (*Necturus*); E, hellbender (*Cryptobranchus*).

as witnessed by the positions of the bullfrog, mudpuppy, and hellbender points on the same curve. Third, the present "lower vertebrate" level of brain evolution, as represented in Fig. 2.4, appears to be representative of the initial level of the earliest vertebrate radiation. This reinforces the conclusion stated earlier in this book that the brain has been a rather conservative organ in its evolution. The fundamental vertebrate pattern was probably achieved with the evolution of the gnathostomes, or jawed vertebrates, and it has been retained in the face of the varied selection pressures faced by many of the fish and by all their amphibian descendants. The bony fish departed somewhat from it with respect to moderate developments of midbrain and hindbrain specializations, but these did not result in significant increases in the total brain size for a particular body size beyond the primitive adaptive level. It is shown in the next chapter that this conservatism of brain evolution persisted through the evolution of the reptiles.

Chapter 7

The Radiation of the Reptiles

The earliest reptiles descended from a group of labyrinthodont amphibians and became clearly differentiated about 325 m.y. ago, in the Carboniferous period. The line between amphibians and reptiles is uncertain in these strata. From a zoological standpoint there is some difficulty in defining the transition because it is impossible to apply to fossil forms the criterion that most clearly specifies modern reptiles, the amniote egg, a development that completely freed the reptiles from dependence on a watery environment. According to Romer (1966), the earliest reptiles can be distinguished from some of their contemporary amphibians only by small differences in the skull, and the distinction is really maintained on the basis of the identification of lines of descent from these ancient reptiles in which more clearly reptilian skeletal features could be seen.

Although it would be helpful to develop some sense of the evolutionary progress of all the reptilian lines, we can appreciate the evolution of their brains by considering two reptilian groups, the synapsids, or mammallike reptiles, and the archosaurs, or ruling reptiles, which included dinosaurs and flying reptiles. The restriction can be justified not only because of the limited material available on other reptiles for the analysis of their fossil endocasts but also because of their relevance for the subsequent evolution of the brain. The mammallike reptiles are our ancestors; their brains are the source of our brain, and it is natural for us, as humans, to be concerned with our ancestors. Our interest in the ruling reptiles can be justified because modern birds are probably descendants of one group of dinosaurs. But do we really need that excuse? If there is any universal love of knowledge in one's experience with science, it is shown in a child's enjoyment of mounted skeletons of dinosaurs in the world's museums. Curiosity about the brains of these "dragons" is natural, and it is a pleasure to try to satisfy that curiosity.

My concern with the endocasts of flying reptiles, the pterosaurs, while also a natural development of ordinary curiosity, is of special importance in developing one of the themes of this book. The pterosaurs invaded a niche that involved unusual demands on the capacity of organisms to process information and to control their bodily movements. The response to the selection pressures of an aerial niche may include the development of an enlarged and somewhat specialized brain. It is discussed in a separate chapter (Chapter 8) when the adaptations by vertebrates to aerial niches are introduced.

In presenting the data on the evolution of the brain as manifested by endocasts from mammallike and ruling reptiles, it will be convenient to compare the endocasts with those of modern crocodiles, lizards, and turtles, and, therefore, the history of these groups is included in this brief exposition of the history of reptiles.

EVOLUTIONARY HISTORY

The stem reptiles (Cotylosaurs), which descended directly from labyrinthodont amphibians, probably fathered all the other reptilian orders. The radiation (Fig. 7.1) occurred during the Carboniferous period and the succeeding Permian period, first with the appearance of synapsids, or mammallike reptiles, and then archosaurs, or ruling reptiles. The earliest synapsids are among the oldest reptile groups, according to the fossil record, with fossil pelycosaurs dating to the mid-Carboniferous period, perhaps 325 m.y. ago. The more immediate reptilian ancestors of the mammals, the therapsids, appeared somewhat later in early Permian times and had an extensive adaptive radiation in the Permian period. They were the only surviving synapsid order after the Permian. Therapsid reptiles persisted in identifiable form through the early and mid-Mesozoic era (Triassic and early Jurassic periods) and probably became extinct during the mid-Jurassic period about 150 m.y. ago. I say "probably" because there is a persistent opinion among some paleontologists, which Simpson (1959, 1960b) has reviewed critically, that the monotremes of present times, the platypus and the spiny anteater, are really surviving therapsid reptiles. The brains of monotremes are generally like those of other mammals, however, and are more like those of marsupials than of placentals (Abbie, 1940; Lende, 1964).

The earliest archosaurs, or ruling reptiles, appeared at the beginning of the Triassic period, about 230 m.y. ago. This group had a tremendous radiation and was clearly a dominant fauna through most of the Mesozoic era. The radiation produced huge creatures that stir the world's imagina-

Fig. 7.1. Phylogeny of reptiles. A, Overall phylogenetic tree; B, the phylogeny of ruling reptiles. Reprinted from "The Vertebrate Story," by A. S. Romer, The University of Chicago Press, Chicago, Illinois, copyright 1959.

Evolutionary History

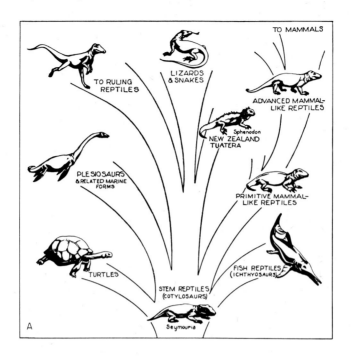

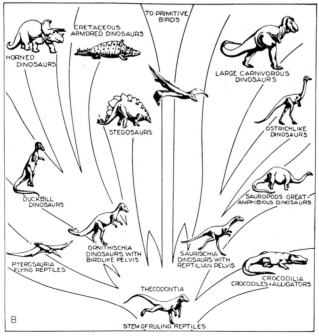

tion, including the largest land vertebrates that have been known. It also included flying reptiles and the ancestors of the birds. The ruling reptiles, except the crocodilians, became extinct in a still unexplained way (Colbert, 1965; but see Axelrod and Bailey, 1968) at the end of the Mesozoic era, about 70 m.y. ago. The most fantastic and largest forms, such as *Triceratops, Tyrannosaurus,* and *Brachiosaurus,* are from the Cretaceous or late Jurassic periods. *Brachiosaurus* was probably the largest land animal that ever lived, weighing, according to Colbert (1962), over 75 metric tons.

Contemporary reptiles have varied histories. The crocodilians, which are members of the subclass of ruling reptiles (Archosauria), date back to the Triassic period and are very little changed skeletally from their ancestors. Turtles are even more conservative forms in most respects, since the present chelonians are essentially identical with their Triassic ancestors. As Romer (1966) points out, their fantastic armor provided an unusually successful adaptation in turtles to the problems of defense against predators, and, once achieved, the specialized adaptation could be maintained.

A number of chelonian endocasts are known, including Cretaceous specimens described by Zangerl (1960). Were it not for their peculiar and successful specializations, reptiles of this group would be of unique interest to comparative neurology and psychology because they have remained isolated and biologically remote from other reptiles and other vertebrates. They are, perhaps, the most truly "living fossils" available among all the vertebrates. In spite of their specialized adaptations they are worth far more analysis by students of the behavior and nervous systems of living vertebrates than they have received. I have not analyzed their endocasts in terms of relative size because of the difficulty of estimating their body sizes. (Much better estimates are available for the dinosaurs.) From the illustrations of their endocasts I judge their brains to have been much like those of living turtles.

The Squamata, which include the contemporary snakes and lizards, date back to late Triassic times, and the snakes, in particular, have a history about as ancient as that of the more advanced mammals. Among the other reptiles of present times, the tuatara (*Sphenodon,* order Rhynchocephalia) of New Zealand is a nearly ideal relict, essentially unchanged from its earliest unspecialized ancestors. It is characterized by an at least moderately functional pineal-parietal eye, and its brain, thus, meets the one morphological criterion that would label an undifferentiated brain as primitive. Perhaps more than the tiger salamander, this form would qualify for detailed comparative anatomical study and behavioral analysis, if one seeks a contemporary tetrapod that represents a primitive and unspecialized condition. Of course, like most "living fossils," *Sphenodon* comes from a branch of the evolutionary tree that differs in its direction from the one that led to the mammals and man. Yet, it validly represents a Permian level of reptilian evolution and is really only moderately beyond the point

Relative Brain Size in Dinosaurs

at which a branch of the stem reptiles evolved into the synapsid mammal-like species.

It is worth noting, finally, that mammals had probably already appeared in the late Triassic period and were certainly present in the Jurassic period, perhaps as modified therapsids. (A paleontologist in a time machine would undoubtedly have classified the mammals of those days as therapsids with slightly aberrant jaws and ears.) The birds had appeared in the form of *Archaeopteryx* in Upper Jurassic times. Thus, an orthogenetic approach to the evolution of the brain, in which one looks for the lines that led to the "higher vertebrates," the birds and mammals, would have to abandon the reptiles at those strata, representing a time 150–200 m.y. ago, and move on to the analysis of mammalian and avian remains.

RELATIVE BRAIN SIZE IN DINOSAURS

In keeping with most of the analysis presented thus far, I will emphasize size relationships in the dinosaurs and other reptiles. However, a few general morphological points about the central nervous systems of these animals are worth making.

"Sacral Brain" and Other Morphological Issues

I have written relatively little about spinal organization of the nervous system beyond the statement (Chapter 4) that, except for the jawless fish (Agnatha), the spinal cord is organized essentially the same way in all living vertebrates. The evidence from casts of the spinal cord in fossil animals is consistent with the view that there is a fundamental identity in the spinal cord of vertebrates, although few casts have actually been taken for study. The most celebrated feature in casts of the spinal cavity occurs in the dinosaurs, which regularly show a massive lumbar and sacral enlargement. The volume of the spinal canal in the sacrum of *Stegosaurus*, for example, is perhaps 20 times as great as that of the endocranial cavity of this animal.

This point seems to me a minor one for understanding the nature of neural control by the brain. In no sense should one expect the sacral expansion of the spinal cord of *Stegosaurus* and other dinosaurs to signify that a "rear brain" had evolved there. Some expansion of the cord occurs in all tetrapods, and in man it is called the "lumbar enlargement." It may be compared to that in some large birds such as ostriches (Fig. 7.2). It is associated with the control of muscles and the response of receptor cells in muscles, tendons, and joints in the rear, or lower half, of the body, and the receptor and effector cells are simply more numerous in larger animals.

The second significant morphological feature in dinosaurs is also peripheral to our concern with the brain per se. It is the impressive enlarge-

142   7: *The Radiation of the Reptiles*

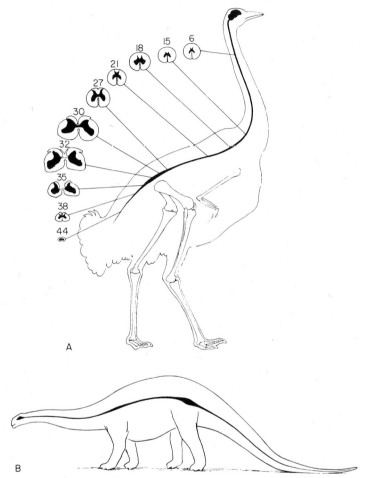

Fig. 7.2. A, Brain and spinal cord in ostrich, which is about 2 m in height, sections of cord numbered by vertebrae, from Streeter (1904). B, Endocast and spinal canal in *Brontosaurus,* which has a head and body length of about 20 m, from Moodie (1915).

ment of the region in which the pituitary body lay, an enlargement that certainly must reflect the fact that the large dinosaurs had large pituitaries. This is a typical condition of gigantism in animals, and its significance has been reviewed by Edinger (1942).

Finally, the difference between brain and endocast limits our ability to analyze the brains in dinosaurs just as it does in the other "lower" vertebrates. The most casual look at endocasts from dinosaurs (Fig. 7.3) shows the absence of distinct brainlike features. It is clear that the brain must have been supported by connective tissue in the brain cavity and was

Relative Brain Size in Dinosaurs 143

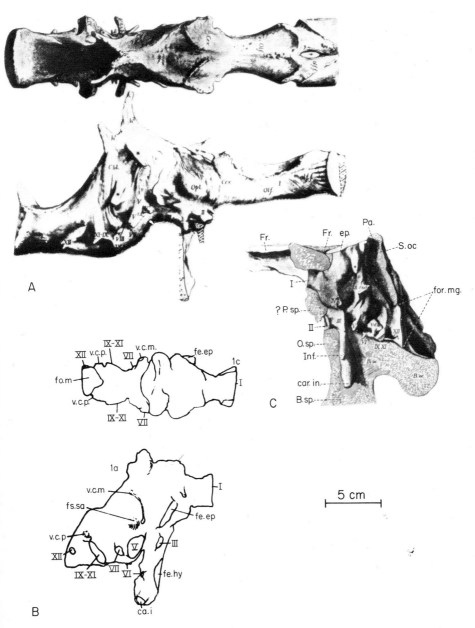

Fig. 7.3. Endocasts from various dinosaurs to illustrate typical patterns. A, Large carnivore, *Tyrannosaurus,* lateral and dorsal views, medulla at left, from Osborn (1912). B, Large herbivore, *Brachiosaurus,* lateral and dorsal views, medulla at left, from Janensch (1935). C, Large herbivore, *Diplodocus,* lateral view only, medulla at right, from Gilmore (1920).

considerably smaller than an endocast of the endocranial cavity. The point was made clearly in the relationship between brain and endocast in iguanid lizards as described in Chapter 2 (Fig. 2.1). It has also been illustrated in turtles (Edinger, 1929) and tuatara (Dendy, 1911). In each instance the volume of the brain is significantly less than that of the endocast. The discrepancies in gross appearance of the brain and endocast, when compared with the gross appearances of most dinosaur endocasts (Fig. 7.3), indicate that the dinosaur brain was probably analogously reduced in size compared to the endocast.

Brain and Body Size

The discussion of the dinosaur brain:body size relationships is made much easier because of a brief report by Colbert in which he estimated the sizes of 14 species of dinosaurs from models of reconstructions and measures on mounted skeletons. As Colbert (1962) put it, "Various other aspects of the problems of giantism in the dinosaurs might be discussed at this place. It is, however, a large subject . . ." (p. 15). His solution of even so small a part of this weighty problem eases the burdens on the present analysis.

There is a popular misconception about the small size of dinosaur brains. Like many ideas about size, this misconception is due to the lack of a sense of the expected proportions of the brain for a body of a given size. Such proportions are presented quantitatively by Eq. (5.2) and even better by the data of Fig. 2.3, in which one may sense expected variations in size. Since the dinosaurs were reptiles, they would be expected to fall within a reptilian or "lower vertebrate" brain size range as suggested in Figs. 2.4 or 2.6. If they were small brained, they would fall below that range. We can find out whether they were unusually small brained.

It is possible to estimate volumes of endocasts such as those illustrated in Fig. 7.3. Following the assumption of Osborn (1912), Swinton (1958), and Colbert (1965), one may assume that the brain volume in these reptiles was half of the endocranial volume. I have done this (Fig. 7.4), although the assumption is not entirely satisfactory. It is more likely that the fraction of the endocranial cavity occupied by the brain is, itself, a (negative) function of body size, as suggested by Aronson (1963) for fish. Data are lacking, however, and it certainly is appropriate to assume that large fossil reptiles had no more tightly packed skulls than do moderately sized reptiles of present times, such as *Sphenodon* (Dendy, 1911) or turtles, alligators (Edinger, 1929), and iguanids (Fig. 2.1). The relationship between brain size and endocranial volume as a function of age and body size could be easily resolved by a quantitative morphological study. This has not yet been done.

I present in Table 7.1 estimates or direct measures of the volumes of the endocasts of 10 dinosaurs in which body weight and volume estimates are available in Colbert's (1962) report. These estimations are based on graphic double integration, a numerical method in which the endocast is sketched in lateral and dorsal views, and measures of the cross-sectional lengths are taken at regular intervals. I have illustrated the method in Chapter 2 (Fig. 2.7) with outlines of the endocast of *Tyrannosaurus* as reported by Osborn (1912). Osborn reported the volume of the endocast, measured by water displacement, as 530 ml. This should be compared with the estimate of 536 ml by the method of graphic integration. The result is gratifying, and, like other geometric models used for quantitative analysis, this method appears to be amply justified.

In Table 7.1 the endocranial volumes of 10 dinosaurs, excluding the olfactory bulbs and the area posterior to the exit of N.XII, are listed along with Colbert's estimates of their body volumes and their predicted brain

Table 7.1
Quantitative Brain and Body Relations in 10 Dinosaurs

Genus[a]	Body[b] volume (P)	Endo- cast[c] volume	Expected[d] brain volume (E)	References
1. *Allosaurus*	2.3	335	120	Osborn (1912)
2. *Anatosaurus*	3.4	300	160	Lull and Wright (1942)
3. *Brachiosaurus*	87.0	309*	1400	Janensch (1935)
4. *Camptosaurus*	0.4	46	38	Gilmore (1909)
5. *Diplodocus*	11.7	100	360	Osborn (1912)
6. *Iguanodon*	5.0	250	200	Andrews (1897)
7. *Protoceratops*	0.2	30	24	Brown and Schlaikjer (1940)
8. *Stegosaurus*	2.0	56*	110	Gilmore (1920)
9. *Triceratops*	9.4	140	310	Hay (1909)
10. *Tyrannosaurus*	7.7	404	270	Osborn (1912)

[a] Genera numbered as in Fig. 7.4.
[b] Colbert (1962); volume in 10^6 ml (metric tons).
[c] By method of Fig. 2.7 or (asterisk) by reported direct measurement; volume in ml.
[d] From Eq. (5.2): $E = 0.007\ P^{2/3}$; volume in ml.

weights according to Eq. (5.2), assuming that their specific gravities were 1.0. Brain size may be estimated as about half the size of the endocast. Table 7.1 is a useful example of the advantages and of the disadvantages of tabular presentation of data and of the limits of Eq. (5.2) as a predictive instrument. Endocranial volume and predicted brain size are incompatible in 4 of the 10 dinosaurs: *Brachiosaurus, Diplodocus, Stegosaurus,* and *Triceratops*. The error is fourfold in the largest of these dinosaurs, with a predicted brain size of 1400 ml that would have to have been packed into a measured endocranial cavity of 309 ml. Since one would expect the brain to be smaller than the endocranial cavity, the actual error of prediction in this animal must be even larger.

Rather than treat these as errors, an alternative view is to use Eq. (5.2) as a predictor of expected brain size. When the evidence shows clearly that the animal to which the prediction is being applied could not possibly have had a brain as large as predicted, then it is reasonable to consider that animal as having been "small brained." I believe that this is at least partly correct and is comparable to the situation in the modern crocodile, which has a brain weight of 15 g (Crile and Quiring, 1940) despite the predicted weight of about 25 g for a 200 kg reptile.

There is probably a relatively simple morphological explanation of the "error" in estimating the brain size of several of the dinosaurs, and the explanation should be apparent upon inspection of several of the endocasts illustrated in Fig. 7.3. In those instances in which the pituitary appears to be near the anterior edge of the endocast, it must have been the case that the forebrain was encased in a cartilaginous portion of the cranium. One can, therefore, measure an endocast of only the midbrain and hindbrain. My estimates in these instances were almost certainly too low, unless the brain occupied a very small fraction of the cranial cavity.

It is also correct to consider the "error" as evidence of the limitations of Eq. (5.2) as an instrument in the analysis. Whenever one uses simple equations that produce rigid deterministic statements, one is faced with the choice of considering the equations as fundamental, reflecting real or natural relationships, or treating them as empirical statements having some ad hoc value for describing real relationships. It is only in the descriptive sense that Eq. (5.2) should be used. If the fundamental processes governing the growth of brains and bodies were known and could be used as the source of a mathematical axiom system, they would be unlikely to generate descriptive statements of the form of Eq. (5.2) as theorems (Chapter 3), although such equations should be reasonable approximations.

Brain:Body Analysis

Table 7.1 presents the usable data on the approximate sizes of the brains, endocasts, and bodies of dinosaurs, but a clearer picture of the relationships among these measures can be developed from a graphic analysis (Fig. 7.4) with the method of minimum convex polygons (Chapter 2). The points representing dinosaur brain volumes as functions of body volumes are numbered in Fig. 7.4 in the same way as in Table 7.1, and the brain volumes have been estimated as one-half the endocast volumes.

Relative Brain Size in Dinosaurs

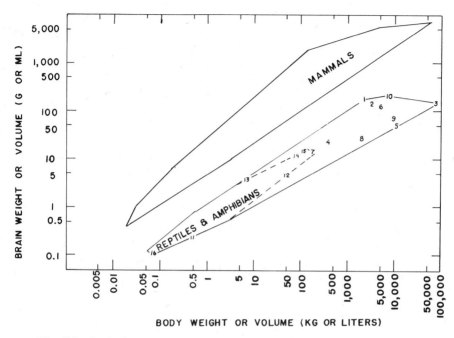

Fig. 7.4. Brain:body size relations in fossil lower vertebrates superimposed on data of living species (cf. Fig. 2.6). Fossil reptiles (1–13 in Tables 7.1 and 7.2); fossil amphibians, *Eryops* (14), *Edops* (15) (Table 6.2); fossil fish, *Boreosomus* (16) (Chapter 5).

(Data on fossil fish of Chapter 5, fossil amphibians from Fig. 6.5, and mammallike reptiles to be described later in Table 7.2 are also shown in Fig. 7.4). One value of the graphic analysis by convex polygons is in providing perspective about the meaning of "error" in these measurements. The apparently large discrepancy between predicted brain size and measured endocast volume in *Brachiosaurus*, for example, is seen to be much less unusual, and, in fact quite normal, in the graphic analysis. If we enclose the space on the graph within which "lower vertebrates" lie, the resulting polygon enveloping the measures appears to yield a truly simple picture of brain:body relationships.

Several inferences can be developed from Fig. 7.4. First, the dinosaurs, including the largest, are typically reptilian with respect to gross brain size. Their large body sizes extend the range of the brain:body map for the lower vertebrates and make the map comparable to the one for the higher vertebrates. Second, the parallel orientation of these two maps suggests a common generating process in the vertebrates with respect to the size of the brain required to govern a body of a particular size.

The discrepancies between estimated brain sizes and calculated endocast volumes fade into insignificance in this kind of analysis. This should not lead us into the error of assuming that either these discrepancies or the differences among the dinosaurs with respect to their brain:body size relationships are not important. The differences are insignificant only in terms of the general results of a graphic analysis as presented in Fig. 7.4. Figure 7.4 demonstrates important similarities among the endocasts; the graph does not imply that there are no important differences.

MORPHOLOGICAL ANALYSIS

The precision possible in an allometric analysis, for example, by Eq. (5.2), cannot be maintained in morphological studies that would interpret endocasts as fossil brains. Such morphological studies are possible in birds and mammals because their endocasts are probably almost perfect models of the external configurations of their brains. In lower vertebrates, however, the poor shaping of the endocast by the brain restricts the analysis to fairly gross features. I have tried to indicate the nature of the data for morphological analysis by illustrating some of the endocasts. It is apparent, for example, that the cranial nerves are arranged and spaced normally, and one can, with some certainty, identify the points of entry of major nerves, such as the trigeminal or vagus, into the brain case. Some authors have remarked on the extensive development of the olfactory nerve (N.I) as indicating adaptations emphasizing the sense of smell. The development does not appear particularly striking to me in any of the endocasts I have seen. Other authors have commented on forebrain development as being beyond the amphibian or fish levels. This, too, seems to me gratuitous; it is hardly possible to ascertain the placement of the forebrain in either photographs or actual preparations of the endocasts. To the extent that one can guess, however, forebrain development is no greater than that in lungfish (Chapter 5) and is hardly noteworthy.

In several of the dinosaur endocasts there is, as mentioned earlier, space for a pineal-parietal eye. In others, such as *Tyrannosaurus,* this space is covered by the skull. Another noteworthy feature, also mentioned earlier, is the enlargement of the pituitaries of these gigantic animals. In instances in which enlarged pituitary fossae are not indicated in a figure, the fault is generally with the preparation (e.g., the *Tyrannosaurus* endocast), in which the casting was not extended into the pituitary region, and one sees only a stump that once housed the infundibulum.

The surface of the dinosaur endocasts indicates that they represent an expected level of reptilian brain evolution, consistent with the assumption that the dinosaurs did not have specialized brains. Their brains were as large as reptilian, amphibian, or fish brains of the present time, relative to their body sizes. Specializations in the form of regional enlargements in the brain were quite absent in dinosaurs, unless one considers the enlarged sacral area of the spinal canal as evidence of such a specialization. This specialization would be without behavioral significance and could be interpreted simply as a neural adaptation to a large body size that minimized demands on the brain. An appropriate analogy might be the adaptation for flight in insects in spite of the absence of extensive neural control systems in the small insect brain. The adaptation in insects is based on the development of intrinsic control mechanisms in the insect's wing musculature that produce appropriate rhythmic response. These are controlled by the central nervous system, which releases or switches off the neuromuscular units (Pringle, 1957; Wilson, 1964). The dinosaur's sacral enlargement could be viewed as a peripheral reflex control system, enabling the animal to carry out appropriate muscular movements of the rear end of its body, with only minimal central switching at the level of the brain to turn the peripheral system on or off. This is, in fact, an aspect of the control mechanism for many body movements in higher as well as lower vertebrates: the brain acts as a central excitatory and inhibitory system to release more or less rhythmic movements of peripheral neuromuscular control systems (Gray, 1968; Lloyd, 1960).

Attempts at reconstruction of the behavior patterns of dinosaurs must rely on the uniformitarian principle. It should, therefore, be assumed that patterns of behavior in dinosaurs were similar in significant ways to those of living animals at a comparable level of brain and skeletal evolution. For example, alligators or crocodiles would be appropriate species as models for such speculations, and I know of no evidence that would suggest that the uniformitarian hypothesis would be misapplied in this case. One should expect that the ancient dinosaurs had the behavioral capacities of the same order as modern crocodiles; an ethological analysis of the behavior of alligators or crocodiles would suggest the kinds of behavior patterns (though not the patterns themselves of course) that are likely to have formed the behavior repertoire of the dinosaurs.

THE MAMMALLIKE REPTILES

Despite our great interest in the antecedents of the mammalian brain in the synapsid reptiles of the Permian and Triassic periods, very few complete endocasts from this group have been described. We are most interested in the therapsid (mammallike) reptiles in this group. Artificial casts are difficult to make for most species because large parts of their cranial walls were cartilaginous. In one group of dicynodonts, represented by *Placerias gigas* (Fig. 7.5), it is clearly impossible to conclude much about the gross size of the endocast because the ventral surface of the braincase is ossified only to the posterior border of the pituitary. Some of the lateral and ventral surface of the braincase of *Placerias* posterior to

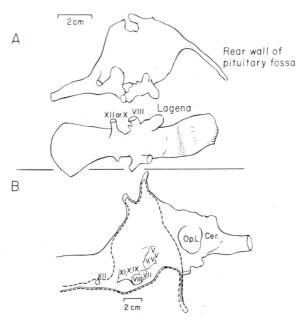

Fig. 7.5. A, Endocast of *Placerias gigas,* upper view, lateral; lower view, ventral. B, Lateral view of endocast (dashed line) superimposed on sketch of endocast of *Tyrannosaurus* (Osborn, 1912) to illustrate portion of endocast of *Placerias* retrievable from fossilized skull. Remainder of endocranium, including forebrain area, in *Placerias* was presumably cartilaginous (cf. Fig. 2.1). *Placerias* specimen is UC27864 (Berkeley).

that point is sufficiently ossified to enable one to identify the probable points of egress of some branches of the trigeminal nerve (N.V.) and more posterior cranial nerves, but only a small bridge of fossilized bone occurs dorsally. The endocast of *Placerias* is, therefore, very incomplete when compared with that of a dinosaur like *Tyrannosaurus*. In other therapsids, such as the cynodonts, the endocranial cavity is more completely ossified. A limitation also exists in interpreting the cynodont endocasts because there was clearly a good deal of cartilage interposed between the bones at the interior cranial wall and the supporting tissues of the brain.

It may be useful to present a more detailed exposition of the structure of the cranial wall in one cynodont, *Thrinaxodon,* which is well described

The Mammallike Reptiles

in the paleontological literature (Brink, 1956, 1958; Estes, 1961; Olson, 1944).

ENDOCAST OF *Thrinaxodon*

The fossil skull of *Thrinaxodon* that I examined is about 6 cm long, 2.5 cm high, and 4 cm wide at its broadest point. In order to examine the cranial cavity, the right posterior quarter has been separated from the rest of the skull by making a midsaggital cut to a point about 2.5 cm forward from the posterior tip at a separated foramen magnum and then cutting coronally to the right from that point. The quartered fragment and the remainder of the skull were, thus, divided by a saggital slice through the endocranial cavity. Several views of the preparation have been sketched (Fig. 7.6). The basisphenoid is missing from the larger

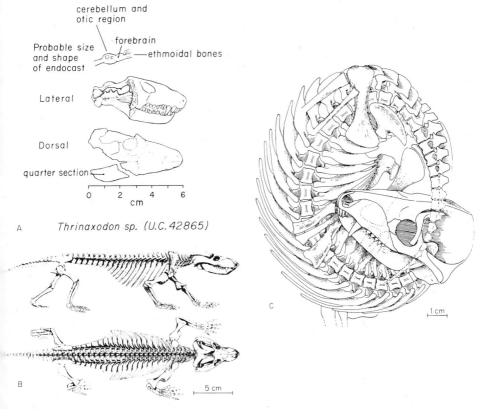

Fig. 7.6. A, Skull of *Thrinaxodon* prepared to show the endocranial cavity; B, skeleton of *Thrinaxodon,* side and dorsal views (Brink, 1956). C, Skeleton of a *Thrinaxodon* specimen as found (Brink, 1958).

fragment, and there is a well-preserved exoccipital. The proötic is partially covered by the epipterygoid, which has two winglike processes covering relatively large openings, probably for several branches of N.V. Basioccipital and, possibly, basisphenoid are identifiable in the smaller fragment. Anterior to this the orbitosphenoid is identifiable in the larger fragment. Superior to the winglike processes of the epipterygoid, the parietal bone has a ridge in it that appears to be suitable for a cartilaginous connection. Anterior to this region the excavated endocranial cavity is pear-shaped, as it would be if it accurately molded the cerebral hemispheres of *Thrinaxodon*. Still farther anterior there is an uncertain region, like a mammalian ethmoidal area, in which olfactory bulbs could extend.

The most important uncertainties in this *Thrinaxodon* endocranium are due to the unknown extent to which the walls were lined by cartilage. It seems unlikely that the visible epipterygoid was directly in contact with the brain; I would guess that a cartilaginous wall was interposed between bone and brain. This summary of the endocranial skeletal anatomy can be used as an exercise to suggest the difficulties of interpretation of the material. If one restricts the interpretation to features of the bony matrix within which the brain lay, then reasonably positive and unequivocal statements are possible. But on the basis of this material the only conclusion about the brain as approaching the mammalian state that I would support is that there may have been some lateral expansion of the cerebellum as described by Olson (1944). Even this statement is based on the undemonstrable assumption that the bone molded the brain without the intervention of substantial amounts of cartilage.

Olson's (1944) discussion of synapsid endocasts is presented by figures of the posterior halves of the endocasts, ending near the origin of N.V. Since forebrains are obviously not represented in these relatively small endocasts, I have made no effort to estimate the volumes of the originals. Olson used serial sections as the basis for his reconstructions. Thus, although not made by a casting process, they are equivalent to endocasts. His most important conclusion, for our purposes, was with respect to the presence of a flocculus on the cerebellum of synapsids (including *Thrinaxodon*). He considered it impossible to get information on the form of the forebrain area of the endocast in any of these forms because of the incomplete ossification of that area. However, the specimen that I examined (Fig. 7.5) appeared to be completely ossified in the forebrain region, as were the specimens studied by Estes (1961).

Olson concluded that the "brains" of synapsids were at an intermediate grade between reptiles and mammals. I have some difficulties with this conclusion because of my uncertainty about how to interpret these reptilian endocasts as brains. The argument based on the form, or shape, of

The Mammallike Reptiles

portions of the cast seems to me more open to preconceived biases and expectancies than an argument based on purely numerical quantitative considerations. I will analyze the problem from a different perspective by developing a quantitative analysis similar to that used thus far for our other reptiles, for our two amphibians, and for fish.

Brain:Body Analysis

There are two endocasts known in sufficient detail to permit an estimation of their volumes by the method used with *Tyrannosaurus* (Fig. 2.7). These are of *Lystrosaurus,* as figured by Kuehne (cf. Piveteau, 1961, p. 346), and of *Diademodon,* as figured by Watson (1913). I also include a volumetric analysis of the *Thrinaxodon* endocast just described, although that analysis should be recognized as provisional, because the cranial bones have not been described with certainty. Two of these mammallike reptiles, *Diademodon* and *Thrinaxodon,* are cynodonts, the group that is generally considered to have fathered the mammals. The third, *Lystrosaurus,* is a dicynodont, a less direct relative of the mammals.

The results of the volumetric analysis are summarized in Table 7.2.

Table 7.2
Quantitative Analysis of Brains and Bodies in Mammallike Reptiles[a]

Genus	Body volume (ml)	Endo-cast volume (ml)	Expected brain weight (g)	References
11. *Thrinaxodon*	500	0.4	0.45	Body, Brink (1958)
12. *Lystrosaurus*	50,000	8	9.6	Endocast and body, Piveteau, 1961)
13. *Diademodon*	7,000	8	2.6	Endocast, Watson (1913) Body, Brink (1956)

[a] Specimens numbered for Fig. 7.4; footnotes for Table 7.1 also apply here.

I have taken into account that the *Thrinaxodon* specimen that I examined was smaller than the one illustrated by Brink (1958), whose figure of the skeletal reconstruction was used to estimate this animal's body size. An interesting contribution of the volumetric analysis is in suggesting that brain and endocast sizes and, therefore, also shapes may have been similar for these mammallike reptiles. This conclusion must follow if these animals had brains of typical lower vertebrate size as estimated by Eq. (5.2). Since their endocasts were all very near the volume of these expected brain sizes and since the endocasts present maximum limits on their brain sizes, the mammallike reptiles could not have had brains that approached a mammalian size. This is certainly the most important contribution of the quantitative analysis and is well illustrated in Fig. 7.4, in which the data of Table 7.2 are included. Although there may have been some expansion of the cerebellum, as suggested by Olson (1944), such expansion could not have been large enough to produce the order-of-magnitude change in brain size needed to affect the quantitative analysis. The mammallike reptiles, in short, were reptilian and not mammalian with respect to the evolution of their brains. In that respect the quantitative analysis is in agreement with the conclusion of most workers who have limited themselves to the examination of the form rather than the size of the brain and endocast of reptiles.

CONCLUSIONS

On the evidence, the brain's size has been determined by approximately the same rule in the bony fish, amphibians, and reptiles, throughout their radiation in time and in space. For a period of at least 350 m.y., and perhaps as long as 425 m.y., this basic rule has been followed, and it continues to be the rule for the living lower vertebrates. The rule is best stated graphically by the polygon enclosing brain and body sizes in lower vertebrates, that is, the polygon shown in Fig. 7.4 and in Fig. 2.6 for modern vertebrates. The earliest positive evidence that this rule was being followed came from recovered endocasts and bodies of certain paleoniscid fish (Chapter 5) in which brainlike natural endocasts had been found. A kind of negative evidence favoring this rule lies in the fact that none of the recovered fossils of lower vertebrates from any strata have yielded endocasts that were so small as to place the animal clearly below the polygon, or map, of brain:body weight relations, shown in Fig. 2.4 for the living descendants of these vertebrate classes. In fact, the data on dinosaurs enable us to extend the reptilian polygon to a body size range comparable to that for the higher vertebrates of Fig. 2.4, and a beautifully regular view of brain:body size relations emerges from that extension.

I must reiterate, here, that the kind of brain:body analysis that I prefer for the analysis of the evolution of the vertebrate brain tends to emphasize regularities and obscure differences. There are, of course, very real differences in the brain pattern both between and within the vertebrate classes that I have discussed. But I would consider these, not in terms of the efficiency of the brains with respect to the amount of information that they can process, but with respect to the kind of information that they did process (Chapter 3). I have invoked a second important rule about the brain, the principle of proper mass: the brains of animal

Conclusions

species are always expanded in the regions that control specialized behaviors for specialized adaptive niches (Chapter 1). The niches occupied by fish, amphibians, and reptiles must never have required enough enlargement of regions of the brain to have had a significant effect on total brain size.

In short, in the early evolution of the vertebrates the brain, as a system, probably achieved the general information-processing capacity that it has presently in the lower vertebrates. Evolution of the brain within these groups has since been, essentially, a matter of specialization for processing different types of information at that level of capacity, and there is no evidence that there has been increased excellence in the quality of such specializations in the evolution of the lower vertebrates.

Although I am very hesitant about drawing morphological conclusions about the brains of fossil lower vertebrates, because of their inadequate representation in endocasts, a few conclusions are suggested by both the endocasts and the quantitative analysis. The spinal cord and cranial nerves in the lower vertebrates followed a generalized vertebrate pattern with the only specialization being the massive enlargement of the spinal canal in the sacral region of large herbivorous reptiles. This specialization is neither unusual nor truly specialized; it is the conservative solution to the problem of a large body size. The pituitary fossa was enlarged in large species, just as it is in modern large animals, and this fact indicates a similar conservatism with regard to endocranial morphology. The only clearly primitive feature in the fossil endocasts is the presence of a channel, the dorsal parietal opening, which must have enabled the pineal (epiphysis) to reach the surface. In modern animals, like *Sphenodon,* this is recognized as a primitive feature of the brain, although it also occurs in many living lizards. The parietal opening had disappeared in some dinosaurs, but it was present in the mammallike reptiles.

There are few suggestions of mammalian features in the brains of the mammallike reptiles (see pp. 204–205), such as the beginnings of the enlargement of the cerebellum. The forebrain, to the extent that its position is identifiable, was of reptilian size and shape. This was not the case in the earliest known fossil mammals. The brains in fossil lower vertebrates were probably similar to the brains of modern lower vertebrates, and all are primitive; only bony fish show significant specializations. A somewhat different story emerges for the flying reptiles, but that is deferred to the next chapter.

Chapter 8

Flying Reptiles and Aerial Niches

The first vertebrates to evolve body shapes that permitted sustained, directed, and controlled flight, powered either by muscular effort or by air currents, were the pterosaurs, or flying reptiles. Such flight evolved independently in many other animal species, and in reviewing the evolution of the brain and of brain size in the pterosaurs, it will be instructive to contrast it with some other groups. In particular, I will contrast it with the neural control of insect flight because of the evidence of neural "parsimony" in insect flight—the insect's control system must be the smallest possible for life in an aerial niche.

THE CONTROL OF FLIGHT

The small size of the insect's neural control system is evidence that the aerial niche, per se, does not necessarily produce selection pressures toward the specialized neural structures of the vertebrate brain. It is only if we recognize that the basic vertebrate organization of brain, muscle, and skeleton is to be the instrument for survival in an aerial niche, that we can appreciate the effect of selection pressures for that niche on the evolving vertebrate brain.

Among the vertebrates, controlled flight has evolved in flying fish, flying reptiles, birds, and mammals. In fish, flight is merely a brief glide, with power provided by ordinary swimming movements of the tail in water and lift provided by the pectoral fins. It is probably nothing more than a means of rapid forward motion through less resistant air rather than the normally resistant medium of water (Gray, 1968). Essentially, no novel neural control and information processing are required by the brain of "flying" fish beyond that required in any other surface fish. The ecological niche of the flying fish is basically in water; its brief excursions in air are no more than an extension of its normal locomotor patterns in water.

When the aerial niche was really central to adaptation, it was a drastic enough challenge to vertebrate genetic systems, a source of sufficiently strong selection pressures, to require clear and unmistakable modifications

The Control of Flight

of the brain and skull. The changes in the vertebrate brain, as reflected in the cranium, have been of three types. First, there was expansion of the cerebellum, presumably as a central neural flight control organ. Second, in birds, bats, and probably in the Mesozoic flying reptiles, there was expansion of the midbrain and forebrain structures, which are the projection areas in living animals for the finely discriminative distance receptors that they characteristically use, that is, the visual system in birds and reptiles and the auditory system in bats. Finally, in the birds and reptiles there was either expansion of the brain or constriction of the cranial bones that form its walls, which resulted in a tight fit for the brain in the cranial cavity, the normal mammalian condition. This resulted in endocasts that are brainlike in their external configurations, even in the flying reptiles, and the endocasts can be used as the basis for deductions about morphology of some special structures in the brain.

Evolutionary developments for life in the air did not necessarily result in notably enlarged brains. The brains of birds did evolve to larger size relative to body weight, whereas the brains of pterosaurs probably did not. Bats, fossil and living, also had enlarged brains, but this can hardly be considered as an evolutionary adaptation by bats to their aerial niches. An enlarged brain is part of the heritage of all mammals, and bats are fairly typical mammals with respect to relative brain size (see pp. 352–354).

The brain of flying mammals is involved in adjustments to aerodynamics and in the coordination of powered and gliding flight, but the aerodynamic issues are secondary for our purpose because the basic aerodynamics of flight are essentially the same for all animals using wings of any type to provide lift (and for flying machines as well). There are some interesting problems of powered flight in animals, which have been most adequately analyzed in insects, and we will consider these briefly for their implications for vertebrate neural control systems. These are still poorly understood in vertebrates, especially with regard to their demands on nervous systems and neuromuscular control.

A Minimum System for Neural Control of Flight: The Insects

The flight control mechanism in insects is like that in vertebrates in many respects, differing fundamentally in a few ways, such as the availability of a specialized detent type operation of the wings (Pringle, 1957). This enables the wings of advanced insects to beat more rapidly than would be possible if each wing required the continuous control of specific muscle groups and individual muscle fibers as it does in birds.

The extent of reflex control is remarkable in insects. Wilson (1964) has described some of its elements in grasshoppers as follows.

> Flight probably begins ordinarily with a jump following a random stimulus, such as a moving shadow or gust of wind, to the head. This not only results in a release of inhibition from the leg proprioceptors but also causes an air stream over the head hairs, which excites the flight system. Once flight is established, the forward movement of the animal maintains this wind stimulus, which maintains flight. Acting similarly are inputs from receptors located about the wings . . . . These wing sense organs respond during flight and help to maintain flight once it is started [pp. 335–336].

Although a more detailed exposition is out of place in a book on the vertebrate brain, the completeness with which the feedback loop that controls insect flight has been described by Wilson is worth our attention since an analogous loop is effective in birds and other flying vertebrates. There is even an analogy between the morphological components of the loop in insects and in vertebrates. The insect's exquisitely preprogrammed control is effected via the thoracic ganglion, corresponding to possibly similar control by the cerebellospinal system in birds. The insect's head ganglion, with its generally excitatory role, releases or disinhibits patterned movements and corresponds to the cerebral control exercised in birds. Thus, if the cerebrum is removed in birds, they may still be capable of flight, provided they are thrown into the air; the familiar demonstration of the decerebrate pigeon flying until it lands on the first roost it happens to reach is part of many first courses in physiology. This is quite analogous to headless flight, which can be demonstrated in insects. Without cerebellum, however, the pigeon is an impossibly gyrating animal, moving randomly from position to position, with each position grotesquely impossible for a normal bird (Dow and Moruzzi, 1958). Ablation of the cerebellum in birds may be comparable to removal of the thoracic ganglia, or much of these ganglia, in insects, although to be exactly comparable the ablation in birds would have to include cervical spinal nuclei as well.

In insects the feedback loop, as outlined in the locust by Wilson (1964, 1968) and summarized for other insects (as well as locusts) by Pringle (1957), is a recursive reflex system, in which the thoracic ganglion is a center for producing and maintaining an exogenous rhythm (analogously to the action of the heart muscle in vertebrates); this rhythm is modulated by signals from sensory cells in the wing muscles and on the wings, the legs, and the head. The control system is automatic and could be simulated by a relatively small number of components in a digital computer. The thoracic ganglion does its work with about 30 microscopically visible nerve cells per wing, instead of the many thousands of neurons that must be involved in the control of the flight musculature of birds. [No counts are available, to my knowledge, of the neurons actually involved in that work in birds, but since the basic spinal innervation of the wing musculature in birds is the same as the innervation of pectoral, back, and forelimb muscles of other vertebrates (Gray, 1968), one can assume that the numerical relations of nerve to muscle cells (Aitken and Bridger, 1961; Cooper, 1966) are also comparable.] These facts of insect flight are important because they illustrate how small the flight control system may be while still being able to do its work.

Neural Control of Flight in Birds

The vertebrate solution to the problem of flight is inefficient in that much more than a minimal amount of biological material is devoted to the job. The vertebrate solution is limited, however, by the fundamental plan of neuromuscular control in vertebrates, with the development of essentially 1:1 relations between muscles and nerves. [The actual relations between number of muscle cell nuclei and their controlling nerve cell nuclei in vertebrates is much higher, of the order of 100:1 or more, for some muscle groups, as pointed out by DuBrul (1967), but the 1:1 ratio correctly describes the physiological, as opposed to the anatomical, situation.] Thus, in order to produce motion in a wing, a more or less continuous contraction and relaxation of opposing muscle groups must be produced, and there must be a correspondingly continuous modulation of the nerve signals to these muscle groups.

The complete control of flight in birds may involve not only flapping the wings, or keeping them steady for gliding, but also changing the angle of attack of the wing, "feathering," and the various changes in the wing and tail that result in momentary stability in an inherently unstable system. Unlike the neuromuscular basis of insect flight, the flight of birds (and bats) is understood only with respect to its aerodynamics (Gray, 1968). We, therefore, know the type of information that birds must process, but we can only speculate about how they process that information. From the better understood, but comparable, information processing in flying insects we can make some quantitative estimations to set a framework for speculations.

The effector system and its control in insects are based on only a few dozen neurons, but the full system with all its feedback loops requires more elements. Nevertheless, the entire central nervous system in insects certainly has fewer than 100,000 nerve cells (Bullock and Horridge, 1965, p. 807), and educated guesses would limit the number to 10,000 or so. The vertebrate solution to the problem of flight involves processing the same kind of information with hundreds of thousands or, more likely, millions of neurons, and one is inclined to believe that rather different mechanisms are used when so many more elements are available for doing the processing.

Consciousness, Reflexes, and the Control of Flight

I would like to speculate about these differences by devoting the following paragraphs to the question of what the millions of cells do. This is really the fundamental problem of psychobiology. Knowledge about the functions of individual nerve cells is now so good that essentially complete histories of the chemistry and physiology of the action potential and other information processing at the cell membrane are introduced in beginning texts (Brazier, 1968; Stevens, 1966). We are rapidly approaching a similarly satisfactory state with respect to the action of small aggregates of neurons, such as those controlling the flight of insects or the movements of the legs of crustaceans (Kennedy, 1967). And it is possible to recover a good deal of information about functional relationships among peripheral and central parts of the nervous systems of vertebrates by examining the patterns of electrical response by receptor fields in the brain to patterns of stimulation of receptor cells in the periphery (Eccles, 1966). But, there is still a gap between our understanding of the functions of these "part systems," or miniature systems, and understanding the functions of systems of cells in very large numbers, as they occur in most vertebrate brains.

Following the general argument presented in the introductory chapter (pp. 16–23), I suggest that the phenomenon of consciousness, which forms so central a part of the human experience, provides a clue to the functions of large aggregates of neurons. One may simulate the functioning of an insect's wing on a modern computer by using about as many information-processing elements in the computer system as the insect actually uses in working its wings (Reiss, 1964). It is, therefore, easy to assume that insect flight is controlled in a way appropriately modeled by our understanding of computer techology. By extension, we may consider the insect to operate as a Cartesian reflex machine, in which it is possible to describe the system deterministically, much as one describes many working machines. The vertebrate system is not adequately modeled by digital machines thus far developed, and I would suggest that, until a machine is specified with something comparable to consciousness as part of its mechanism, a machine model of the vertebrate brain as a whole will be faulty. The essentials of consciousness for such a machine may be the production of analogies or models that become a "real world," within which action is possible.

Craik's (1943) description of thought as model building to handle the information of the senses may be the best and most complete statement of this view of the working of a nervous system. I have reviewed this point of view in the introductory chapter, in which I suggested that the work of vertebrate brains, beyond the level of simple reflexes, involves at least a primitive type of consciousness in which "maps" of a "real" world are created, and an animal's behavior is developed with respect to these maps (Tolman, 1948). The human animal, in creating such a map, also creates the "real" world (the "really" real world!), the world of everyday experience. This is the world filled with solid objects, rather than the empty space that the physicists proclaim; it includes living plants and animals, more or less predictable actions (which occur in similarly created "real" time) by living and nonliving matter, and socially desirable and undesirable people and things.

The bird or bat (or pterodactyl) in flight may be thought of as a more or less conscious machine, following instructions to move through a space defined by a map in its head, rather than as a machine preprogrammed and thus instructed to activate muscle-controlling systems when specific patterns of stimuli occur at selected points on its anatomy. The latter could result in the kind of movement that we perceive when an insect flies through our real-world space (our "map") from one point to another along a particular path. Is an insect's reflex system fundamentally different, in any way, from the hypothetical vertebrate system that creates maps of possible worlds? I suspect that it is. Whether all vertebrates have had brains large enough to contain stored "maps" with "objects" placed against "background" is an open issue, and, in general, I have assumed that extensive conscious perceptual worlds appeared only with the evolution of mammals and, to a lesser extent, birds.

I have presented these speculative paragraphs to suggest the nature of the problem of understanding the function of neurons in large numbers and to contrast the kind of solution to the problem of flight that may have occurred in some vertebrate species with an equally successful solution developed by insects. Let us, now, return to the discussion of the available data on the evolution of the vertebrate brain for life in aerial niches.

FLYING REPTILES

Among the spectacular archosaurians, or ruling reptiles, the dinosaurs discussed in the last chapter are rivaled only by the pterosaurs in their appeal to the imagination. These flying reptiles, which are mainly from the middle and later Mesozoic era, are remarkable in paralleling birds in some ways and bats in others with respect to their structural adaptations for flight. Excellent photographs, drawings, and paintings, which are both esthetically pleasing and scientifically accurate, are available in Augusta and Burian (1961), a popular work on Mesozoic flying vertebrates. These pictures effectively catalog the adaptations. There is also much charm in Seeley's (1901) classic account, recently reprinted in 1967. The pterosaurs' efficiency as flying machines and their probable methods of flight have recently been reanalyzed by Bramwell (1970a,b), Bramwell and Whitfield (1970), and in great detail by Heptonstall (1971a). The adaptations of the pterosaurs are discussed mainly for their implications for body size estimation, an important problem for the present approach in which the measure of relative brain development depends on brain:body relationships.

The common ancestors of pterosaurs, birds and bats, come, of course, from older reptile species and are to be found among the stem reptiles of the Permian period. The earliest recovered pterosaur fossils are from Lower Jurassic levels, 175 m.y. old. These are the long-tailed rhamphorhynchids, which did not survive the Jurassic period. The pterosaurs that are familiar from popular writings and fiction are the pterodactyloids, which include small species such as *Pterodactylus elegans,* the size of a sparrow or bat, as well as the largest of the flying reptiles, *Pteranodon,* which had a wingspan of as much as 5 m and probably weighed about 30 kg. The pterodactyloids, which were almost or entirely tailless and had more elongated wings than the rhamphorhynchids, survived into Cretaceous times. The body configurations and relative sizes of representatives pterosaurs are illustrated in Fig. 8.1.

The endocasts of pterosaurs look like brains, and, like the endocasts of the small paleoniscids among the fish (Chapter 5), they can be analyzed as direct representations of the brain. They are genuinely fossil brains with respect to external configuration, and there is no reason to assume that the living brains of these animals, as they existed during the 100 m.y. duration of the order Pterosauria, were different from the endocast in gross appearance. One reason for such excellent molding of the brain by the endocast is to be found, at least in pterosaurs, in the general loss of excess bone as part of the adaptation to flight. The long bones of pterosaurs were pneumatic, as are the long bones of living birds. Reduction in weight is also readily effected by minimizing the surfaces of the cranial bones, and this is most easily done by having thin bones closely fitted to the tissue they enclose.

Another likely reason for the reduction of the cranial bones and the development of a cranial cavity that molds the brain can be derived from diurnal habits, which can be deduced for the pterosaurs. These flying reptiles were all characterized by great enlargement of the orbits and the development of the ossified sclerotic ring of the eye (Fig. 8.2). This feature, and the fact that the adaptations of all the pterosaurs suggest that their habits were analogous to those of fishing birds such as albatrosses

Flying Reptiles

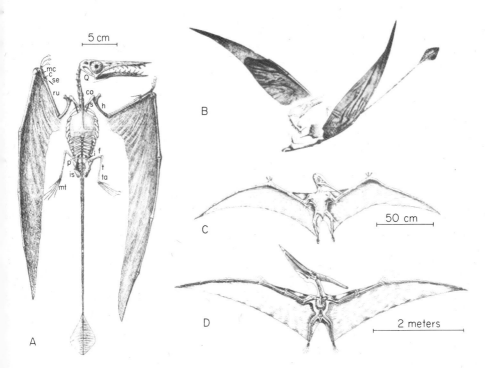

Fig. 8.1. Restorations and fossil pterosaurs to illustrate body and wing configurations and relative size. A, *Rhamphorhynchus*; B, lithographic slate of *Rhamphorhynchus* as found; C, *Nyctosaurus*; D, *Pteranodon*. From Seeley (1901); Heilmann (1927); and Piveteau (1955).

and pelicans, identify the pterosaurs as animals that were active during the day and as animals whose circadian and other rhythms were controlled by light as well as temperature. Thinning of the cranial bones would have allowed variations in ambient light to affect the control centers in the brain (e.g., pineal body, hypophysis, or hypothalamus), as they are known to affect these hidden centers in living birds such as sparrows (Binkley *et al.*, 1971).

PTEROSAUR BRAINS

Speculations aside, the evidence is unequivocal that the endocasts of pterosaurs are brainlike in appearance. The several that are illustrated (Fig. 8.3) clearly reveal the rounded cerebral hemispheres of the forebrain. The optic lobes, which form the visible portion of the midbrain, are in positions similar to their place in the brains of living birds, although they seem relatively smaller than those of birds. The endocasts reflect a

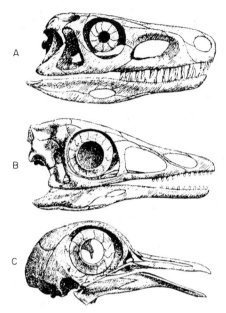

Fig. 8.2. Skulls of reptiles and birds to illustrate the sclerotic ring. A, *Euparkeria capensis*, a Triassic reptile; B, *Archaeopteryx*, the Jurassic bird; C, *Columba domestica*, the living pigeon. From Heilmann (1927).

brain that appears avian also in the region of the cerebellum. Edinger (1927, 1941) described the brains of flying reptiles and in particular that of *Pterodactylus*. Her conclusions follow.

> The brain of *Pterodactylus* is quite un-reptilian, almost entirely bird-like in form.
>
> Avian were the nervous centers of the olfactory and optic senses. The olfactory bulbs, shifted about the orbits, were as much reduced as in the later Pterodactyloidea. The large optic lobes were situated latero-basally; they were removed even farther downward in Cretaceous forms—in the Rhamphorhynchoidea, similar development can be followed from the Lower to the Upper Jurassic—.
>
> The cerebellum, center of the statotonus, was placed as in birds adjoining the forebrain; but it did not extend as far downward as in birds though farther than in a Liassic late Triassic rhamphorhynchoid.
>
> The upright direction of the medulla oblongata is avian.
>
> The forebrain seems to have possessed both the fissures of the bird forebrain; but in acquiring avian outlines it was slower to follow the trend towards bird-likeness than the other brain parts. Slender in the Liassic rhamphorhynchoid and the Tithonian *Pterodactylus*, it achieved avian breadth in the Tithonian Rhamphorhynchoidea and in the Cretaceous Pterodactyloidea [Edinger, 1941, p. 681].

Flying Reptiles

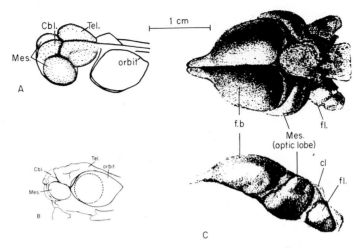

Fig. 8.3. Endocasts of pterosaurs. A, *Pterodactylus kochi*; B, *Pterodactylus elegans*; C, *Scaphognathus purdoni*. From Edinger (1941); Newton (1888).

Edinger's conclusions are notable both with respect to their detail and to their certainty in identifying the pterosaurian brain as avian. I am generally less impressed than Edinger, however, by the geometrical relations among the brain's structures. It seems almost irrelevant to me whether the medulla is oriented vertically or horizontally, whether the cerebellum touches the cerebrum, or whether the optic lobes are oriented laterally or laterobasally. These facts are probably determined by mechanical considerations such as the space available for the brain in the skull and the orientation of the head on the neck and body. A vertically oriented medulla is typical of living birds and man, but the horizontally oriented medulla of most mammals is oriented parallel to, and in line with, the normal orientation of the body, as in the "lower" vertebrates, and is hardly a measure of primitiveness. The position of the optic lobes and the degree of contiguity of cerebellum and cerebrum seem to raise issues that are mechanical in the same sense as the problems of the source of fissurization in the brains of mammals, as discussed by LeGros Clark (1945).

In short, I would reserve judgment about the avian aspects of the brains of pterosaurs until I could add an analysis of their sizes to the analysis of their shapes. The avian form of the pterosaurian brain may be nothing more than evidence that the cranial cavity molded the brain and not that the brain contained sufficiently expanded neural tissue to function at the advanced levels achieved by birds. To determine the functional level of pterosaurian brains I will rely on volumetric data of the type used in previous chapters.

Brain and Body Size in Pterosaurs

The method of numerical double integration used in other chapters (see Chapter 2) will also be used for the analysis of the size of the brains of pterosaurs. It will be recalled that the method of double integration requires estimates of the averages of both the height and the breadth of the brain (or endocast—or any irregular solid) and can be made from sketches of lateral and dorsal views. The measurements define the lengths of the two principal axes of an elliptical cylinder, the length of which is the length of the brain. The volume of the cylinder is then given by the formula $\frac{1}{4}(w \times h \times l \times \pi)$, where w, h, and l are the average width and height and the measured length of the brain, respectively. For those endocasts in which there is only a lateral or dorsal view, we assume that the relationship between width and height is sufficiently stable across species that each can be predicted from the other. With that assumption one can estimate brain size in *Pterodactylus elegans* by considering it as similar to that of *P. kochi,* and Newton's (1888) figures of the brain of *Scaphognathus* serve as the model for the ratio between breadth and height of the endocast of *Rhamphorhynchus gemmingi*.

Reconstructions as figured in the literature for the endocasts of two species of *Pterodactylus,* two rhamphorhynchids, and *Pteranodon,* shown here in Fig. 8.3, enable us to estimate the brain sizes of these specimens. These are presented in Table 8.1. Body size estimates in this table were made by developing length:weight and lift-surface:weight formulas.

A weight:length formula was developed first by assuming that living bats are similar to pterosaurs in this relationship. The weight:length function in living bats was derived by taking a sample of measurements from

Table 8.1
Estimated Brain and Body Sizes in Pterosaurs[a]

Species	Brain weight (g)	Body[b] weight (g)	Head and body[c] length (cm)
1. *Pterodactylus elegans*	0.14	60	11
2. *P. kochi*	0.42	450	30
3. *Ramphorhynchus gemmingi*	0.70	310	25
4. *Scaphognathus purdoni*	1.7	1,500	55
5. *Pteranodon* sp.	4.8	20,000	200

[a] Edinger (1941) has a full bibliography of these endocasts.

[b] Estimated from body length by use of Eq. (8.1): $P = 0.5 \, L^2$.

[c] Head and body length of *Rhamphorhynchus* and *Scaphognathus* shown as snout to cloaca (base of tail). Body length of *Scaphognathus* estimated from skull length as reported by Newton (1888), assuming its relation to torso and tail as in Seeley's (1901) reconstruction of *S. crassirostris*.

Walker (1964). The procedure differed from that used in Chapter 2 because it was important to weight the sample toward larger specimens. My actual procedure was ad hoc: I found an approximate centroid in the most populous region of weight:length points—among the smaller bats. I then drew a line on log–log paper from that centroid to a point that included the largest living bat, the "flying fox," *Pteropus giganteus,* using a length (head and body) of 41 cm and a weight of 900 g, as given by Walker for the heaviest specimen that he had recorded. I then wrote an approximate equation for that line, an equation that could serve as a kind of nomogram for estimating the weights of large pterosaurs. The equation was:

$$P = 0.5\, L^2 \tag{8.1}$$

with the body weight P in grams and the length L in centimeters.

It is instructive to compare Eq. (8.1) with the fitted equation to the large sample of carnivores and ungulates as given in Chapter 2, that is, Eq. (2.4), $P = 0.025\, L^3$. The values of P and L in the two equations are the same at $L = 20$ cm and $P = 200$ g. As a curve-fitting exercise, in fact, the data on a large sample of bats do fit Eq. (2.4) quite well. It is only the larger specimens, such as *Pteropus,* that show major shifts from the more general mammalian equation, Eq. (2.4). According to that equation a 41-cm mammal should weigh 1700 g, almost twice the weight of *Pteropus.* The correct interpretation of this result is that bats, like many flying vertebrates, responded to selection pressures by reducing body weight, and this is most apparent in the larger forms. Eq. (8.1) is, therefore, the best approach to estimating body size in large flying vertebrates such as the pterosaurs, as a "quick" method. An analysis based on aerodynamic considerations is, of course, superior (cf. Heptonstall, 1971a, and our discussion below).

With the help of Eq. (8.1) it was possible to complete the analysis of brain:body data on the pterosaurs. The analysis is summarized in Table 8.1, including comments on the measures presented in the footnotes to the table. On the whole the estimates seem reasonable, but some special comment on the estimate for *Pteranodon* is in order in view of the recent debate about this (see Heptonstall, 1971a, for the most complete discussion).

Analysis of Body Size in Flying Vertebrates

An alternate and more rational approach to the analysis of body size in flying vertebrates would emphasize aerodynamic considerations—the requirements for powered flight and the requirements for gliding and performing controlled maneuvers in the air. A full analysis is beyond our present scope (see Heptonstall, 1971a), but a partial analysis is appropriate, both to suggest the style of thinking that is required and to correct some errors that may persist in textbook accounts. I will limit this discussion to the weight of the largest of the specimens, *Pteranodon,* and to the probable style of life of this largest of all animals in an aerial niche. *Pteranodon* certainly glided and probably was capable of powered flight; it had adapted to a niche similar to that of the living albatross. One can estimate the efficiency of this reptile as a flyer and determine the ease (or difficulty) with which it could move its assumed 20-kg weight about as a glider. The procedure for making the calculations is described by Gray (1968) and in the works he cites. Two points are worth mentioning. First, powered flight by a bird, that is, flight that results from flapping wings and expending muscular energy rather than from simple gliding, is considered effectively impossible in winged vertebrates weighing more than 14 kg (Gray, 1968, p. 235). Thus, if my estimate of the weight of *Pteranodon* is correct, this animal must have been a glider rather than a flier. Second, in a consideration of the aerodynamics of gliders the weight of *Pteranodon* was estimated to be 12 kg, possibly in order to be under the 14-kg limit of muscle-powered flight (Wilkie, 1959). In a review of the capability of *Pteranodon* as a flier (Piveteau, 1955, p. 980), this reptile was described as an unusually efficient glider, comparable to the very best man-made machines available in the early 1930's. Its capacity for powered (flapping) flight was not considered.

To improve the estimate, data on the surface area of the wings of *Pteranodon,* 4.65 m² (Piveteau, 1955), were used for further computations. The efficiency of a glider and its speed of descent are determined by the ratio of the wing surface to the body weight, and one would expect such a ratio to be fairly stable among animals adapted for gliding flight or for flight in general. Hartman (1961) presents extensive data on the relationship between wing surface area and body weight in many birds. The relationship is, in fact, rather good and is approximated by the equation

$$S = 8.2\, P^{0.82} \tag{8.2}$$

(cgs system) in which wing surface S is in centimeters squared and body weight P in grams. Applying this equation to *Pteranodon* to determine its "body weight" from its wing surface, we find that its body weight is estimated as 37 kg. It is apparent that a bird with the wing surface of *Pteranodon* would have had to weigh even more than my estimated weight for that pterosaur, and if I erred, it was probably in the direction of estimating too low. Since the brain size estimates of Table 8.1 will tend to place the flying reptiles with other reptiles with respect to the

Flying Reptiles

level of development of the size of the brain for a given body size and since this controverts the "classic" position held by Edinger, it is more appropriate to err in the direction of underestimating the body weight, as mentioned before, because that minimizes the risk of erroneously rejecting Edinger's generally accepted views. I regard my estimate of 20 kg for *Pteranodon* as minimal, however, and the suggested 12 kg seems markedly low and unrealistic. When Heptonstall (1971a) reviewed this issue in detail, he concluded, independently, that 38 kg was the most likely weight of *Pteranodon*.

The decision to regard the pterosaurs as reptilian rather than avian with respect to relative brain size rests on the relationship between brain size and body size in reptiles and in birds. In Fig. 8.4 I have graphed the

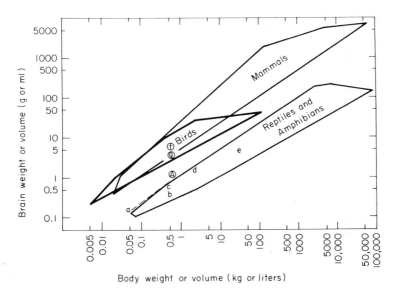

Fig. 8.4. Brain:body relations in pterosaurs and in fossil birds, compared with brain:body maps (minimum convex polygons) of reptiles, birds, and mammals, as shown in Figs. 2.6 and 7.4. Pterosaurs are labeled by letter to correspond to Table 8.1. Birds, to be discussed in Chapter 9, are A, *Archaeopteryx lithographica* (Upper Jurassic); g, *Numenius gypsorum* (Eocene); t, *Numenius tahitiensis* (Recent).

relevant data on contemporary vertebrates from Crile and Quiring (1940), as I have done in previous chapters in making decisions on this kind of question. I have added to the graph the results presented in Table 8.1. The unique data on *Archaeopteryx* and on *Numenius gypsorum*, discussed in the next chapter on birds, are also included in Fig. 8.4. The decision regarding the pterosaurs is easily reached with the help of Fig. 8.4. The pterosaurs were reptilian in brain size, although they were all near or even at the upper border of the convex polygon that contains the data on living reptiles. The contrast with *Archaeopteryx* is also notable because this earliest of the birds, which was the contemporary of some of the pterosaurs shown in the graph, had apparently departed from the reptilian level, though it was not yet at the level of living birds. The Eocene *Numenius,* however, was well within the polygon of living birds.

SOME RESERVATIONS

The data in Fig. 8.4 are impressive, but I have to provide a few caveats, lest they convince too easily. I have reservations, in this case, about both the brain size estimates and the method of estimating body size. The best method for estimating brain size is the direct method used for some of the dinosaur endocasts discussed in the previous chapter and which will be used for many determinations of mammalian brain (endocast) sizes. This is the Archimedean method of water displacement. Since the method of double numerical integration depends heavily on accurate dorsal and lateral views, I am uncertain of the accuracy of the very limited data on fossil flying reptiles. The lateral projections in *Pterodactylus kochi* and *Pteranodon* involve guesses from incomplete or distorted photographs and drawings. The dorsal and lateral projections of *Scaphognathus* are entirely accurate, and I have been able to verify this by comparing them with the originals in London. The lateral projection of *Rhamphorhynchus gemmingi* had to be guessed by analogy with the uncertain *Scaphognathus,* and the dorsal projection of *Pterodactylus elegans* had to be guessed by an uncertain estimation from *P. kochi,* which is illustrated only in three-quarter view.

My major reservation is about the body size estimates in the analysis of the tailed pterosaurs. The tails of these animals were so long that it is impossible to ignore the likelihood that they contributed significantly to the body weight, and I have almost certainly underestimated the body weights of these animals by using the weight-length function derived for bats. The best approach to correct the problem would be to estimate the volumes of the whole animals by constructing scale models, as Colbert (1962) did with dinosaurs. I might note, parenthetically, that I am, generally, more at ease with volumetric data than with weight data and that the information that I have presented on living animals, which was almost always information on weight, would be easier to integrate with fossil data if it were given by volume. This is especially true for flying animals, which are under selection pressures toward lower body weights and specific gravities.

CONCLUSIONS

Given these reservations, it seems to me, nevertheless, that the pterosaurs, though having brains that looked like the brains of birds, had not departed from the reptilian level of brain development. As I indicated earlier, they had responded to selection pressures toward lower weights and specific gravities by reducing the weight of the bone and molding the soft tissues of their bodies more exactly by their surrounding bones. The system that best illustrates the response to these selection pressures is, of course, bone itself, but I argue (following Heilmann, 1927, and many others) that the cranium/brain system also responded to such pressures by thinning the bone around the brain and fitting it more closely to the brain. The endocasts of the flying reptiles are, thus, brainlike, and they give the appearance of having reached an avian level. But this is appearance, in my judgment, not reality.

In one respect the form of the brain in pterosaurs is clearly avian, and this requires special discussion. As Edinger (1941) remarked in her summary of the brain of *Pterodactylus elegans,* there was a great expansion of the optic lobes (midbrain) in the pterosaurs and a reduction in the olfactory bulbs. This is a clear indication that the life habits of pterosaurs were like those of modern birds, as indicated earlier. The pterosaurs must have been active during the day, living by their visual sense, and they probably made little use of olfaction as a distance-sensing system. Within a particular level of brain evolution, as measured by the gross brain size, there will always be specialized adaptations reflected in the relative enlargement of specialized structures. The same kind of specialization occurred in the paleoniscid fish (Chapter 5), which antedated the pterosaurs by 100 m.y. or more; these fish also had markedly enlarged midbrains, specifically, optic lobes, and their brains were similar in form to the brains of certain living fish. But this is not an indication of a jump in the level of brain evolution. It is a measure of the similarity of specializations in the ancient and living fish. Only from the fact that living fish have brains similar in size as well as in shape to those of their Carboniferous ancestors do I judge that the level of brain evolution has not advanced in the fish since those times.

One might, thus, conclude from the similarity in the shape of the pterosaur brain to that of living birds that there was a convergence of response to selection pressures, a kind of response presumably inherent in the vertebrate genetic systems that control the growth of the brain. To be a flying visual vertebrate demands that the control systems for the activity include brain centers, such as the optic lobes, cerebellum, and forebrain, that are involved in processing visual information. And these brain centers will necessarily become enlarged relative to other parts of the brain. Within the reptilian level of brain evolution, the flying reptiles were at the upper edge of development rather than scattered throughout the reptilian polygon of Fig. 8.4. But if the reptilian level of brain evolution is defined by the reptilian polygon of Fig. 8.4, the flying reptiles were still clearly reptilian. The response to selection pressures by an overall enlargement of the brain, or by a sufficient enlargement of parts of the brain to result in an increase in its total size above the lower vertebrate level, had not occurred in the pterosaurs.

Part III

Brain Enlargement and the Basic Vertebrate Radiation

It is my thesis that the evolution of mammals and birds was the result of the invasion of practically empty adaptive zones rather than the replacement of reptilian occupants that were well established in a significant number of niches within such zones. The basis of the argument is twofold. First, the earliest mammals were contemporaries of reptiles that were only beginning their major Mesozoic radiation, and the first birds, which appeared shortly thereafter, were contemporaries of an extremely successful (according to evolutionary criteria) order of flying reptiles. This indicates that mammals and birds did not compete successfully for niches invaded by the reptiles during the great adaptive radiation of the Age of Reptiles. Second, the characteristic adaptations of mammals and birds would be suitable for new niches closed to the reptiles, which lacked these adaptations. The adaptations were homoiothermy (or reasonably stable body temperature controlled by reflex neural mechanisms), enlargement of the brain, and, in the mammals, the specific enlargement of the olfactory and auditory systems and the development of the rod system in vision.

I will argue that the earliest mammals were "reptiles" that had found a niche as nocturnal animals. A nocturnal niche would exert strong selection pressures for developing mechanisms to control temperature in the absence of sunlight and to use distance information from modalities other than fine vision. I will suggest that the niche of the earliest birds was similar to that of many living primates, as tree dwellers in woodlands, often deprived of much of the sun's warmth, which have to make unusually fine visual discriminations to detect well-camouflaged targets against the mottled and colorful background of a typical woodland.

This thesis is related to older viewpoints, such as Marsh's (1886) and

Osborn's (1910), that the relative size of the brain was instrumental in determining the survival and replacement of vertebrate species. Evolutionary thought of the present time is incompatible with the details of statements of that vintage. A modern statement must specify the selective advantages of enlarged brains, especially in the light of the successful survival and evolution of so many small-brained species of "lower" vertebrates.

Prior to the appearance of mammals and birds, the demands of vertebrate niches on capacities for sensorimotor integration and neural control could be met by relatively small additions or reorganizations of nervous tissue in the brain. Thus, as shown in earlier chapters, the overall map of brain:body relations for lower vertebrates is a polygon that has been unchanged in size and shape since the earliest vertebrate radiations, except as it has accommodated the evolution of larger bodies. The replacement of species within these niches did not involve the evolution of measurably enlarged brains (relative to body size) in the successor species. The niches occupied by birds and mammals, however, did provide a selective advantage associated with some increase in relative brain size. A mammalian or avian succession is, therefore, characterized to some extent by enlargement of the brain.

The evolution of the brain in birds is described in the first of the ensuing chapters. Although the evidence is admittedly scanty, based as it is on endocasts of a single Mesozoic bird, an early Tertiary bird, and Quaternary species that are essentially modern, it is, nevertheless, crucial and compelling. The Mesozoic bird, *Archaeopteryx,* had a brain larger than in any comparable living reptile, though smaller than in any comparable living bird. There must, therefore, have been an adaptive response by the birds in their evolution after *Archaeopteryx,* which resulted in the continued enlargement of the brain. The late Eocene species *Numenius gypsorum* provides evidence of that enlargement in progress, in the evolutionary sense.

When the mammals first appeared, there must have been a rapid evolution of sensorimotor and neural control apparatus. The earliest mammals in their adaptive zone probably had brains that were already enlarged compared to the brains of their immediate ancestors among the reptiles. As the zone remained occupied by a succession of mammalian species, the selection pressures for enlarged brains did not change, and there were no more major changes in relative brain size, except as required by an early Cenozoic radiation in which many species evolved large bodies.

This theme is developed in the following three chapters. In Chapter 9 I describe the evolution of the avian brain. In Chapter 10 I outline the evolution of the mammals in the Mesozoic era, as well as some of the subsequent occupation by Cenozoic species of what I will call "Mesozoic niches." The endocasts, brains, and bodies of animals occupying these niches are described, and I discuss the selection pressures associated with those niches. In Chapter 11 I discuss aspects of the first major radiation of mammals in the Cenozoic era, during the Paleocene and Eocene epochs, most of which was accomplished without further significant relative enlargement of the brain beyond a basal mammalian level. The final chapter of this section gives a speculative theoretical analysis of the reasons for the early enlargement of the brain in birds and "archaic" mammals, that is, mammals from orders or families that were completely replaced in their niches. In the chapters of Part IV I will discuss the adaptive radiation of the mammals, in which further brain enlargement, which may be correlated with the evolution of "intelligence," is particularly important.

Chapter 9

Evolution of the Brain in Birds

As in the pterosaurs reviewed in the previous chapter and the mammals considered later in this book, the endocranial cavity of birds is a tightly fitted capsule about the brain. Endocasts of birds can truly be visualized as "fossil brains." Only two important fossils are considered in this chapter in detail, but these will enable us to get a reasonably good picture of the way in which brain size evolved in birds.

The most important avian endocast, and one of the most interesting of all vertebrate endocasts, is that of *Archaeopteryx lithographica* of the Upper Jurassic period. This species is a genuine "missing link" in evolution, classified with the birds, yet retaining enough reptilian features to have been considered a reptile had there not been evidence of fossilized feathers. The endocast is of special interest because in the most complete monograph on *Archaeopteryx* it was considered to be one of the reptilian rather than avian features of this bird (de Beer, 1954). I have recently shown that this was probably an erroneous assignment (Jerison, 1968), although there is still some question about the correct view of the endocast (Dechaseaux, 1968).

Of other fossil avian endocasts, we analyze only that of *Numenius gypsorum,* a Tertiary species that had been exhibited at the Museum d'Historie Naturelle in Paris for many years but was only recently described in detail (Dechaseaux, 1970). This cast is of special interest because it represents an early member of an extant genus and, therefore, provides evidence about the evolution of the bird brain during the past 40 m.y. or so.

The extensive reports published by Marsh (1880) on the "toothed birds" of the late Cretaceous are considered only in passing. Several generations ago these figured prominently in most accounts of the evolution of the birds. Marsh presented illustrations of their endocasts and made various conjectures about the evolution of their brains. Much of that material is now questioned, and his reconstructions of the brains have been very severely criticized by Edinger (1951). To the extent that it is appropriate, I will review the problem and the controversy surrounding Marsh's material. The remaining known endocasts of extinct birds are

Pleistocene or Recent, and these are indistinguishable from those of living representatives of related groups.

EVOLUTIONARY BACKGROUND

The evolution of the birds is described, briefly and popularly, by Swinton (1965). A less detailed but beautifully illustrated review has been presented by Augusta and Burian (1961). Heilmann's (1927) classic work is still useful, although often fanciful; it has been superseded in the case of the description and analysis of *Archaeopteryx* by de Beer's (1954) description of the specimen at the British Museum. Stingelin's (1958) monograph is an excellent source of comparative data on living birds. The standard texts like Romer's (1966) and Piveteau's (1955) are quite useful, and the brief, readable descriptions by Young (1962) are models of exposition of the zoological framework within which to appreciate the place of *Archaeopteryx* and other fossil birds. There are also several relevant articles in Marshall (1961), including Portmann and Stingelin's (1961) discussion of the comparative anatomy of the visible portions of the brains of living birds, based in part on Stingelin's (1958) monograph mentioned above.

It is a familiar fact that among the flying vertebrates birds, pterosaurs, and bats represent very different lineages. It is worth reviewing these lineages, however, to note when in history these groups diverged from one another. All can be traced to the stem reptiles (cotylosaurs) of the Carboniferous period, and their common ancestors date back perhaps 300 m.y. The first divergence was between the reptilian ancestors of the mammals and those that were common ancestors of both birds and pterosaurs. That divergence (between the synapsids, from which therapsid, or mammallike, reptiles were derived, and the thecodonts, which were the earliest of the ruling reptiles and the ancestral group from which both birds and pterosaurs descended) is, thus, perhaps 300 m.y. old.

The divergence between birds and pterosaurs must have occurred much later, and their common (thecodont) ancestor lived at the end of the Permian period, about 230 m.y. ago. The earliest members of these groups are known only long after the divergence of their ancestral lines. *Archaeopteryx,* as noted earlier, is a late Jurassic fossil of about 150 m.y. ago. The earliest pterosaurs are known from the Lower Jurassic, about 180 m.y. ago, and it is noteworthy that their major adaptive radiation was contemporaneous with the first appearance of birds. Although mammals are known from that entire period (and from Upper Triassic beds as well), the earliest known flying mammals, the bats, are Cenozoic species.

Evolutionary Background 179

For our purpose, this outline should be used to suggest the duration of the independent histories of the several groups and to suggest the extent to which common adaptations are to be attributed to common ancestry (homology) as opposed to convergent or parallel evolution. The enlargement of the brain in birds and all mammals (including bats), which will be a major theme of the discussion in this and the following chapters, is an instance of parallel evolution that occurred independently in two divergent "reptilian" lineages (the lineages are identified by their original reptilian ancestors). The persistent, relatively small brained, status of the flying reptiles is evidence that entry into an avian niche did not, of itself, insure selection pressures toward an enlarged brain.

Archaeopteryx AND OTHER MESOZOIC BIRDS

The first fossil birds that have been recovered are several skeletons and a single feather of *Archaeopteryx* (de Beer, 1954; Ostrom, 1970). These are all Upper Jurassic and are thus about 150 m.y. old. *Archaeopteryx* had many reptilian features typical of land vertebrates and of its thecodont ancestors, such as solid rather than pneumatic bones, true teeth, and perhaps a dozen characters, in all, that de Beer describes as reptilian. The major avian feature of this species, and the only one that really identified it as a bird, is a full set of feathers (Fig. 9.1). There are, in addition, finer osteological features of the jaws and pelvis, which de Beer considered as approaching the avian form and departing from the reptilian form.

Among taxonomists the line between the class Reptilia and the class Aves is often considered to be less sharp than the lines among other living classes of vertebrates that are in an ancestor–descendant relationship. Romer (1966), for example, began his chapter on birds by stating:

> Although the birds are grouped as a separate vertebrate class—Aves—they are, apart from the power of flight and features connected with it, structurally similar to reptiles. Indeed, they are so close to the archosaurians that we are tempted to include them in that group . . . [p. 164].

The closeness of reptiles and birds may have led to a number of unwarranted judgments about reptilian features in a group of later Mesozoic avian fossils, that is, the Upper Cretaceous toothed birds, Odontognathae. As mentioned earlier, these fossils were first described and popularized by the great impressario of nineteenth-century American paleontology, O. C. Marsh (1880; for the flavor of Marsh's personality see Schuchert and LeVene, 1940). The enormous range and importance of Marsh's work, in which he supervised the discovery, excavation, preparation, and classification of thousands of vertebrate specimens recovered from the fossil beds

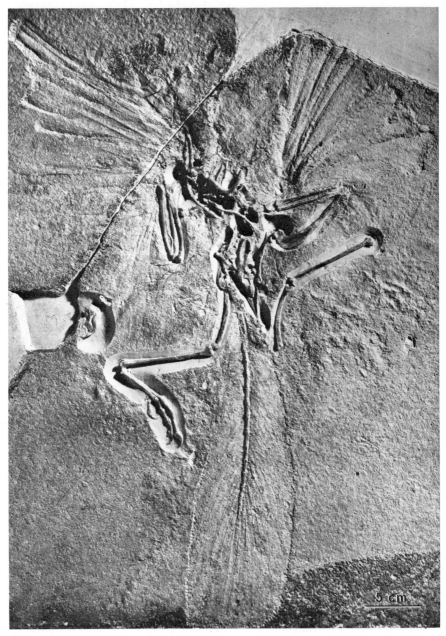

Fig. 9.1. Photograph of fossil remains of *Archaeopteryx lithographica*, British Museum specimen. Note endocast in excavated area at left center. From de Beer (1954); courtesy of the British Museum.

Evolutionary Background 181

of the great American western deserts, is marred by his occasional scientific flamboyance, which led him to reconstruct specimens to conform to his occasionally faulty imagination. When he attempted imaginative reconstructions, he rarely warned his audience, and one would suspect that he may have himself been imperceptive in distinguishing between an imaginative and realistic reconstruction. If one has managed a reasonable reconstruction from a set of incomplete bones, often scattered and distorted, and has actually mounted them into a realistic skeleton (or partial skeleton) of a once living animal, one may perhaps be excused for failing to recognize the tentative and hypothetical nature of the construction.

There is a very limited amount of evidence on the brain available from partial endocasts of the toothed birds of the Cretaceous, but in the presently published literature an unwary reader may find reference to the brains or endocasts of *Hesperornis* and *Ichthyornis* as described by Marsh. These should be ignored. The account to be presented later is based on Edinger's (1951) work. The remarkable thing, nevertheless, is that, on the whole, Marsh's reconstructions, projections, and guesses have been supported by more careful analysis. His picture of the odontognathan brain is actually a reasonable reconstruction for *Archaeopteryx* (cf. Fig. 9.5 and 9.7).

Cenozoic Birds

The evolution of the birds during the last 65 m.y. is still considered an uncertain business by paleontologists. Some very large nonflying Tertiary birds are known skeletally, but nothing has been reported about their endocasts. In fact, there is an almost complete gap in information about the brain during the Tertiary period, with a single fossil reported sufficiently well to be used in this account. This is *Numenius gypsorum,* a congener of living species of Curlews such as *Numenius americanus,* which is a familiar shore bird in California and elsewhere. The major orders of living birds are known from most of the Tertiary period, and the paucity of the fossil record is generally attributed to the low probability of fossilization of birds, which are rarely in ecological niches that produce good interments for fossilization.

A significant number of Quaternary fossil birds are known, and a fair number of subfossil extinct species have been described as well. These descriptions include good pictures of the endocasts (Piveteau, 1955; Stager, 1964), and a few are mentioned in this chapter. From their endocasts one can readily see that Pleistocene birds and extinct Recent birds are not to be differentiated from their living descendants with respect to relative brain development.

BRAIN AND BODY IN *ARCHAEOPTERYX*

Edinger (1926) had concluded, and de Beer concurred, that the brain of *Archaeopteryx* was "similar in type and structure to that of reptiles and different from that of all other known birds" [de Beer, 1954, p. 13]. I have examined a newly molded copy (Fig. 9.2) of the endocast excavated by de Beer (Jerison, 1968), and I agree that the brain differed from that of known birds, but it seems to me clearly avian in external form and intermediate between bird and reptile brains with respect to size.

In presenting these conclusions I must emphasize that there is an important element of personal and unverifiable judgment in statements about brains that depend on evidence from incomplete endocasts. The material studied is stone, not flesh, and ambiguities cannot be resolved by microscopic analysis. The uniqueness of the *Archaeopteryx* specimen prohibits a "crucial experiment" in which serial sections are prepared because the experiment would destroy the specimen. But the evidence that is presently available is sufficient to prove that the orientation of the endocast embedded in its matrix of lithographic limestone has been incorrectly perceived, and this has led to erroneous conclusions about the morphology of the brain. Before reviewing the evidence, it is worthwhile to devote a few paragraphs to the interesting history of the endocast. The history, quoted from de Beer's (1954) monograph on the British Museum's specimen, follows.

> In his paper on *Archaeopteryx* read before the Royal Society on 20 November 1862 Richard Owen stated that it lacked the skull, but towards the end of of the same year John Evans discovered on the counter-slab part of a jaw-bone with teeth, and on the main slab a nodule which he recognized as a natural cast of the brain-cavity. When Owen's paper was published [Owen, 1863] Evans's discoveries were mentioned in the explanation of the plate.
>
> . . . The first publication of Evans's discovery of the brain-cast was made without his permission by S. J. Mackie in *The Geologist* of 1 January 1863. It was also referred to by Sir Charles Lyell in the first edition of *The Geological Evidence of the Antiquity of Man* which was published in February 1863. In order to test his interpretation of his discovery, Evans made casts of the brains of magpies and rooks and found them to resemble the nodule on the main slab of *Archaeopteryx*. Unfortunately he mistook the orientation of the brain and imagined that the object exposed was a cast of the brain as seen from in front. Adopting his interpretation, S. J. Mackie and Carter Blake claimed to identify the object as the olfactory lobes.
>
> Tilly Edinger in 1926 recognized that the brain-cast represents the right cerebral hemisphere and succeeding portions of the brain as seen in lateral view from the right side. The corresponding left portions of the brain-cast are covered over by bone. Dr. Edinger pointed out the probability that when fully investigated the brain-case of *Archaeopteryx* would show marked rep-

tilian features, a consideration whose importance is strongly emphasized by her demonstration in 1951 that the brain of Cretaceous birds already shows avian characters in the relations of the cerebellum to the midbrain. In the present investigations, observation by ultra-violet fluorescence revealed various bones surrounding the cast of the brain-cavity on the main slab, and has confirmed that the cavity on the counter-slab is lined with the bones of the right side of the brain-case [de Beer, 1954, pp. 5–6].

As indicated earlier, the conclusions first reached by Edinger (1926) and supported by de Beer, above, must be revised. Let us, therefore, review the endocast of *Archaeopteryx lithographica,* the earliest known bird. The review will be more detailed, for a number of reasons, than that of any other endocast that is described in this book. First, the endocast is unique, showing the transition between two classes of vertebrates, the reptiles and birds. Second, the view is so strongly ingrained among evolutionists that the brain of this fossil bird was reptilian (e.g., de Beer, 1956; Portmann and Stingelin, 1961; Rensch, 1959; Swinton, 1965) that only a detailed discussion can enable one to shake that view and show why it is mistaken.

Endocast and Brain

Dorsal and lateral views of the exposed portion of the unique natural endocast of *Archaeopteryx* displayed at the British Museum are shown in Fig. 9.2. The dorsal view is a photograph of a copy of the endocast; the lateral view is a photograph of the original. A dorsal photograph of the original cannot easily be taken because only enough of the overlying matrix of limestone has been excavated to expose some of the dorsal surface. The exposure has the dorsal surface as a shelf embedded in a half shell (Fig. 9.3; see also Fig. 9.1), and it is impossible to gain a clear view of the inner edge of the "shelf" on the original specimen. The excellent copy of the specimen provided by the British Museum could be appropriately prepared by removing the copied overlying matrix and thus exposing the dorsal area. This was done by chipping away the portion of the copy shown in Fig. 9.3 as the dashed line, indicating the "plane of the limestone" matrix dorsal to the endocast.

Since this analysis challenges the long held view that the brain of *Archaeopteryx* was reptilian, it will be best to contrast it with the classic analysis. The main difference between the two interpretations is that de Beer (1954), following Edinger, perceived the skull as having been embedded in the matrix with the midsaggital plane of the brain exactly at the exposed plane of the lithographic limestone. He described the fractures in the cranial bones as edges of the left frontal and parietal bones. The present view (Jerison, 1968) is that the midsaggital plane of the brain is

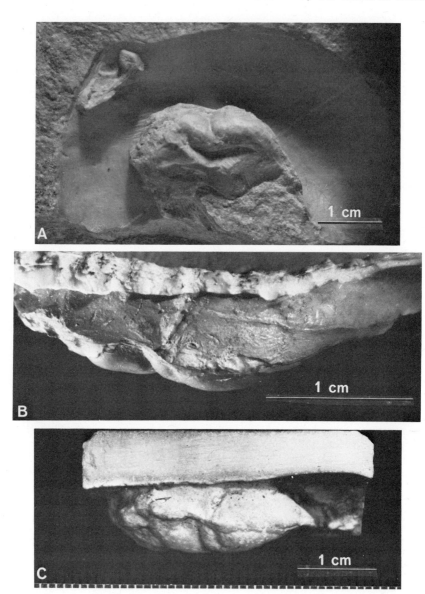

Fig. 9.2. Photographs of dorsal and lateral views of the endocast of *Archaeopteryx*. (See Fig. 9.4 for labels of landmarks.) A, Lateral view from de Beer (1954, Plate VII, Fig. 1); B, dorsal view of new copy of endocast, showing clear midline; C, de Beer's Plate VIII, Fig. 2 with poorly reproduced midline. In proper dorsal view (B) forebrain is clearly more expanded than midbrain; apparently equal expansion of midbrain (optic lobes) occurs in orientations of C because this is a dorsolateral rather than dorsal view, as explained in Fig. 9.3. Figures 9.2A and 9.2C courtesy of British Museum (Natural History), published in de Beer (1954).

Brain and Body in Archaeopteryx

rotated about its anteroposterior axis to form an angle of about 40° with the exposed plane of the limestone; the right ventrolateral edge of the natural endocast protrudes as a "nodule," and the fracture in the cranium is to the right of the midline suture. The endocast is also rotated about its dorsoventral axis to make an angle of about 5° to the exposed plane of the limestone; the anterior tip is closer to the observer. The first rotation is, of course, more likely to distort one's analysis, and it is the one that I emphasize in this discussion.

Two orientations about the anteroposterior axis, in coronal section, are shown in Fig. 9.3 to help one visualize the rotation and its effect on

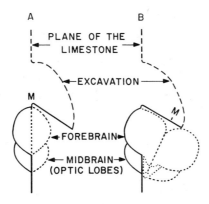

Fig. 9.3. Schematic view illustrating rotation of the endocast from the normal orientation. These are schematic coronal sections (end-on views) through the endocast and matrix. A, Interpretation of Edinger (1926) and de Beer (1954), assuming that midline (M) of brain is in plane of limestone matrix and that midsaggital section is also in that plane. B, Present interpretation, showing midline (M) as revealed in de Beer's excavation, and dorsal surface inclined at angle to plane of limestone; midsaggital section forms angle of 30°–50° to plane of limestone. Dotted lines indicate probable outlines of brain in coronal section. Note relation of forebrain to optic lobes as well as difference in brain size under the two interpretations. Portion of excavation shown as dashed line was removed in copy of endocast and surrounding matrix in order to make a photograph, Fig. 9.2B, from an orientation perpendicular to dorsal plane of skull.

the interpretation of the brain. This is a schematic view of a cross section of the endocast shown embedded in the partially excavated matrix. The cross section is drawn at about the maximum width of the forebrain. I have tried to suggest the relationship between forebrain and midbrain (optic lobes) in this end-on view by projecting the maximum width of the midbrain into the plane of this coronal section, orienting the endocast to have its anterior portion raised relative to the posterior portion. Figure 9.3A illustrates the effect of assuming with Edinger and de Beer that the midline (M) was at the point of fracture and in the exposed plane of the limestone. The symmetry of the brain about its midline then demands a projection of the cross section as indicated by the dotted arc in Fig. 9.3A. In addition to resulting in a rather narrow brain, this also aligns the forebrain and midbrain as if they were equally wide. In lateral view (see Fig. 9.4) the orientation of Fig. 9.3A leaves no space for cerebellum dorsal to the midbrain and near the forebrain. The view of Fig. 9.3A demands that the brain be relatively long and narrow, with the forebrain, midbrain, and cerebellum aligned like railroad cars, a situation reminiscent of the reptilian brain. Were the orientation of Fig. 9.3A correct, the conclusion offered by Edinger (1926) and "confirmed" by de Beer (1954), who consulted with Edinger, would be inescapable.

The correct midline (M), which had been embedded in the limestone matrix until exposed by de Beer, but not noticed, and the angle of the skull relative to the limestone matrix are both indicated in the schematic coronal section Fig. 9.3B. The midline is visible only as an irregularity near the inside angle formed by the dorsal surface of the skull and the excavated shell of the matrix. It is quite difficult to see this in the original endocast because the exposed region is only a small excavation overlying the endocast. The midline is readily identifiable as a part of the structure of the frontal and parietal bones only in a good copy of the endocast, when the impression made by the overlying shell of the matrix is removed, and it is easy to understand why the midline was overlooked. Figure 9.4 is a labeled outline drawing of the photographs of the endocasts, showing the new interpretation (Figs. 9.4A and 9.4B) and de Beer's (and Edinger's) interpretation (Figs. 9.4C and 9.4D).

The reorientation of one's view of the brain that results from the correct placement of the midline (M) and dorsal surface of the endocast is also indicated in Fig. 9.3B. The orientation of the midsaggital section is shown as a perpendicular (dotted line) to the frontal and parietal bones, running through the midline. In this orientation the forebrain may be reconstructed as two nearly spherical lateral expansions of the cerebral hemispheres, and the optic lobes are oriented more or less as in living birds, ventromedial to the hemispheres. There is less information about the orbital and orbitomedial surface of the forebrain. These are illustrated as dotted lines, assuming that, as in modern birds, there is a considerable fraction of the forebrain in that region. This is also suggested in the sketch of the lateral view of the endocast (Fig. 9.4) and in the reconstruction of the brain (Fig. 9.5). In Fig. 9.5 a dashed line is drawn to suggest the position and the extent of the orbitomedial surface. The matrix around the endocast must be excavated further to explore its ventral limits more thoroughly. The exposed portion of the forebrain, Evans' nodule, appears

Brain and Body in Archaeopteryx

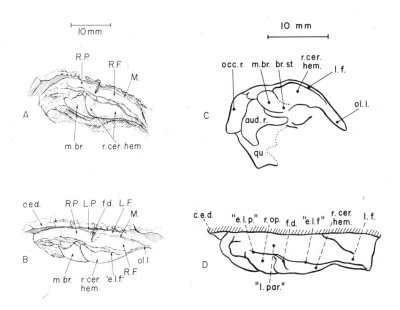

Fig. 9.4. Dorsal and lateral views of endocast of *Archaeopteryx lithographica*, corresponding to photographs in Fig. 9.2. A, Lateral view; B, dorsal view (note newly identified midline M). C, Lateral view; D, dorsal view. After de Beer (1954, Plate VIII); scale not exact for these, which are de Beer's identifications of landmarks in Figs. 9.2A and 9.2C. Following are abbreviations (lower case from de Beer; those in quotes are considered erroneous) for all figures: aud. r., auditory region; br. st., brain stem; c.e.d., cut edge of block (matrix); "e.l.f.," de Beer's median edge of left frontal; "e.l.p.," de Beer's median edge of left parietal; f.d., depression in surface of cerebral hemisphere caused by fracture; "l.f.," de Beer's left frontal; L.F., left frontal according to present view; "l.par.," de Beer's left parietal; L.P., left parietal according to present view; m.br., midbrain (optic lobe); occ.r., occipital region; ol.l., olfactory lobe (bulb); qu., quadrate region; r.cer.hem., right cerebral hemisphere; R.F., right frontal bone according to present view; r.op., right optic lobe (midbrain); R.P., right parietal bone according to present view.

to be part, but not all, of the lateral expansion of the cerebral hemispheres of *Archaeopteryx,* a harbinger of the expansion that is even more marked in living birds.

Another important result of the reorientation is the more birdlike placement of the cerebellum. There is a slight hump at the midline near the suture between the frontal and parietal bones. The hump probably covered the anterior end of the cerebellum, which must have overlaid the midbrain and touched the hemispheres. Further exposure of the endocast, with the orientation of Fig. 9.3B as a guide, may clarify the position of the cerebellum as well as the orbital and medial extents of the brain.

Rather than being oriented to provide a lateral view of the brain in the limestone, as suggested by Edinger's and de Beer's drawings and photographs, the fossil has been rotated about its anteroposterior axis, and that axis is, itself, at a small angle to the plane of the limestone. The observer has a ventrolateral view of the brain of *Archaeopteryx* when looking directly at the limestone.

The endocast is presently only about half-exposed. It was less than one-quarter exposed prior to the excavation of the shell above the frontal and parietal bones. The anterior dorsal surface is now slightly more than half exposed. Most of the posterior dorsal and all the ventral surface is still largely hidden, and the left side is almost unknown. The left side of the endocast, if exposed, might reveal the auditory capsule, the cerebellum, the orbitomedial portion of the forebrain, the pituitary, and the medulla. It would clearly be an important contribution to paleoneurology to complete the preparation of the brain of *Archaeopteryx,* removing it entirely from the matrix and exposing both sides. Such a preparation should also result in the exposure of the stapes and the entire tympanic region as well as the cerebellum.

On the basis of present information, and with the orientation of Fig. 9.3B, restorations of the endocast suggesting the brain of *Archaeopteryx* are sketched in dorsal and lateral views (Fig. 9.5). The lateral view is almost as in normal brains of living birds because the extensive lateral expansion of the cerebrum is not visible in that projection. The dorsal and lateral views also show (dotted lines) the dimensions of the endocast of a typical living bird, such as a pigeon. The dorsal view shows the lateral expansion of the forebrain region, the lesser expansion of the optic lobes, and the likely similarity of the cerebellum in living birds compared to *Archaeopteryx*.

Brain and Body Size

The orientations presented in Fig. 9.5 are the basis for a quantitative analysis of the brain of *Archaeopteryx* to determine its position relative to reptilian and avian brains with respect to size. Brain size was estimated by "numerical double integration" in which the brain is modeled by an elliptical cylinder, constructed by averaging successive heights and widths measured at regular intervals in the lateral and dorsal projections in Fig. 9.5. The average height was found to be 7.4 mm, and the average width was 7.2 mm. The endocast is 22 mm long in the orientation of Fig. 9.5. Its length from the olfactory lobe to foramen magnum is about 25 mm. This leads to an estimate of the volume of the endocast as 0.92 ml.

Brain and Body in Archaeopteryx

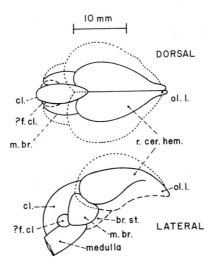

Fig. 9.5. Reconstruction of endocast of *Archaeopteryx* based on view of orientation of Fig. 9.3B. Dashed line reconstructs orbitomedial and ventral limits not visible in specimen. Dotted lines indicate outline of endocast of living bird, the curlew (*Numenius tahitiensis*), to show expansion of forebrain (r.cer.hem.) and midbrain. Additional abbreviations: cl., cerebellum; f.cl., flocculus of cerebellum; other abbreviations as in Fig. 9.4 (cf. Fig. 9.9).

The endocast of *Archaeopteryx*, like that of living birds, is brainlike in configuration, indicating that the brain filled the endocranial cavity and was the same size as the endocast. One may assume the specific gravity of the brain to be 1.0, permitting us to estimate its weight as 0.92 g. Is this an avian or reptilian size? The comparison can be made only if we take body size into account. Let us assume for this purpose that *Archaeopteryx*, which according to Heilmann (1927) was comparable in size to a pigeon (275 g) or a crow (430 g), was slightly more heavily built, since its bones were solid. This suggests that its maximum weight was 500 g.

Alternate approaches to make this estimate can follow the procedures used in Chapter 8 for pterosaurs. On the basis of the length:weight relation Eq. (8.1) we estimate a body weight of 310 g. The wing-surface:body-weight relations in birds may also be used, but one should consider the total glide surface, including tail and body in *Archaeopteryx*, because of the obviously well-feathered tail in the fossil bird. The total glide surface of *Archaeopteryx* estimated from Heilmann's drawing (Fig. 9.6) is approximately 1200 cm^2. From Eq. (8.2), which is the same for both wing surface and glide surface (the latter leads to a higher body weight estimate), the estimated body weight of a birdlike *Archaeopteryx* is 425 g.

Fig. 9.6. Restoration of *Archaeopteryx* to estimate body dimensions. Sizes may be estimated from Fig. 9.1. From Heilmann (1927).

An estimate of 500 g for the solid-boned fossil bird, therefore, seems to be a realistic maximum.[1] The analyses of body size and shape and of brain size are summarized in Table 9.1.

IMPLICATIONS OF THE BRAIN:BODY ANALYSIS

Was the brain of *Archaeopteryx* reptilian or avian in size? We can now answer this question as we did for the pterosaurs by comparing its brain:body data with those on reptiles and birds, as described in previous chapters. The analysis was included in Fig. 8.4.

One of the upper convex polygons in Fig. 8.4 encloses points representing brain and body weights in the base sample of living birds, as pre-

[1] This issue was thoroughly discussed in a spirited correspondence in *Nature*, in which it was suggested that *Archaeopteryx* may have weighed about 250 g rather than the 500 g indicated here. If the lower value is accepted, it would place *Archaeopteryx* further from the assemblage of reptiles though still not among the birds (Fig. 8.4). My estimate of the heavier body size, which is close to Heptonstall's (1970), was intended to represent a maximum likely size. The point was to be conservative about the decision that would place *Archaeopteryx* outside the range of reptiles with respect to brain evolution (Bramwell, 1971; Heptonstall, 1971b,c; Yalden, 1971a,b).

Brain and Body in Archaeopteryx

Table 9.1
Endocast and Body Dimensions of *Archaeopteryx lithographica*

Endocast[a]	
Volume	0.92 ml
Length (foramen magnum to olfactory tract)	2.5 cm
Maximum width, forebrain (cerebral hemispheres)	1.1 cm
Maximum width, midbrain (optic lobes)	0.95 cm
Body[b]	
Volume	500 ml
Body length (snout to base of tail)	25 cm
Length of tail	21 cm
Wing span	56 cm
Surface area of wings only	730 cm^2
Total glide surface (wings, body, and tail)	1200 cm^2

[a] Endocast length measured on copy; width determined by assuming similar right and left halves of the endocast.

[b] Body lengths measured from Fig. 9.1; surfaces based on reconstruction by Heilmann (1927) as shown in Fig. 9.6.

sented in Chapter 2. The lower polygon is the complete reptile polygon including data on living reptiles from Chapter 2 and the fossil data from Chapters 7 and 8. Extreme points at the angles of the polygons represent maximum and minimum measurements: for example, the smallest bird was a hummingbird (brain = 0.2 g; body = 4.8 g), and the heaviest reptile was *Brachiosaurus* (brain = 150 g; body = 85,000 kg).

The position of *Archaeopteryx* on this map, the circled "A" in Fig. 8.4, is significant. *Archaeopteryx* was intermediate between reptiles and living birds in brain size. As we see in Fig. 9.5, its brain approached, but did not reach, the form of living birds. The analysis of Fig. 8.4 shows that this was also true of the relative size of the brain.

We know from earlier chapters that the vertebrate classes that appeared before the mammals and birds had achieved a "basal" common level of relative brain size. This level was essentially as shown in the lower polygon in Fig. 8.4. The evidence from the brain:body data of *Archaeopteryx* indicates that a major new development had taken place, paralleling that which appeared in living and fossil mammals. *Archaeopteryx* apparently made the evolutionary response to its niche in the then uncharacteristic way of evolving an enlarged brain. That evolutionary response had at that time (about 150 m.y. ago) occurred in only one other vertebrate class—the mammals.[2] The question is not whether *Archaeopteryx* had

---

[2] Evolutionary enlargement of the brain may also have occurred in some species of the class Chondrichthyes as mentioned in Chapter 5, although fossil evidence is presently lacking.

achieved the present avian level; it had not. But had it shown a trend toward that level and away from the basal or reptilian level? The answer in Fig. 8.4 is that it had.

Another question that may be raised but cannot be answered yet has to do with the rate of evolution of the brain. There is a clear disparity between *Archaeopteryx* and living birds, shown by the dorsal view of the reconstruction of the brain in Fig. 9.5, as well as by the brain:body analysis of Fig. 8.4. There was still a considerable lateral expansion of the forebrain to occur in bird evolution in the 150 m.y. after the late Jurassic period in which *Archaeopteryx* lived. Some of that expansion is well documented, especially as it has appeared in the differentiation of living forms (Portmann and Stingelin, 1961). Between the late Jurassic period and the present, however, there is only one reliable fossil endocast to provide evidence on evolutionary trends of the type to be considered in the chapters on the mammals. Reports by Marsh on brain casts of the Odontognathae, or toothed birds, of the late Upper Cretaceous period have been discredited (Edinger, 1951), and our ignorance is, thus, almost complete until the appearance of the Upper Eocene *Numenius gypsorum* of about 40 m.y. ago, a gap of 100 m.y. There must have been a progression because the avian brain has evolved to its present size and shape. *Archaeopteryx* had left the reptilian "lower vertebrate" level and is an example of a point in the progression; it was on its way but had not yet achieved the lower limits reached by living birds.

CRETACEOUS AND CENOZOIC BIRDS

In presenting a review of the evolution of the avian brain there is an embarrassing problem of what to do with Marsh's reconstructions of the brains of Odontognathae, the toothed birds of the Cretaceous, *Hesperornis* and *Ichthyornis*. I have reproduced Marsh's (1880) figure as Fig. 9.7, although according to Edinger (1951) this was really a work of the imagination. She considered it evidence of Marsh's well-disciplined creativity rather than evidence about the brains of these animals of perhaps 80 m.y. ago. The skeletal remains are clear enough: the size and habits of *Hesperornis* were probably as in living large penguins, and those of *Ichthyornis* were similar to small terns.

Edinger (1951) was able to find no convincing evidence for Marsh's reconstructions of the skulls and brains in his specimens. Her examination of the original material that had been available to Marsh indicated that direct evidence in *Hesperornis* from endocasts exists only for the midbrain and cerebellar region and in *Ichthyornis* only for the olfactory bulbs and a

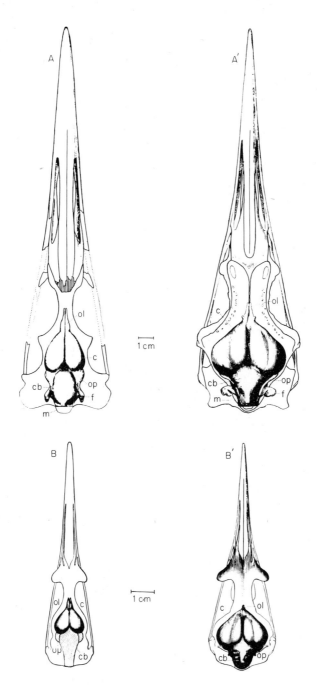

Fig. 9.7. Marsh's "reconstructions," now known to be based on inadequate data, of endocasts of *Hesperornis* (A) and *Ichthyornis* (B) compared to correct reconstructions (also from Marsh) of endocasts of the diver *Colymbus* (A') and the tern *Sterna* (B'). Compare Marsh's imaginative restorations with Fig. 9.5.

bit of forebrain. These she characterizes as avian rather than reptilian (in contrast to Marsh who had described them as reptilian).

The really interesting thing to me is the meaning of "reptilian" to Marsh. Accepting Edinger's view that Marsh was exercising his imagination in seeking reptilian features in his Cretaceous birds, we can compare the results of his imagination with the endocast of *Archaeopteryx*. Compared to endocasts of living birds, as shown in Marsh's figure (Fig. 9.7), specifically the diver and the tern, Marsh's picture of a "reptilian" bird brain is startlingly similar to our photograph of the endocast and reconstructions of the brain of *Archaeopteryx* (Figs. 9.2 and 9.5). One must applaud Marsh, despite stern critics such as Edinger (1962), for the excellence of his imagination. He pictured the brains of his "missing links," which he believed his Cretaceous toothed birds to be, more or less as we have found the brain of *Archaeopteryx,* the most truly missing link in the evolution of birds.

Numenius gypsorum: An Early Tertiary Bird

In her critique of Marsh's work Edinger (1951) mentioned two fossil early Tertiary birds, in the Muséum d'Histoire Naturelle in Paris, which had not yet been adequately described. Her impression was that the fossils were the same as living birds with respect to the size and configuration of the brain. When I visited the museum a number of years ago, I was able to find the specimens, which were, apparently, two fragments of a single fossil. They have now been reanalyzed from a paleoneurological and paleontological point of view by Dechaseaux (1970), who described the fossil as an Upper Eocene *Numenius gypsorum*. In comparing its brain to that of a living species of approximately the same body size, *N. tahitiensis,* she observed that the forebrain of the fossil was smaller and the optic lobes larger than in the living species. This conclusion was based on a comparison of the endocranial casts (Fig. 9.8).

Dr. Dechaseaux kindly provided me with copies of these endocasts, and photographs of lateral views of these endocasts are included in Fig. 9.9. To her conclusions with respect to the features visible in the casts I can add that the Eocene species' endocast had a volume of 3.1 ml and the living species' volume was 3.7 ml. Since the fossil endocast is only slightly distorted in shape, we may consider both measures to be reasonable estimates of brain size. There was, therefore, an increase in total brain size of about 20% during the 40 m.y. history of the brain of these species, from the late Eocene epoch to the present.

The only point on which Dechaseaux's conclusions should be questioned is the relative size of the optic lobes. Inspection of Fig. 9.8 and the photographs of these endocasts (Fig. 9.9) do not really permit one to make a

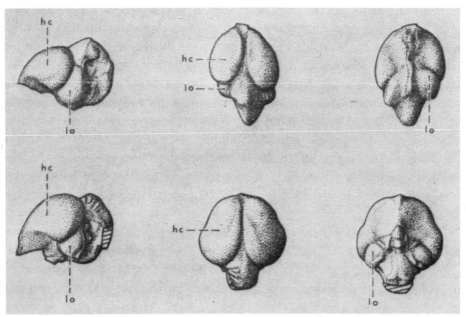

Fig. 9.8. Endocasts of the Eocene fossil *Numenius gypsorum* (top) compared with its living relative of about the same body size, *N. tahitiensis* (bottom). From left to right: left-lateral, dorsal, and ventral views; hc, cerebral hemispheres; lo, optic lobes. From Dechaseaux (1970).

definitive statement about the total size of the optic lobes. The portion of the optic lobes visible in the endocast of the living species is obviously less than in the fossil, but this is probably due to the expansion of the forebrain to cover more midbrain area. One would conjecture that the actual extent of the optic lobes was not smaller in the living species, merely that these lobes were less completely exposed.

The result on *Numenius* is important because it demonstrates that contrary to Edinger's (1951) suggestions the brain of birds had not completed its expansion during the Cretaceous. Quantitatively, considering the average level of brain size in "lower vetebrates" as the base line, the history of the bird's brain involved an approximate doubling in size with the appearance of *Archaeopteryx* during the Upper Jurassic period. By the Upper Eocene, if *Numenius* represented the typical condition of the birds, the size of the brain was further increased by a factor of about 3.5, that is, the bird brain was about seven times as large as that of a "typical" reptile or fish of comparable body weight. The endocast (or brain) of the living species of *Numenius* indicates a relatively small post-Eocene increase to about 8 times the basal size, to achieve a typical brain size for living birds.

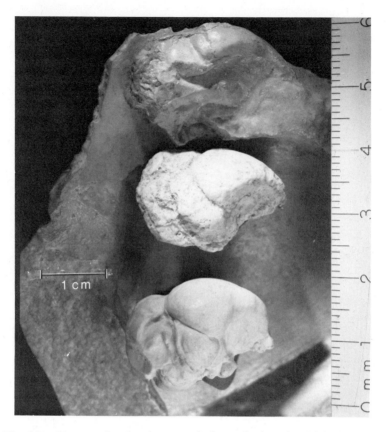

Fig. 9.9. Photographs of endocasts to indicate absolute size (right lateral views) of (top to bottom) *Archaeopteryx, Numenius gypsorum,* and *Numenius tahitiensis.*

We will see in later chapters that this pattern in the rate of evolution of brain size may be contrasted to that in mammals, although one hesitates to generalize from such limited data. If *Numenius* is typical of the birds, then we would conclude that, by the Upper Eocene, birds were closer to their present level of evolution of brain size than were the mammals because the mammals had not yet achieved half their present "typical" brain size (Fig. 13.12).

For purposes of a graphic analysis we may consider *Numenius* to have weighed about the same as *Archaeopteryx*. (Weights for living species of *Numenius* are between about 300 and 700 g.) The appropriate brain: body data were included in Fig. 8.4, with *N. gypsorum* labeled as g and *N. tahitiensis* labeled as t. Both are well within the convex polygon of living birds, and the difference between them in relative brain size is obviously small on the scale of the graphic analysis. The illustrations of the endocasts, however, clearly indicate that the enlargement of the brain during the evolution of *Numenius* was a real phenomenon.

QUATERNARY BIRDS

A number of endocasts are available for late Cenozoic (Quaternary) fossil birds such as the Madagascar ratites (Piveteau, 1955) and the La Brea giant vulture (Stager, 1964). These are similar in size and shape to those of their living descendants and relatives, differing only to the extent that one would expect from the differences in body size. The avian brain had certainly evolved to its present size before the Pleistocene epoch.

BRAINS FOR AERIAL NICHES

Vertebrate survival in aerial niches, at least as shown by the birds and pterosaurs, was associated with some shifts in the brain away from the earlier vertebrate pattern. In the case of the pterosaurs the shift was of the same type that occurred in specialization for other niches—enlargement of some parts of the brain (optic lobes) and reduction of other parts (olfactory bulbs). If there was an increase in the total quantity of neural tissue, it was not great enough to appear as an overall enlargement beyond the reptilian or "lower vertebrate" range in this admittedly gross analysis. Although the same kinds of specializations occurred in birds, reflecting the importance of vision for life in a diurnal aerial niche, they were accompanied by a net enlargement of the brain as a whole. Thus, the birds (like the mammals) responded to a selection pressure by an enlargement of the brain, as a whole, that was sufficiently great to appear even in the gross brain:body analysis.

If the difference between *Archaeopteryx* and the pterosaurs in relative brain size was real, as I believe it was, this implies that the earliest birds had invaded a somewhat different aerial niche than had the flying reptiles. We saw in the previous chapter that flight alone cannot account for the increase in neural tissue because the minimum necessary amount of tissue is really quite small, as shown by the evolution of the insects. Of course, a particular evolutionary problem would be solved in different ways by lineages with different histories, and birds as vertebrates had to solve the problem of flight with a vertebrate nervous and muscular system. The unusual wing specializations of insects were not available, and the system to be adapted to flight had to be the control mechanisms, both neural and muscular, associated with the movement of the arms and digits. The interesting speculations by Maynard Smith (1952), who discussed the enlarged avian brain as an adaptation to permit continuous control in flight of an aerodynamically unstable system, can be considered in the light of the present approach. The problems faced by birds would also have been faced by pterosaurs, and since the great enlargement of the brain did not occur in pterosaurs, despite the evidence that they were excellent flying machines (Bramwell, 1970a,b; Bramwell and Whitfield, 1969; Heptonstall, 1971), the reasons for the enlargement should not be sought in the vertebrate solution of the problem of flight control.

It is important to try to identify the selection pressures toward brain enlargement in *Archaeopteryx* and to try to reconstruct its habits. Because of its reptilian body and relatively small brain (for birds), it has been considered an awkward or clumsy flier. Heilmann's (1927) fanciful reconstruction of an afternoon in the life of *Archaeopteryx* illustrates the point. He wrote:

> Suddenly a feathered creature launches itself from the top of a tree-fern, sails gently through the air, borne by the expanded wings, and tries to catch the dragonfly. But this one eludes it by a deft turn, and the bird glides on a little, before it is able to stop. At last it succeeds in turning; it flaps its small wings and tries to rise a little higher in pursuit of its glittering quarry. Its movements look feeble and awkward, evidently requiring considerable effort and prove unavailing [p. 36].

If we examine this picture more closely, we find it inadequate in important respects. It is true that the absence of a large sternum ruled out the possibility of heavy wing musculature typical of living birds that can sustain powered flight by flapping their wings. And it seems likely that *Archaeopteryx* was a relatively poor glider even when compared with the pigeon (Heptonstall, 1970). When *Archaeopteryx* did fly, however, it could have flown accurately, gracefully, and with good control. It was a well-constructed animal aerodynamically if one considers the lift provided by its body and tail in addition to that from its wings, and it had strong enough bones to flap its wings at least a few times for brief flight within a woodland niche and amid primitive flowering plants.

The detailed argument by Heptonstall (1971a) assesses *Archaeopteryx* as inferior to modern birds in every aspect of flight on the basis of aerodynamic considerations, but his argument is based on the most conservative of the possible assumptions about the flight design of this bird. I have constructed simple gliders following the approximate outline of the body shown in Fig. 9.6 and have found them to be excellent designs, neither tail heavy nor particularly unstable. *Archaeopteryx* was adequately adapted for very short powered flight, of the type considered by Heilmann as quoted earlier. I would modify his description by suggesting that the "feathered creature" did not launch itself from the top of a fern to "sail gently through the air." Rather, it probably more or less jumped from a more modern-looking tree trunk or branch, maneuvered by one or two quick flaps of its wings and arched tail to catch its prey, and caught another branch or trunk in a four-handed grasp made possible by the claws on its wings as well as its feet. Its motion may have been analogous to take-off and landing procedures of living monkeys or prosimians as they jump among branches, but added to these was the ability to control the course of motion while moving through the air.

The flight motions and the control of flight present no significant problem for the control system of the central nervous mechanism, and it seems doubtful that the slight (but probably significant) expansion of the brain in *Archaeopteryx* was necessary in order to exercise such control. To that extent, I would not accept Maynard Smith's argument as relevant for the problem of the expansion of the brain in birds.

The clue to the expansion of the brain in *Archaeopteryx* may lie in the suggestion that the new niche for birds was like that for primates. The selection pressures associated with such a niche are considered in more detail in later chapters (Chapters 12 and 16). They lead us to consider the possibility that the birds were a successful evolutionary "experiment" by the class Reptilia of the same type that was later "performed" by the class Mammalia with the evolution of the primates. Just as the earliest primates required enlarged brains compared to the brains of other mammals of their time (Chapter 16), so did the earliest birds require enlarged brains compared to the brains of "other" reptiles of their time. In both cases the enlargement of the brain may be correlated with the unusually difficult problems in the analysis and use of visual information by active, diurnal, tree-dwelling animals that live in a mottled world of branches, leaves, and other foliage.

Chapter 10

Mammalian Brains for Mesozoic Niches

In mid- or late Triassic times, perhaps 200 m.y. ago or more, one or several species of mammallike reptiles evolved into mammals. Hopson and Crompton (1969), Olson (1959, 1961, 1970), Parrington (1967), Romer (1961, 1968), Simpson (1959b, 1960, 1961), and Van Valen (1960) are among recent authors who have discussed aspects of that evolution, and brief popular discussions have been presented by Olson (1965) and by Simpson (1967). Only one endocast of a Mesozoic mammal has been described in sufficient detail to contribute to the analysis of the evolution of the brain and its size, and only two other endocasts are known sufficiently well to be directly pertinent for an analysis of the evolution of the brain in Mesozoic mammals. But from this limited Mesozoic material, supplemented by later brains and endocasts from related species, important principles can be derived about the expansion of the brain for a given body size as a characteristic of the mammals. I will develop the hypothesis that the Mesozoic mammals were the nocturnal "reptiles" of that era and that their brains evolved to accommodate life in nocturnal niches in which hearing and smell, rather than the then normal reptilian vision, could be the characteristic sense modalities for information about events at a distance.[1]

THE EMERGENCE OF THE MAMMALS

The early history of the mammals and their emergence from among the mammallike reptiles has recently become a somewhat controversial topic (Hopson and Crompton, 1969). The main issue is whether the mammals were monophyletic or polyphyletic in their origin, an issue that is, fortunately, peripheral to our problem because there is general agreement on the points that are most important about the evolution of the brain and sensory systems. Although the earliest mammals are insufficiently

---

[1] The hypothesis that the earliest mammals were nocturnal is, of course, not new. The full implications of this hypothesis for the evolution of the brain have, apparently, not been recognized before.

known for us to be certain that they had a fully developed set of middle ear bones, this was probably a very early feature of the evolution of the mammals (Olson, 1970, p. 710). There is also agreement that the braincases of Mesozoic mammals as a group were similar to one another and were relatively larger than those of their reptilian ancestors (Kermack, 1963; Olson, 1970, p. 718 *et seq.*). Finally, there is general agreement that the earliest mammals must have been like their immediate ancestors among the mammallike reptiles (perhaps a *Thrinaxodon*-like species of therapsid) in having unusually well-developed (for reptiles) palatal and nasal bones, including ethmoidal bones that could support an extensive olfactory epithelium.

The critical selection pressures and the probable adaptive zone of the Mesozoic mammals may be hypothesized by correlating these facts with the evolutionary history of the several reptilian groups that dominated the land before and during the emergence of true mammals. In particular one may contrast the subclass of ruling reptiles (archosaurs) with the subclass of mammallike reptiles (synapsids) and, among the synapsids, the therapsid reptiles from which the true mammals evolved.

The present view (Romer, 1966) is that the relative numbers of archosaurs and synapsids changed after the transition between the Permian period at the end of the Paleozoic era and the Triassic period early in the Mesozoic era. It is reasonable to think of these great subclasses of reptiles as having been in a kind of competition for major adaptive zones; from the fact that the synapsids became extinct by the Middle Jurassic (175–150 m.y. ago), while the archosaurs became increasingly dominant as the land animals of the Mesozoic, we can judge that the archosaurs won the competition. We may then view the mammals, which first appeared when the mammallike reptiles as a group were approaching extinction, as a surviving and further evolved remnant of an otherwise nonadaptive group. And those peculiar "reptiles" that we retrospectively recognize as the first mammals could survive because they could invade adaptive zones for which the archosaurs were not as well adapted.

These zones, I suggest, were for nocturnal land animals; they were zones within which "reptiles" derived from therapsid stock could survive if there were further evolution of the brain that was consistent with the evolution of major skeletal features and functional adaptations. Trends among the therapsids toward the reduction in the number and size of the jaw bones, for example, may have been fortuitously correlated with functional trends toward a more efficient sound transmission system. The miniaturized bones no longer used in articulating jaw to skull could readily interact with the reptilian stapes as middle ear bones and could produce a sound transmission system that would be adaptive in species that made use of hearing as a distance sense to replace vision. To use auditory information for this purpose a somewhat larger brain is necessary in order to package the neural networks that would be needed to analyze sound for information about the location of objects at a distance. Thus, the elaboration of new functions for a sensory system had, as a necessary correlate, the evolution of relatively larger brains. Trends in the development of the olfactory system also imply the development of enlarged structures in the brain itself because all the neural elements of that system are contained in the brain.

These developments may be contrasted with the minimum selection pressures toward brain enlargement in diurnal Mesozoic reptiles, in which vision was retained as the fine distance sense. In such animals many of the neural elements of the information-processing system do not have to be packaged in the brain because of the elaborate neural organization of the retina as a "peripheral brain." The argument is developed in more detail in a later chapter (Chapter 12).

The point I would emphasize is that the earliest mammals were not merely an accidental development in the history of life and of animals that lived on land. Their evolution was a natural consequence of the special adaptations for the niches open to them, and the history of the mammals followed its peculiar course because of the adaptive zone that they first invaded. The consequences of their adaptations to that zone were to be especially important for their later evolution during the Cenozoic era, when they could reenter diurnal niches as a result of the extinction of many reptilian groups (Jerison, 1971a, discussed also in Chapters 12 and 17).

The mammals were able to survive as an only moderately diversified class throughout the "Age of Reptiles," for over 100 m.y. As species, the Mesozoic mammals were almost all quite small, and none would be considered large animals by present standards. The largest was perhaps 60 cm long and 5 kg in weight, whereas most were in the size range of living shrews, moles, mice, and rats. They were animals that had developed sensory capacities in adapting to their peculiar niches, which one may retrospectively consider as preadaptive for the further increase in relative brain size and for the evolution of intelligence (Chapter 1). It is, nevertheless, possible to consider the earliest mammals as nothing more than specialized "reptiles" living out a relatively normal reptilian existence in an environment for which other reptiles were ill adapted. Adaptations to that environment naturally had consequences for structures and their functions, but it was only incidental that those consequences had some utility for later adaptations to other niches by the descendants of the earliest mammals.

History of Mesozoic Mammals

In summarizing the history of the earliest mammals one can paint a picture with broad strokes; the reader may refer to Olson's (1970) latest discussion of this topic, as well as that of Hopson and Crompton (1969), cited earlier, for a detailed analysis.

The general picture is that of an evolutionary trend among several lineages of therapsid reptiles toward reduction of the jaw bones, differentiation of the teeth, and the adoption of limb suspensions that are, in retrospect, identifiable as movements toward the mammalian condition. According to Hopson and Crompton, only one of these lineages was actually ancestral to the mammals, and by paleontological criteria (which have to exclude information about soft tissue) the mammalian level was achieved among one group (eozostrodontids = morganucodontids) in late Triassic times, more than 200 m.y. ago, when the earliest specimens are identifiable. Other early mammalian groups included the triconodonts, which are the best known of the Mesozoic mammals with respect to the brain because of the endocast of *Triconodon*. An even earlier specimen, of the genus *Sinoconodon*, has been assigned to this group (Patterson and Olson, 1961), although Hopson and Crompton (1969) consider it as perhaps closer to *Eozostrodon*.

These very early mammals were at least related to early and remote ancestors of the Mesozoic mammals, the pantotheres, from which the marsupials and placentals probably evolved. Of the pantothere stock, one species is known at least in part by its endocast, and this is the Upper Jurassic *Amblotherium* (Edinger, 1964a), which was probably a contemporary of *Triconodon*. The endocast is too incomplete for a quantitative analysis (Fig. 10.1C).

Among the longest "lived" of all mammalian orders, the Mesozoic order Multituberculata persisted into the early Tertiary, and although no endocasts are known from Mesozoic representatives of the order, there is one of *Ptilodus,* a Paleocene form, that has been described in detail and is discussed in this chapter. The multituberculates were a really separate mammalian lineage and were only distantly related to others described in this book by the time they had reached the level of evolution represented by *Ptilodus*. They are of some interest because they occupied niches that have probably since become characteristic of living rodents, lagomorphs, and some primates (Van Valen and Sloan, 1966).

To conclude this brief account, it may be repeated that, with the exception of the multituberculates, the Mesozoic mammals as a group were on the whole undifferentiated skeletally. They were small animals, and the giants among them were hardly larger than an alley cat. The multituberculates represented the outstanding variations from the simple mammalian pattern, but most other Mesozoic mammals were probably ecologically like small species of "primitive" living groups such as the hedgehogs or opossums.

In late Cretaceous times, about 75 m.y. ago, there is fossil evidence of the evolution of a few of the archaic Cenozoic orders[2] and of the first marsupials and insectivores (Clemens, 1966; Sloan and Van Valen, 1965). The continuing survival of undifferentiated insectivores such as the hedgehogs and undifferentiated marsupials such as the opossum suggests that the Mesozoic niches that these groups first invaded have remained occupiable by mammals not much different from the earliest occupants. I will refer to such niches as "Mesozoic" despite their existence in present times.

There is quite a bit of information about the radiation of the insectivores with respect to relative brain development, largely as a result of the efforts of Bauchot, Stephan, and their colleagues; for example, Bauchot and Stephan (1967). I will report their data in this chapter to indicate how some of the mammals that have persisted in these ancient niches have fared.

Brains of Mesozoic Mammals

Most present knowledge of the brains of Mesozoic fossil mammals was summarized by Simpson (1927) when he described an endocast of *Triconodon,* fragments of endocasts of an otherwise unidentified American Jurassic mammal from Como Bluff, Wyoming, and a pantothere, *Amblotherium nanum.* A later paper by Simpson (1937) describes the endocast of the Paleocene multituberculate, *Ptilodus,* which is also included in this analysis. Except for the discovery of the cranial bones of *Sinoconodon* (Patterson and Olson, 1961) and a figure of the fragmentary *Amblotherium* endocast published by Edinger (1964a), there have been no additions to the data on Mesozoic mammalian brains beyond those presented by Simpson. There are, apparently, some as yet undescribed late Cretaceous endocasts that have been uncovered (Van Valen and Sloan, 1966), but they are not available for analysis at this writing.

At a qualitative level Simpson (1927, pp. 265–268) reviewed the major features of the endocast of *Triconodon* by contrasting it with one of the therapsid *Nythrosaurus,* and his main points are illustrated by the comparison of the two endocasts in Fig. 10.1. We may summarize the comparisons as follows, paraphrasing and updating Simpson.

> 1. Anteriorly, both have large nasal chambers and cartilaginous ethmoturbinals and relatively large olfactory bulbs compared to the size of the rest of their brains.
> 2. The cerebral hemispheres (and possibly the olfactory tracts as indistinguishable anterior extensions) are long and narrow in the reptile whereas

---

[2] An order (and less often a family or other taxon) will be referred to as archaic when it has become entirely extinct and has, presumably, been entirely replaced in its niches or adaptive zone by later forms.

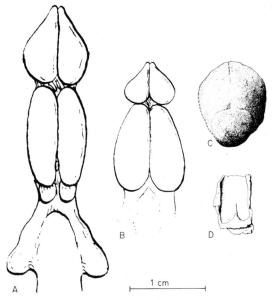

Fig. 10.1. Endocasts of Mesozoic mammallike reptile *Nythrosaurus* (A) and mammals *Triconodon* (B), *Amblotherium* (C), and cranium of *Sinoconodon rigneyi* (D). A and B are from Simpson (1927); C is from Edinger (1964a); and D, from Patterson and Olson (1961).

the hemispheres are laterally and dorsally expanded in *Triconodon* with no representation, dorsally, of the olfactory tracts.

3. The brain is serially arranged, railroad-car style, in *Nythrosaurus*, whereas there is evidence that the midbrain is overlapped at least by the cerebrum and possibly also by the anterior tip of the cerebellum in *Triconodon*.

4. The relative size of the brain compared to the skull is much greater in *Triconodon* than in *Nythrosaurus*. Parrington (1936) indicated the skull length of *Nythrosaurus larvatus* as 7.6 cm, whereas *Triconodon's* skull was about 4 cm long.

5. Simpson pointed out that *Nythrosaurus* had an unusually developed floccular portion of the cerebellum, for a reptile; comparable information was not available for *Triconodon*. Olson's (1944) detailed review of therapsid endocasts (see Chapter 7) confirmed Simpson's observation.

These contrasts between a mammallike reptile (Chapter 7) and a true mammal may be validly extended to the Mesozoic mammals and their Cenozoic relatives in similar niches. It should, perhaps, be restated that the larger mammallike reptiles were probably like other reptiles in having brains smaller than their endocranial cavities, whereas this was not the case for true mammals.

RELATIVE BRAIN SIZE

The quantitative analysis to be presented now is based on comparisons of the brain and body size in *Triconodon* and the Paleocene multituberculate *Ptilodus* with a number of later Cenozoic fossil and living species. Most of the comparison species have probably been in comparable "Mesozoic" ecological niches. Another set of such species constitutes the archaic ungulate sample to be analyzed in detail later (Chapter 11). The methods of analysis described earlier (Chapter 2) are used, and I present the computations for *Triconodon* and *Ptilodus* to indicate some special assumptions required in order to take full advantage of the limited data on these fossils. The estimates of body size were particularly difficult because there are data only on the lower jaw and fragments of the cranial bones of *Triconodon* and on the skull of *Ptilodus* but no useful data on the postcranial skeleton of either animal. I, therefore, consider the possible effects of very large errors in the estimates of body size on our conclusions with respect to the evolution of brain size. The analysis of the other species, in particular of relative brain size in living insectivores, is based on much more complete information.

Three fundamental questions about the evolution of the brain are addressed with the help of this analysis. First, did the earliest mammals really have brains larger than their reptilian ancestors, as suggested by the qualitative observations of braincases? (They almost certainly did.) Second, was there progressive enlargement of the brain within the mammals as they evolved during the Mesozoic era? (There probably was not.) Third, was there progressive enlargement of the brains of mammals that replaced the Mesozoic species within similar "Mesozoic" ecological niches of the Cenozoic era? (There was, but not to the same extent as the enlargement in the more "progressive" orders of the Cenozoic.)

In discussing the first question, I will compare the new data of the present chapter with data on reptiles considered earlier. For the second question we have to anticipate a few of the results to be presented in more detail in later chapters, in particular, the results of the first Cenozoic (Paleocene and Eocene) radiation of the mammals. For the third question I present a detailed analysis of the available data on living Cenozoic species that could be considered to be living in "Mesozoic" niches. This includes the analysis of the distribution of the encephalization quotient, EQ, in fossil and living species from such niches. The living species to be used are insectivores and rodents.

Brain Size in *Triconodon* and *Ptilodus*

The raw data for the analysis of brain size (endocranial volume) are from dorsal and lateral views of the endocast, and the volume is determined by graphic double integration. Only dorsal views are available for the endocasts of *Triconodon* and *Ptilodus*, and, in order to perform the analysis of these two early mammalian endocasts, the lateral dimension must be estimated. This is done with the help of data on other endocasts, illustrated in Fig. 10.2, from fossil and living species of insectivores (the Upper Eocene *Neurogymnurus* and the living *Setifer* and *Echinosorex*), as well as the opossum (*Didelphis marsupialis*) and the Norway rat (*Rattus norvegicus*).

In performing a graphic double integration, it will be recalled (Chapter 2) that one can determine the area of each projection planimetrically, excluding the olfactory bulb anteriorly and considering the brain to end posteriorly at the twelfth cranial nerve. Then, each area is divided by the anteroposterior length to obtain the average width of the dorsal projection and the average height of the lateral projection.

The ratio of height to width in the six specimens of Fig. 10.2 was 0.8, and I assumed this ratio for the endocasts of *Triconodon* and *Ptilodus*. and it was assumed that the same ratio held for *Triconodon* and *Ptilodus*. (This is one of the uncertain assumptions that had to be made because of the paucity of data.) I, then, undertook the analysis of brain size summarized in the upper part of Table 10.1, which is presented in detail to illustrate all the computations.

In the analysis the brain is modeled by an elliptical cylinder, with length and major and minor axes derived from the projections and computations as indicated in Table 10.1. Experience with this method indicates that errors rarely exceed 5% as compared with volume determined by water

Table 10.1
Computations for Estimating Brain (Endocast) and Body Size in *Triconodon* and *Ptilodus*

Computations[a]	*Triconodon mordax*	*Ptilodus montanus*
Endocast		
Mean width ($\overline{w}$)	0.87	1.15
Estimated mean height ($\overline{h} = 0.8\,\overline{w}$)	0.70	0.92
Length[b] (l)	1.53	1.31
Estimated volume ($\frac{1}{4}\,\pi\,l\,\overline{w}\,\overline{h}$)	0.73	1.09
Body		
Skull length (s)	4[c]	4
Body length ($4\,s$)	16	16
Body weight [Eq. (2.4) or (2.5)]	100	200
Habitus[d]	average	heavy

[a] Lengths in centimeters, volumes in milliliters, weights in grams.
[b] Exclusive of olfactory bulbs.
[c] Estimated as equal to *Sinoconodon* on basis of similarity of mandible.
[d] See Chapter 2, p. 54; Chapter 13, p. 239.

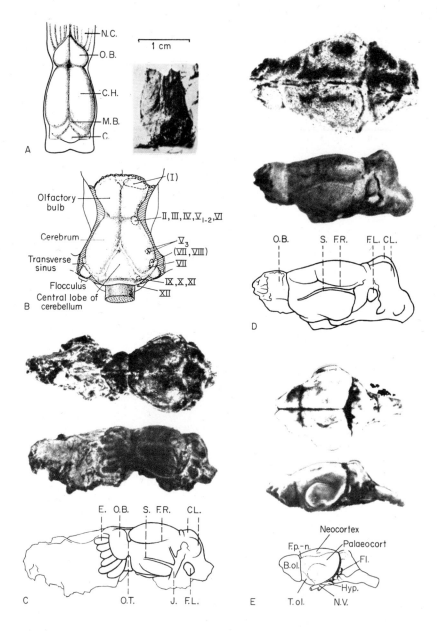

Fig. 10.2. Dorsal (and lateral when available) views of the endocasts or brains of A, *Triconodon*, British Museum, BM 47763 (photograph and labeled reconstruction); B, *Ptilodus*; C, *Neurogymnurus*; D, *Echinosorex*; and E, *Setifer*. From Simpson (1928, 1937), Dechaseaux (1964), and Stephan and Spatz (1962). (See also Fig. 15.5 for comparison with brain and endocast of the tenrec.)

Relative Brain Size

displacement. In the present data, however, since the average height of the endocasts had to be estimated by analogy with those of other mammals, we may expect that greater errors could be introduced. This is especially true for *Ptilodus*, which, as we can see from Fig. 10.2, had an unusually broad brain. Its brain may also have been relatively flattened compared to the brains of the other mammals shown. Brain size may have been overestimated in this multituberculate, because the value for h was too high.

BODY SIZE IN *Triconodon* AND *Ptilodus*

The basic approach in estimating body size is to use analogies to the form of the body in living mammals. I determined the length of the skull either by estimation in the case of *Triconodon* or directly in the case of *Ptilodus*. The fossil data for these determinations (Fig. 10.3) are a set of jaws and skulls. I first estimated body length from the length

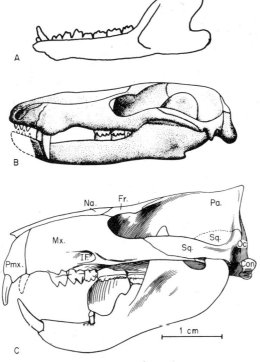

Fig. 10.3. Skulls of Mesozoic and later mammals, used to estimate body size. A, Mandible of *Triconodon mordax* (Upper Jurassic), from Simpson (1928); B, skull of *Sinoconodon rigneyi* (Upper Triassic), lateral view, reprinted from "Vertebrate Paleontology," 3rd ed., by A. S. Romer, The University of Chicago Press, Chicago, Illinois, copyright 1966; C, skull of *Ptilodus montanus* (Paleocene), lateral view, from Simpson (1937).

of the skull, and then estimated the body weight from the body length, using procedures similar to those presented earlier (Chapter 2). The chain of assumptions in these computations is more extensive and less certain than for the estimation of brain size, and we must, therefore, be prepared to cope with much larger errors in the estimates. The results of the computations were summarized in the lower part of Table 10.1.

In the case of *Triconodon*, Simpson's (1928, p. 84) discussion indicated a skull at least 3.8 cm long, or perhaps even longer. He took into account the length of the jaw (Fig. 10.3) and the cranial fragments, which were associated with the endocast. It seems safe to take skull length in this specimen to one significant figure as 4 cm. This is, fortuitously, the same length indicated by Simpson (1937) for the skull of *Ptilodus*, an estimate based on an "average" skull to which his carefully reconstructed endocast (Fig. 10.2) was referred.

In order to estimate the body length from the skull length, data from small living insectivores, rodents, and marsupials (Walker, 1964) were used to permit the regression analysis shown in Fig. 10.4A. It is apparent that the total head and body length of these animals ranged between 3–5 times the length of the skull, and I, therefore, estimated body length at 4 times the length of the skull. This indicated that both specimens were 16 cm long, exclusive of the length of the tail.

The analysis of body weight followed the same procedure indicated for the camelids (Chapter 2), in which Eq. (2.4) was determined for the carnivores and ungulates (Fig. 2.9). The equation is

$$P = 0.025\ L^3 \tag{2.4}$$

As an equation to estimate body weight P from the head and body length L, Eq. (2.4) also seems to fit the data reported by Walker (1964) on insectivores, didelphids, and rodents, in which head and body lengths were between 10–30 cm. (These data were selected to enable us to estimate the weight of a 16-cm long "primitive" mammal.) In Fig. 10.4B the data from Walker mentioned above are shown, and Eq. (2.4) is graphed. It is apparent that the equation gives a reasonable estimate of body weight from the head and body length. According to this equation, a 16-cm "primitive" mammal of average habitus would be expected to weigh 100 g (two significant figures). A similar animal of heavy habitus should weigh 200 g as indicated by the dashed line in Fig. 10.4B (Eq. 2.5).

RELATIVE BRAIN SIZE IN *Triconodon, Ptilodus,* AND OTHER SPECIMENS

The complete brain:body analysis for the animals in this sample is summarized in Table 10.2. The analysis includes an estimate of the volume of the olfactory bulbs as well as of the rest of the brain. Our consideration

Relative Brain Size

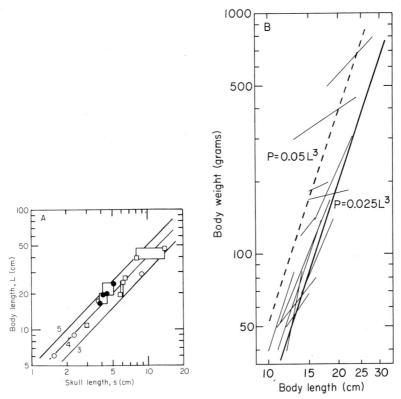

Fig. 10.4. Graphs for estimating body weight from body and skull lengths (cgs system). A, Body length as a function of skull length; open circles, insectivores; filled circles, rodents; squares, marsupials. When skull and body lengths were indicated as ranges, the lower limits were set as lower left corner of rectangle and upper limits as upper right corner. Equations, fitted by eye, have a slope of 1.0, on log–log coordinates, with intercepts at 3, 4, and 5 as noted. Equations are thus, $L = 3s$; $L = 4s$; and $L = 5s$. B, Body weight as a function of head and body length in living insectivores, rodents, and marsupials, between 10 and 30 cm long. Lines are Eqs. (2.4) and (2.5). Solid line is for "average" habitus, dashed line is for heavy habitus (cf. Fig. 2.9).

of the olfactory bulbs as "nonbrain" is arbitrary from a neurological point of view and is based on the assumption that the neural elements of the bulbs are essentially transmission elements. The bulbs are thought of as performing, functionally, in a way analogous to the action of bipolar cells of the retina or of the spiral ganglion of the cochlea. This elimination of the olfactory bulbs from the analysis of brain size is also based on the observation that the volume of the bulbs is frequently overestimated by the volume of their representation on an endocast, whereas the endocast represents the volume of the rest of the brain accurately.

Table 10.2
Brain, Olfactory Bulb, and Body Size in Mammals from "Mesozoic" Niches

	Endocranial "brain" volume (ml)	Endocranial "olfactory bulbs" volume (ml)	Body weight (g)	Encephalization quotient (EQ)
Triconodon mordax	0.73	0.09	100	0.28
Ptilodus montanus	1.09	0.27	200	0.26
Neurogymnurus cayluxi	1.57	0.15	200	0.38
Setifer setosus	1.36–1.70	0.23[a]	187–300	0.35–0.32
Echinosorex gymnurus	3.52	0.29	850	0.32
Didelphis marsupialis	7.65	0.83	5000	0.22
Rattus norvegicus (wild)	1.92–2.55	0.06[b]	177–507	0.51–0.33

[a] $P = 190$ g for this specimen.
[b] $P = 300$ g for this specimen.

The relationship of the specimens described in Table 10.2 to other mammals and to the reptiles from which they evolved is indicated with the method of minimum convex polygons in Fig. 10.5. The brain:body data of the specimens are presented graphically against the background of the brain:body maps of living and fossil reptiles and of living mammals as discussed in previous chapters (Chapters 2 and 7). It should be kept in mind that the solid-lined mammalian polygon in Fig. 10.5 is based on a large, though not complete, sampling of the mammals by Crile and Quiring (1940) and that the several smaller-brained living species included in Table 10.2 are species not included in their sample of mammals. Had they included these (and related) species, the polygon for the mammals would be shaped somewhat differently at its lower end, as indicated by the dashed extension of the mammalian polygon. [The new species for extending the polygon are described by Bauchot and Stephan (1966) and are included in the data of Table 10.3. The archaic ungulates shown in Fig. 10.5 are discussed in Chapter 11.]

It is apparent that the extended polygon for living mammals is broad enough to include the fossil data, although the mean estimate for *Triconodon* and values for the Virginia opossum may have fallen somewhat below the lower margin of that polygon. Given the uncertainty of our estimates, however, it would be difficult to argue that either *Triconodon* or the opossum is completely out of the range of this sample. Like the other specimens of Table 10.2, they may be considered as lying near the small-brained edge of the living range.

Evolutionary Implications

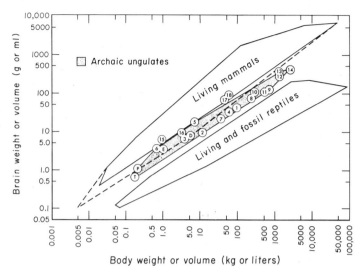

Fig. 10.5. Brain:body relations in living and fossil mammals in "Mesozoic" niches. Four specimens described in Table 10.2 are shown by generic initials: T, *Triconodon;* P, *Ptilodus;* D, *Didelphis;* E, *Echinosorex.* Numbered points are of archaic early Tertiary mammals described in Chapter 11. Minimum convex polygons shown for archaic ungulates (Chapter 11), living and fossil reptiles (Chapters 7 and 8), and living mammals (Chapter 2). Extension of polygon of living mammals (dashed line) is to add data of Bauchot and Stephan (1966) and of *Didelphis* to those used in Chapter 2 to define the polygon. The extended polygon, therefore, includes mammals presently living in "Mesozoic" niches.

EVOLUTIONARY IMPLICATIONS

We can now consider more detailed answers to the evolutionary questions raised at the beginning of the previous section. These were all on the rate of the evolution of brain size, and in particular about periods of rapid evolution of relative brain size. Such periods could produce a noticeable upward shift of a brain:body map (minimum convex polygon) or, at least, measurable increments in distribution of indices of relative brain size, such as the encephalization quotient EQ.

The first question that was raised, about the reptile–mammal boundary, has now been answered by the data in Fig. 10.5. The earliest mammal for which there is reasonable evidence, *Triconodon* of the Upper Jurassic period, was apparently already at or near the level of living "primitive" mammals such as the insectivores or the Virginia opossum. It was certainly larger brained than its reptilian ancestors of comparable body size.

Although less adequate from the point of view of precision, essentially the same statement has been implied for the much older Upper Triassic mammal *Sinoconodon* (Figs. 10.1 and 10.3), known by a braincase from which an endocast has not yet been made and by a reasonably complete skull. Qualitative observation of that braincase by Patterson and Olson (1961) led them to conclude that this mammal of more than 180 m.y. ago was similar to *Triconodon* in brain size.[3] Its skull, like that of *Triconodon*, was 4-cm long, indicating an animal of about the same body size. Hence, the qualitative analysis would indicate that the expansion of the brain evident in *Triconodon* had probably occurred as part of the transition between the mammallike reptiles and the true mammals, before the appearance of *Sinoconodon*. This seems less certain from our view of Fig. 10.1, since the older braincase was obviously much smaller. It is, perhaps, wisest to defer judgment on *Sinoconodon* until more evidence is available.

Evolution of Brain Size during the Mesozoic Era

The subsequent evolution of brain size in Mesozoic mammals cannot be studied directly because of the small number of specimens—only *Triconodon* is really sufficiently well known—that can be examined. We have noted (Table 10.2 and Fig. 10.5) that the Paleocene multituberculate *Ptilodus* was probably much larger brained than *Triconodon*. I have assumed that the relatively massive skull of *Ptilodus* was associated with heavy habitus, that is, a more massive body than *Triconodon*, and estimated *Ptilodus* to have been twice as heavy as *Triconodon*. That is, the body weight of *Ptilodus* was estimated with Eq. (2.5), that of *Triconodon* with Eq. (2.4). The index EQ, with *Triconodon* weighing 100 g and *Ptilodus* weighing 200 g, would have values of $EQ = 0.28$ and 0.26, respectively. That is, both would be describable as having one-quarter the brain size of an "average" living mammal of their body sizes.

Even if the two animals were the same body size and if *Ptilodus*, though Paleocene, represented the persistence of a late Cretaceous level

---

[3] This observation should be reviewed and verified quantitatively; it does not seem to be supported by the illustrations of the *Triconodon* endocast and the *Sinoconodon* braincase in Figs. 10.1B and 10.1D. Figure 10.1 is an excellent illustration of problems of scaling that are faced in an analysis of relative size. With some effort each of the parts of this figure was enlarged or reduced in order to fit a single scale. The result displayed several unsuspected similarities and differences. Most dramatic was the contrast between the endocast of the mammallike reptile, *Nythrosaurus* (A), and *Triconodon* (B). In Simpson's (1927) originals of these drawings the two endocasts were shown equal in anteroposterior dimension in order to emphasize relative size of the parts of the brain. The fact that the total brain size of the reptile was almost certainly greater than that of the mammal was obscured, however, as was the fact that their forebrains were about the same size. It is only because *Triconodon* was a much smaller animal that one is able to conclude that its brain was actually larger relative to its body size than was that of *Nythrosaurus*.

Evolutionary Implications

of relative brain development, these data would not be sufficient to answer our second evolutionary question about progressive changes in relative brain size during the Mesozoic era. We would then conclude, of course, that the multituberculate with a recalculated $EQ = 0.42$ had a relatively larger brain than *Triconodon*. But this does not necessarily signify a progression with respect to brain size. It could properly be interpreted as an effect of the different niches occupied by the two animals; *Ptilodus* and related multituberculates could have been in niches in which there were selective advantages for brains enlarged even more than those of the triconodonts. We do not expect identity in relative brain size, and the question about a progression really must be about successive distributions of EQ in sets of species that can represent grades or levels of evolution. For that we must assemble a larger amount of data in order to characterize the appropriate population.

An appropriate population would be a late Mesozoic or early Cenozoic assemblage of mammals. In Chapter 11 I present data of the latter type under "Archaic Tertiary Mammals"; the data were included in Fig. 10.5. These are on 13 species from archaic ungulate orders, Amblypoda, and Condylarthra, whose fossils were recovered from Middle Paleocene to Upper Eocene deposits and which, therefore, may be dated from about 60–40 m.y. ago. The adaptive zones of these species are also considered in the next chapter, but we can be certain that they were broader than those of the Mesozoic mammals. I will argue, in fact, that these early Tertiary species, which were part of the first mammalian radiation of that period, reinvaded a set of niches that became available to mammals as a result of the extinctions of reptilian groups at the end of the Mesozoic era. The radiation was apparently accomplished without increases in relative brain size. The niches included those of large herbivorous land animals as well as of moderately large species that were probably carnivorous (the arctocyonids). The idea that these niches were occupied by persistently small-brained mammals follows from the data on relative brain size in these species as compared with the data on *Triconodon* and the other specimens just discussed.

The values of EQ for the archaic Tertiary ungulates (Chapter 11, also Fig. 10.6) ranged from 0.11 to 0.37 with a modal value of 0.22. This is similar to our best estimate, presented earlier (Table 10.2), for the Mesozoic *Triconodon*. We may note, incidentally, that the estimate of EQ for the multituberculate *Ptilodus* would be relatively high in the context of an archaic early Tertiary assemblage unless it was of heavy habitus as assumed in Table 10.1. In that case, it, too, fits into the range of the archaic early Tertiary level of relative brain development.

On the basis of these comparisons I conclude that there is no evidence for a progression in the development of relative brain size in mammals during the Mesozoic era. On the contrary, the evidence is most consistent with the position that the Mesozoic mammalian radiation involved a restricted range of relative brain sizes, between about 10–30% of the size of the brain in living mammals of comparable body size, because this range includes both the earliest known Mesozoic mammal endocast and most of the first great mammalian radiation, which was completed early in the Tertiary period, or "Age of Mammals."

Cenozoic Evolution of Brain Size in "Mesozoic" Niches

One of the basic themes of this book is that the early evolution of mammals during the Mesozoic era was related to their invasion and successful adaptation to the then relatively open adaptive zone of inconspicuous, nocturnal land animals. Many living vertebrate species, amphibians and reptiles as well as mammals, can be considered to inhabit such an adaptive zone. Can we legitimately characterize the ecological niches within which the Mesozoic mammals evolved as persisting to the present? We shall, at least for heuristic purposes, because it is convenient to compare relative brain size in different groups and to relate changes in brain size both to progressive changes across a span of time and to the adaptive radiation of species into the many possible niches within a broadly defined adaptive zone.

Although an ecological niche can scarcely be defined independently of the adaptations of the animals and plants that interact within it, there are similarities among niches within which different species are in similar roles. The best known examples are those instances of parallel evolution in which almost entirely unrelated species such as the marsupial mole (*Notoryctes*) and the placental mole (*Talpa*) have evolved remarkably similar bodily forms and skeletal adaptations. Less dramatically, one may identify common patterns in marsupial and placental evolution by recognizing the similarity of the ecological niches of the larger Australian kangaroos to those of grazing and browsing ungulates of the rest of the world, despite the many differences in body shape and habits. It is in that sense that one may think about some of the niches of living mammals as similar to those of their Mesozoic ancestors and relatives and describe such niches as "Mesozoic."

Among living orders the insectivores as a group seem to fit most clearly into a pattern that survived relatively unchanged from Mesozoic times. The earliest insectivores are known as fossils from the Cretaceous period, and, to quote Romer, surviving shrews, moles, and hedgehogs "have, despite various specializations, many primitive characters . . . . Many common features were . . . characteristic of a generalized ancestral placental"

Evolutionary Implications

(Romer, 1966, p. 208). Similar statements are often made of the didelphids among the marsupials. In describing the earliest mammals, Romer (1966, p. 202) suggested their probable similarity to rats and mice in size and general appearance and indicated that they occupied an adaptive zone similar to that of many living rodents. We will, therefore, examine data on relative brain size in these groups to determine the course of evolution during the Cenozoic era of mammals in "Mesozoic" niches. Insectivores and didelphids are more "primitive" and are in older mammalian orders; the rodents represent the newest approach to life within "Mesozoic" niches.

Our analysis will, therefore, be performed by comparing the data on "archaic" Tertiary ungulates, which, as we have seen earlier, probably represent a persistence of the typical Mesozoic condition of relative brain size, with the data on living insectivores, didelphids, and rodents. We will, of course, compare *Triconodon* and *Ptilodus* with all these groups, and we will be able to verify the fact that these two animals were entirely within the archaic ungulate range.

If each of these groups could be represented validly by minimum convex polygons that did not overlap, the analysis could follow the same procedures as used in previous chapters. Unfortunately, but inevitably, as our samples approach the same mammalian grade, it is impossible to present the analysis in that simple and easily understood way. Were we to attempt it, we would be faced with confusingly overlapping figures. Instead we will use the index of relative brain size, the encephalization quotient EQ (Chapter 3) and consider its distribution in each of the groups that we will compare. This method indicates the density of points within the space defined by a minimum convex polygon and enables us to use the fact that animals are more likely to be represented by points in the middle of a polygon than at the edges. Thus, the overlap of the edges of two polygons does not have to obscure the fact that the central tendencies and variabilities of two groups may be different.

We have already seen the data on EQ in several individuals (Table 10.2), and we will now examine a set of cumulative frequency distributions of that measure in three of the groups just mentioned (Fig. 10.6), omitting the didelphids because we have data on too few species. The data on archaic ungulates are from later chapters (Chapters 11 and 13), and the raw data for that curve are presented in Chapter 11. The data on living insectivores are based on information in Bauchot and Stephan (1966) and are summarized in Table 10.3. They also serve to illustrate the method of analysis, which is reviewed briefly (see also Chapter 13). The 33 insectivore values of EQ shown in Table 10.3 are represented as individual points on the cumulative probability distribution for that group, graphed in Fig. 10.6. By plotting the points on normal probability paper, one should find

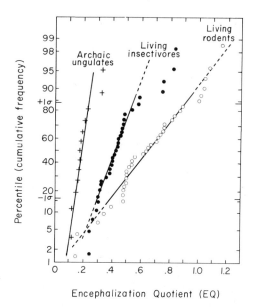

Fig. 10.6. Cumulative frequency distributions of encephalization quotients (*EQ*) in mammals with brain development suitable for "Mesozoic" niches. Note that *Triconodon* and *Ptilodus* (see Table 10.2) would fall at the upper end of the archaic ungulates or at the lower end of living insectivores and rodents. Curves demonstrate progressive increase in brain size in groups of mammals within these niches (cf. Fig. 13.11).

them lying more or less along a straight line if the basic distribution is normal (bell-shaped). We can, thus, linearize a cumulative normal distribution and can read the standard deviation and the mean from the lines fitted to the data. We can, at the same time, determine whether the data do tend to be normally distributed by how well they can be fitted by a straight line.

It is apparent that the insectivore data are reasonably normal until about the eightieth percentile, when they depart from a single straight line in a manner indicating a cluster of relatively large-brained species. None of the insectivores reached the level of an "average" living mammal, that is, in every instance, $EQ < 1.0$. But it does seem to be the case that the living insectivores are made up of at least two populations. One of these is small-brained with mean $EQ = 0.40$, approximately, and with S.D. = 0.09. The larger-brained insectivores cluster about $EQ = 0.7$. In their analysis of insectivores, based on histological criteria and an estimate of degree of neocorticalization, Bauchot and Stephan have described nine of the species in Table 10.3 as "basal" insectivores, suggesting a more primitive

Evolutionary Implications

Table 10.3
Brain and Body Data on Living Insectivores[a]

Specimen[b]	Brain size (g)	Body size (g)	EQ
Solenodon paradoxus	4.67	900	0.42
Tenrec ecaudatus (B)	2.57	832	0.24
Setifer setosus (B)	1.51	248	0.32
Hemicentetes semispinosus (B)	0.83	110	0.30
Echinops telfairi (B)	0.62	87.5	0.26
Oryzorictes talpoides	0.58	44.2	0.39
Microgale cowani	0.42	15.2	0.57
Nesogale dobsoni	0.56	32.6	0.46
Nesogale talazaci	0.79	50.4	0.48
Limnogale mergulus	1.15	92	0.47
Potamogale velox	4.10	660	0.45
Erinaceus europaeus (B)	3.35	860	0.31
Sorex minutus (B)	0.11	5.3	0.30
Sorex araneus (B)	0.20	10.3	0.35
Blarina brevicaudata	0.37	18.5	0.44
Neomys fodiens	0.32	15.2	0.43
Sylvisorex megalura	0.15	5.3	0.41
Sylvisorex lunaris	0.34	18.5	0.40
Suncus murinus	0.38	35.5	0.29
Crocidura hildegardae	0.22	10.6	0.38
Crocidura russula (B)	0.19	11	0.32
Crocidura niobe	0.28	11.5	0.46
Crocidura jacksoni	0.25	12.6	0.38
Crocidura occidentalis (B)	0.44	28	0.40
Crocidura giffardi	0.55	82	0.24
Galemys pyrenaicus	1.33	57.5	0.74
Desmana moschata	4.0	440	0.58
Talpa europaea	1.02	76	0.47
Scalopus aquaticus	1.16	39.6	0.83
Chrysochloris asiatica	0.70	49	0.44
Chlorotalpa stuhlmanni	0.74	39.8	0.53
Elephantulus fuscipes	1.33	57	0.75
Rhynchocyon stuhlmanni	6.10	490	0.82

[a] Data from Bauchot and Stephan (1966).
[b] Species labeled "(B)" are "basal insectivores" as defined by these authors.

level of organization. These were labeled as such in Table 10.3, and it is apparent that they do include the relatively smaller-brained species as determined by our more gross index.

The most important point in Fig. 10.6 is that the living insectivores are clearly relatively larger brained than were the archaic ungulates. If the latter group is representative of a Mesozoic level of relative brain size, then the answer to the question about progressive increase in relative brain size within "Mesozoic" niches during the Cenozoic era must clearly be positive. We may assume that the living insectivores were derived from ancestral forms that would be adequately represented by the data of the archaic ungulates, and Fig. 10.6, therefore, indicates that the living insectivores had advanced in relative brain size within their adaptive zone. The advance may have occurred relatively early in the Cenozoic era, because the Upper Eocene or Lower Oligocene insectivore, *Neurogymnurus* (Table 10.2), as described by Dechaseaux (1969), with $EQ = 0.38$, clearly fits the living insectivore pattern and was larger brained than any of the archaic ungulates in our sample.

The advance within a "Mesozoic" adaptive zone, as suggested by the data in Fig. 10.6 on living rodents, apparently continued. The rodents are also more varied in EQ, implying that they occupy more diverse niches than insectivores, including some for smaller-brained species that actually overlap the archaic forms as well as larger-brained species that are relatively larger brained than "average" living mammals ($EQ > 1.0$). The data on rodents used in constructing their curve were taken from Crile and Quiring (1940) and from Brummelkamp's (1940) summary of earlier measurements, with the restrictions that only the largest specimen of a species was used if data on several were available and that each species was represented by only one point.

IMPLICATIONS OF THE QUANTITATIVE ANALYSIS

The questions on the evolution of relative brain size are answered most precisely by the polygons in Fig. 10.5, with respect to the departure of the earliest mammals from a reptilian level. They are answered by the curves of Fig. 10.6 with respect to evolutionary trends among the mammals during the Mesozoic era. The relatively static situation of the Mesozoic era with respect to brain size was apparently replaced by progressive increases in relative size as the Cenozoic era proceeded.

It is probably easiest to defend the position that during the Cenozoic era a broader range of niches, which were generally similar to, but not necessarily identical with, those of the Mesozoic, was invaded. For example, the place of the shy nocturnal creatures of the Mesozoic era is more or less paralleled by the lives of some living rodents with adaptations for diurnicity. The lives of such animals actually may not be too different from those of nocturnal animals: they could be mainly burrowing creatures or could live out their lives in underbrush along tracks, where little use is made of vision as a sense giving precise information about the location of objects at a distance. Walker (1964, p. 869) pointed out that among the living murid rodents the arboreal forms are usually nocturnal, whereas the ground dwelling or burrowing forms may be diurnal or nocturnal; thus, from our point of view none of these would have to be "visual" animals. Some living rodent groups, such as many squirrels (Polyak, 1957), have redeveloped daylight vision with significant numbers of cones in their retinas, and it is probably no accident that among these one finds species with $EQ > 1.0$.

The dimension of relative brain size, as shown by the abscissa in Fig. 10.6, is, therefore, one in which the orderly variation over a broad range of numerical values probably reflects a broadening range of niches to which adaptations have occurred during speciation. When an evolutionary progression could be identified during the Cenozoic era with respect to increased relative brain size of animals within an adaptive zone, it could generally be associated with the diversification of ecological niches. The major exception to this rule occurred during the first mammalian radiation of the Cenozoic era as represented by the archaic ungulates, when there was diversification in many ways but not in relative brain size. The rule is manifested, in an analysis of the type presented in Fig. 10.6, by the progressive increase in relative brain size as shown by an increase of the mean value of EQ and by the fact that this was correlated with an increase in diversity in relative brain size shown by the shift away from the nearly vertical orientation of the cumulative frequency distribution of the archaic ungulates.

The Olfactory Bulbs

Although I have attempted to maintain a consistent treatment of the role of the olfactory bulbs by excluding them, whenever possible, from the measure of brain size, some of the variations in the measurements taken from other authors are due to inconsistency in this regard. It should be clear, however, from data already presented (e.g., Table 10.2) that even in animals with "large" olfactory bulbs the volume of the bulbs is only a fraction of that of the rest of the brain. Anatomically, the olfactory bulbs are part of the brain and contain the bipolar cells of the olfactory system (Adey, 1959). If one is concerned with the expansion of the brain as an information-processing organ, one should expect the olfactory bulbs to reflect the amount of sensory information received through that modality rather than the amount of analysis of sensory information that is performed by the central nervous system.

In discussions of the evolution of the brain the precise role of olfaction and the way to treat the brain's work in handling olfactory information have not been major concerns. The usual statement is that the progressive evolution of the brain has involved the reduction of the relative size of the olfactory bulbs and the olfactory system and the increase in neocorticalization. The latter statement is certainly correct, but I believe that the statements about the olfactory system have generally been either incorrect or misleading. I review here, and again in Chapter 11, data on absolute and relative size of the olfactory bulbs, and try to interpret these in more appropriate functional terms. The role of smell and of the olfactory bulbs in the evolution of the mammalian brain was probably one of the reasons for its enlargement relative to reptilian brains. A quantitative understanding of the significance of the olfactory bulbs is, therefore, particularly important.

Within the last few years several new quantitative analyses of the size of the olfactory bulbs and associated brain systems have become available. Bauchot and Stephan's long-term studies of the brains of insectivores and their comparisons of these brains with those of primates have been the source of much of that analysis; part of the discussion of the meaning of brain size (Chapter 3) was based on these data. Sacher's (1970) more complete factorial analysis of their data is also used in this section.

In the "basal insectivore" group (Stephan, 1966) the size of the olfactory bulbs is a simple function of body size, well fitted by an allometric function

$$B = 0.015 \; P^{2/3} \tag{10.1}$$

in which B is the volume of the olfactory bulbs in milliliters, and P is body weight in grams. When the estimates of the volume of the olfactory bulbs of fossil and living mammals in Table 10.2 are compared with Stephan's (1966) data on basal insectivores, these mammals fall reasonably close to the level of the basal insectivores (Fig. 10.7). Even *Ptilodus,* with its body weight estimated as about 200 g, and despite its obviously large olfactory bulbs, is seen to be comparable to the living insectivore, *Setifer,* which has similarly large bulbs for its body size. The living rat has smaller bulbs than the living insectivores of comparable body size.

In Fig. 10.7 we can see directly that the sizes of the olfactory bulbs of fossil mammals, including *Ptilodus,* were matched in every instance by those of some living insectivores, and the sizes were orderly functions of body size. We should, therefore, reject the idea of olfactory bulbs that receded in size. Like the brain as a whole, the olfactory bulbs were matched to animals' niches, and this is as true of descendants as it was of their ancestors. A similar result can be demonstrated in "progressive" living orders such as carnivores, as well as "primitive" orders such as insectivores, and we will see that the progressive enlargement of the brain was generally accompanied by some enlargement of the olfactory bulbs (Fig. 11.9).

Let us now consider the significance of the absolute size of the olfactory bulbs. Figure 10.7 includes data on many living primates as reported

Evolutionary Implications

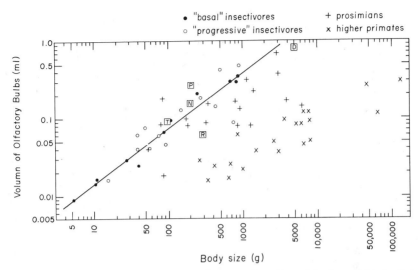

Fig. 10.7. Volume of olfactory bulbs as a function of body weight in living "basal" insectivores, other insectivores, and primates, and mammals in "Mesozoic" niches (D, N, P, T, R in Tables 10.2 and 10.3). Note the similarity of the Mesozoic mammal *Triconodon* (T), living insectivores, and some prosimians and the approximately constant and small sizes of the bulbs in living Anthropoidea (monkeys, apes, and man). The line is Eq. (10.1). Data from Stephan *et al.* (1970).

by Stephan (1966). It appears that within the primates the olfactory bulbs have approximately the same volume regardless of species. The human bulbs, with a volume of about 0.15 ml, are somewhat smaller than average for the primates. If the generally inverse relationship between brain volume and neuronal packing density (Chapter 3) holds for neurons packed in the olfactory bulbs, the human disadvantage would be even more marked, reflecting an unusually nonolfactory niche.

A level of development of olfactory bulbs as indicated by Eq. (10.1), representing the status of living insectivores, is, nevertheless, probably the primitive mammalian condition. The primitive condition was, thus, toward macrosmatic life in niches in which olfaction was an important sense. The reduction of the relative size of the olfactory bulbs below the basal level thus suggests the invasion of new niches in which smell was less important. This is probably the only correct way to phrase the statement that the bulbs became smaller in mammalian evolution. The overall pattern of evolution must always have been associated with the invasion of new niches by mammalian species and the evolution of brain structures appropriate to those niches. Some of the new niches involved reduced dependence on smell; consequently, the olfactory system—or parts of it—were reduced in size and importance. However, other living mammals retain olfaction as a primary distance sense, and their olfactory bulbs approach the predicted size from an extension of Eq. (10.1) to larger mammals.

This, and other issues raised here, are clarified in the next chapter (Chapter 11) in which a larger sample of archaic mammals is analyzed. Late in that chapter the significance of the olfactory bulbs is considered again, in a section titled "Olfaction and the Evolution of the Olfactory Bulbs." Other implications of the evolution of olfaction in mammals are presented in Chapter 12 under the heading "Olfaction as an Analogue of Vision."

Chapter 11

Archaic Tertiary Mammals and Their Brains

Following the great extinction of reptiles at the end of the Cretaceous period, many terrestrial niches were opened to mammals. A number of orders radiated rapidly (by geological standards) at that time, and species evolved that varied in both body size and skeletal structure. Some of these orders are "archaic" in that they were destined to be extinguished early, never having diverged much from the ancestral type, and to be replaced in their niches by modern orders of hoofed and clawed animals. Their brains are known from perhaps a score of endocasts in various museum collections, but these provide a good basis for the analysis of the early evolution of the mammalian brain. We will be able to establish with some certainty the general nature of the response by brain evolution to the selection pressures entailed by the first mammalian invasion of new niches at the end of the Cretaceous and the beginning of the Tertiary period.

One can get a sense of the general appearance of these unusual archaic mammals and of their ecological niches from the excellent paintings reproduced in "The World We Live In" (Barnett, 1955). Some of these are shown in Fig. 11.2.

Although there were many specialized adaptations in the previous Mesozoic radiation of the mammals, reflected, for example, in the teeth of multituberculates, the brain was less clearly modified or specialized than other organs and structures. In order for evolutionary changes in the brain to be clearly reflected by endocasts, there must be either gross evolutionary changes in body size or major behavioral specializations. Changes in body size, which were notable in the early adaptive radiation of the Tertiary period, would affect the size of all organs including the brain. Their effect on the brain (other things being equal) is predictable from the exponent $\frac{2}{3}$ of the brain:body equation [Eq. (2.1)]. Behavioral specializations, such as development of eye–hand coordinations, echolocation, and so on, are also reflected in the enlargement of parts of the brain, and if the enlargement is great enough, it can affect the gross size as well as the appearance of the endocast.

11. Archaic Tertiary Mammals and Their Brains

EVOLUTIONARY BACKGROUND

In this section I will review the positions of the various species that are mentioned in this chapter relative to one another both temporally and ecologically. I also discuss the history of the discovery and identification of a few of the specimens and the construction and interpretation of their endocasts. Despite the inherent oversimplifications, it should be helpful to summarize the probable evolutionary relationships among our specimens by presenting a conventional phylogenetic tree of the mammals, beginning with their origins among the mammallike reptiles and reaching to the present, or Quaternary period. This has been done in Fig. 11.1, in which the progress of time is not shown to scale, but the approximate times of extinction of the last known species of the several archaic orders of the placental mammals are indicated (Hopson and Crompton, 1969; Olson, 1970; Romer, 1966, 1968).

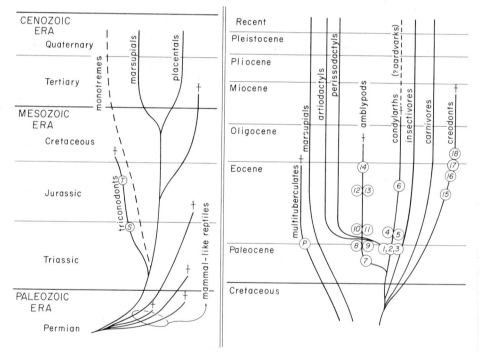

Fig. 11.1. Phylogeny of the mammals. Cenozoic radiation shown in more detail in right half. Specimens discussed in this chapter and Chapter 10 identified by letter and number (cf. Fig. 10.5, Fig. 11.5, and Tables 10.2 and 11.1; S, *Sinoconodon*).

Evolutionary Background 227

THE ARCHAIC ORDERS

The archaic Tertiary mammals are of special interest because of the variety of their body sizes and ecological niches. Among the most primitive of these, the archaic ungulate order Condylarthra, there are some fossils that suggest carnivorous species, although most were omnivorous or herbivorous (Szalay and Gould, 1966); it is from this order that the living ungulates descended. In the later archaic orders Pantodonta and Dinocerata, which are now usually classified as suborders of the Amblypoda, were giant herbivores. And the order Creodonta was an archaic order that included a wide range of definitely carnivorous mammals. The taxonomic issues do not concern us seriously. The main point is that these groups of archaic mammals had invaded many and varied niches, and there was an appropriate opportunity for adaptations of body size, brain, and behavior. The evidence of their endocasts and bodies is the first really extensive information for an analysis of the relationship between the radiation of mammals and the evolution of their brains. The information takes on greater interest because of the broad range of body sizes represented; specimens as small as living squirrels and, by the late Eocene, as large as the rhinoceros have been recovered and their endocasts described.

There were other Paleocene orders that are not discussed in this chapter because they cannot be considered "archaic" as we use the term. There were many Paleocene primates, for example, although the earliest primate endocast is Eocene. We do not call primates "archaic," however, because they were not replaced (as an order) in their niches, and as we all know they continued to evolve and eventually produced the human species. Primates from Eocene or later strata were definitely specialized with respect to their brains (Gazin, 1965b; Radinsky, 1967b). They are discussed in Chapter 16. The living order Carnivora, which shows characteristic specializations of the brain, is also known from fossils from the same Paleocene and Eocene levels as the creodonts. However, these "true" carnivores evolved more slowly, although they eventually replaced the Creodonta in the carnivorous adaptive zone.

Two orders of Paleocene mammals, the marsupials and the insectivores, also well known from the Mesozoic, have survived to the present time, and their brain:body relationships were considered in Chapter 10. Among the living marsupials, the opossums retain many primitive Cretaceous features (but see Clemens, 1968), including the characteristic brain:body relationship. The living insectivores, on the other hand, are slightly advanced in that regard but still much below the average level of living mammals. Though not "archaic" in the sense of having been replaced in their niches, these two orders represent a level of brain evolution in which the brain was relatively small and showed no unusual specializations. In size, their brains were at or only slightly beyond that associated with survival in Mesozoic and Paleocene environments, as we saw in the previous chapter. Data on these primitive, though still surviving, orders are included in some of the graphs and figures to be presented shortly as part of the background for a full description of the brain in nonprogressive Paleocene mammals.

One important archaic order of Paleocene mammals, the order Multituberculata mentioned earlier, also survived from the Mesozoic era and did not become extinct until the middle or late Eocene. This most long-lived of all known mammalian orders, whose oldest fossils are from Upper Jurassic strata (Simpson, 1961), was probably eventually replaced by condylarths, primates, and rodents (Van Valen and Sloan, 1966; Hopson, 1967). Like the late Mesozoic didelphids and insectivores, the multituberculates share the brain size characteristic of the primitive mammalian level. The endocast of the Paleocene multituberculate *Ptilodus* was illustrated in Fig. 10.2.

It should be kept in mind that the three orders of Mesozoic mammals just mentioned, as well as a fourth (Triconodonta) with known endocast, are much more different from one another than are the orders of placentals (Fig. 11.1). They represent major subgroups probably at the level of subclasses (Olson, 1970). We have already reviewed the evolution of their brains (Chapter 10).

The archaic mammalian species of the orders discussed in this chapter, Condylarthra, Amblypoda, and Creodonta, are known from both Paleocene and Eocene strata. The limited available data indicate that the two epochs present a single picture of brain evolution in these groups: no clear advances in grade can be detected within these orders in a succession of archaic species between those epochs of the Tertiary period, 65 and 40 m.y. ago. Endocasts of early and late forms differ only in detail not in gross measures that can lead to judgments about changes in brain size or structure independently of body size.

NICHE AND HABITUS OF THE ARCHAIC FAUNA

There was an important adaptative radiation during the period covered in this chapter, which is dramatically manifested by changes in body configuration and size. During the first few million years of that radiation, perhaps until the Upper Paleocene, the mammals remained within a size range considerably more restricted than in average living mammals. The largest species probably weighed little more than 100 kg and most were considerably smaller. Among the species in our sample the most ancient is

Pantolambda bathmodon, a Middle Paleocene pantodont that weighed about 30 kg (Table 11.1). It was probably among the heaviest species of its time. In general, it is more difficult to obtain data on relative brain size in smaller than in larger species because endocasts are more difficult to prepare, and our sample is probably biased toward the physically larger archaic species of the early Tertiary period. The actual estimates of their body sizes, considered in a later part of this chapter (Table 11.1), range from 630 g for the Middle Eocene condylarth *Hyopsodus miticulus* (Gazin, 1968) to 2500 kg for the Upper Eocene amblypod *Tetheopsis ingens* (Marsh, 1886; Wheeler, 1961).

To provide a sense of the appearance of some of these animals I have presented lifelike reconstructions of some of the species discussed in this chapter (Fig. 11.2). These reconstructions have fascinated me for two reasons. First, they show pretty well that the archaic forms can be visualized as quite similar in body dimensions and skeletal adaptations to living mammals. Second, and as a corollary to the first observation, the archaic forms can readily be visualized as living within niches comparable in many ways to those of living animals. These observations are not entirely unbiased, of course, because the artists and scientists involved in making the reconstructions were obviously influenced by the anatomy and appearance of living mammals. Yet, unless one examines the skeletons of the archaic fossils with some care and with a practiced eye, one may easily miss the fact that they are, indeed, of primitive animals. The illustrated reconstructions were made quite carefully, with the reconstruction moving outward, from muscular sheaths to efforts at indicating skin and sense organs appropriately (Osborn, 1929; Knight, 1947). The only serious errors would arise if a species had developed an unusual adaptation of soft tissue, such as the camel's hump, which is not clearly mirrored in the skeleton. The particular reconstructions in Fig. 11.2 are all based on rather complete material and involve few guesses about the skeleton.

Those who have worked most closely with the remains of these fossils have frequently speculated about their probable habits and appearance in life. Their views are the bases of the sections that follow, and it will probably be useful to refer to the illustrations in Fig. 11.2 as the various species, genera, and higher taxa are introduced.

CONDYLARTHS

The condylarths that have left fossil evidence of the configurations of their brains as well as their bodies for our analysis, and of which the quantitative analysis is summarized in Table 11.1, were probably reasonably representative of the full range of body sizes and ecological niches that characterized this archaic ungulate order. One of the older specimens

Fig. 11.2. Selected reconstructions of archaic mammals numbered as in Table 11.1 and drawn to a common scale: oldest (mid-Paleocene) at top and latest (late Eocene) at bottom.

in our sample, *Arctocyon primaevus* from the Upper Paleocene of France, is representative of an old condylarth family, Arctocyonidae, known in the late Cretaceous fossil record. As an animal it has been described by Russell (1964) as heavy, clumsy, and relatively unspecialized. It was about 45 cm tall at the shoulders, and by analogy to its relative *Mesonyx* it was probably about 120 cm long from its snout to the base of its tail. *Arctocyon* has been described as being "as large as bears and . . . an early parallel to bears in dentition and probably in omnivorous habits as well" (Romer, 1966, p. 243). Alternatively, Piveteau (1961, p. 649) identifies it as almost reaching the size of a wolf. Because of its heavy habitus, I estimate *Arctocyon*'s weight as 84 kg [Chapter 2, Eq. (2.3), and Table 11.1] on the basis of weight:length equations for heavy mammals. This may be compared to the weight range of 27–79 kg for the timber wolf (*Canis lupus*) and 120–150 kg for the black bear (*Euarctos americanus*) as reported by Walker (1964). Our quantitative approach led to a happy compromise between the divergent qualitative judgments of Romer and Piveteau.

Of the other Paleocene condylarths in our sample, *Arctocyonides arenae* (Russell, 1964) was apparently similar in habitus to *Arctocyon primaevus*, though much smaller. Because of their dental adaptations for cutting flesh with carnassiallike molars, these and related forms had been classed as creodonts until only a few years ago. But there is now general agreement that they belong with the condylarths taxonomically (Romer, 1966; Olson, 1970). Behaviorally, they may have been scavengers rather than hunters. Their brains, which are considered later, suggest no specializations either in size or shape beyond a primitive mammalian level. The last Paleocene condylarth in the sample, *Pleuraspidotherium aumonieri* (Russell, 1964; Gazin, 1965a), was a more lightly built animal, probably entirely herbivorous, and on the basis of the size of its skull I have estimated its body weight as no more than about 3.0 kg.

The three Eocene condylarth genera on which we have brain:body data are *Phenocodus* and *Meniscotherium*, which are also known from the Paleocene, and *Hyopsodus*, the earliest remains of which are from the Lower Eocene. The particular specimens in the present sample are all from Eocene strata. *Phenacodus primaevus* was probably longer in body than *Arctocyon*, but *Phenacodus* was probably also more graceful and more like living ungulates in habitus and niche. "Swift-footed, cursorial, small-brained," according to Osborn (1898c), "intermediate in size between a sheep and a tapir," according to Cope (1884), it would weigh 56 kg according to Eq. (2.2), assuming that it was in the middle of the size range of living ungulates and carnivores of similar length. Its probable niche is suggested by the living mammals to which Cope referred, the sheep and tapir, and its appearance in life is suggested in Fig. 11.2.

The habits and habitus of *Meniscotherium robustum* have been analyzed in detail (Gazin, 1965a), and its reconstruction is also included in Fig. 11.2. This animal was the most fully herbivorous of this group, according to Gazin, probably a grazer in savanna country with trees that were the homes of contemporary primates. It was about 64 cm long, and assuming that it was of average build, it would have weighed [Eq. (2.2)] about 6 kg. The remaining species of condylarth was also the smallest, *Hyopsodus miticulus,* which probably weighed about 500 or 600 g and was about as large as a gray squirrel. Its endocast and body reconstructions have been described by Gazin (1968), who suggests a weasellike appearance for the animal and a digging, grubbing mode of life with an omnivorous diet like that of the living hedgehog.

AMBLYPODS

The amblypods, as mentioned earlier, have at times been placed in two orders of mammals (Pantodonta and Dinocerata), although taxonomists presently include them in a single order. They represent the largest mammals of their time; those in our sample probably ranged in weight from about 30 to 2500 kg or more. A total of six genera are represented in our sample: four pantodont genera and two uintathere ("Dinocerata") genera. All were heavily built mammals as indicated in Table 11.1. The oldest species is the Middle Paleocene *Pantolambda bathmodon,* which was about 85 cm long. The two Upper Paleocene pantodonts, *Leptolambda schmidti* and a probable specimen of the genus *Barylambda* or *Haplolambda* were perhaps twice as long or more (see Simons, 1960, for a review).

The Eocene amblypods in our sample include two species of the pantodont *Coryphodon,* two specimens of *Uintatherium anceps,* and a third, somewhat larger and later uintathere, *Tetheopsis ingens. Coryphodon,* which is also known from the Upper Paleocene, has been described as the largest genus of the Lower Eocene, and the uintatheres were the largest genera of the Middle and Upper Eocene. The species of *Coryphodon* are *C. hamatus* (Marsh, 1893) and *C. elephantopus* (Cope, 1877). The latter species was apparently much larger than the former (Table 11.1), and was apparently similar in size to the mounted skeleton of *C. radians* (Osborn, 1898b). The two older uintatheres, originally described by Marsh (1886) as *Uintatherium ingens* and *Dinoceras mirabile,* have recently been reviewed by Wheeler (1961) and placed in the genus *Uintatherium* with the single variable species *U. anceps.* The third uintathere specimen, *Tetheopsis ingens,* has also been reclassified on the basis of Wheeler's revision of the uintatheres. It was the heaviest of the archaic Tertiary mammals that I consider.

The uintatheres are of additional interest because there are three different, more or less independent, estimates of their body size. These can be compared as a way of estimating an "error of measurement." In his monograph on Dinocerata, Marsh (1886) guessed at the weight of *U. anceps* as 1400 kg, and his guess for *T. ingens* was 2750 kg. On the basis of Marsh's reconstructions of these animals, I estimated their respective weights from their body lengths, using Eq. (2.3) for mammals of heavy habitus. My estimates are included in Table 11.1 and are 1400 kg for *U. anceps* and 2500 kg for *T. ingens*. In the case of these particular specimens I also had available a scale model constructed under my supervision and from this model the estimate for *U. anceps* was 1250 liters and for *T. ingens* it was 2160 liters. The variation in these estimates suggests that the data of Table 11.1 are within 20% of a correct figure. This is a tolerable error. In most of the illustrations of brain:body maps of fossil animals an animal is placed on the graph in the form of a circle that encloses a region of about 30% error. It may be noted that if, like man (Webb, 1964), the uintatheres had a specific gravity of 1.06, the weights of the two species to two significant figures, as determined from the scale model, would have been 1300 and 2300 kg, respectively, thus further reducing the differences among these several estimates.

CREODONTS

The creodonts were the major flesh eaters of the Paleocene and Eocene, although some of the condylarths, such as *Arctocyon,* were probably also at least partially carnivorous. If we imagine a balanced ecological community of 50 or 60 odd m.y. ago, it would have been dominated by these archaic forms, the condylarths, amblypods, and creodonts, occupying niches similar to those of the moderate to very large-sized ungulates and carnivores of recent times (Scott, 1937).

The earliest data on both brain and body are from the Middle Eocene creodont *Thinocyon velox,* whose skeleton was described most fully by Matthew (1909). It was reviewed later by Denison (1938) with respect to both skeleton and endocast. It was a small animal, with a skull about the size of a mink's (*Mustela vison*). Its habits were certainly carnivorous, and its niche may have been analogous to the living mink's or genet's. From data on its skull length, and by analogy with its contemporary, though larger, creodont *Sinopa rapax,* in which the head was about one-fifth of the total body length, its most probable body weight was estimated as about 800 g (Table 11.1).

The specimen of *Cynohyaenodon cayluxi* is from a late Eocene stratum in the phosphorites of Quercy in France, a famous fossil bed that has yielded many other specimens reviewed in this and later chapters. Although the fossil is from an animal that lived about 40 m.y. ago, the genus is probably of older vintage, known from Middle as well as Upper Eocene strata. It is considered to be a more primitive genus than *Pterodon,* which is described presently.

Cynohyaenodon was a moderately small carnivorous or carrion-feeding creodont, with its teeth well adapted for crushing bone as well as tearing flesh. Its skull was about 11 cm long, and if its body proportions conformed to those of other older hyaenodonts, its head and body would have been about 4.5 times as long as its skull, giving it a body length of about 50 cm. Its most probable body weight (Table 11.1), assuming it to have been "average" in relative bulkiness is, therefore, about 3.0 kg.

Piveteau (1961) described the species *Pterodon dasyuroides* from Upper Eocene or Lower Oligocene strata in the phosphorites of Quercy as having the dimensions of a large wolf. He considered it to be particularly well adapted by its teeth and jaw musculature to the niche of the living hyena, feeding as a scavenger on dead prey and consuming bones as well as flesh. As Scott (1937) stated, "So far as they are known, skull and skeleton [of *Pterodon*] are like those of *Hyaenodon*" (p. 637), a Lower Oligocene creodont well known in the fossil record of North America. By analogy to *Hyaenodon horridus,* the skeleton of which is well known, we may estimate the body weight of *Pterodon* as 42 kg. This is the weight entered in Table 11.1 and is in the middle of the range of Walker's (1964) summary of data on the timber wolf *Canis lupus* (body length 107–138 cm; body weight 27–79 kg). Piveteau's original judgment of *P. dasyuroides* as being the size of a large wolf is, therefore, consistent with the result of this estimate of body weight derived from weight:length equations.

The last species to be considered is the Oligocene *Hyaenodon horridus.* The various species of *Hyaenodon* lived in Upper Eocene times in Europe and Lower to Middle Oligocene times in North America. They ranged in skull length from about 15 to 35 cm and probably occupied niches comparable to living minks, civets, and hyenas. Sufficient data for quantitative analysis of the evolution of the brain exist only for the largest of the species, the Lower Oligocene *H. horridus* from the White River beds of South Dakota. As indicated earlier, this species is similar to and probably is a collateral (rather than direct) descendant of *Pterodon*. The entire genus *Hyaenodon* appears to have achieved major advances in the evolution of its brain over that of the related hyaenodonts and other creodonts. For the brain:body analysis I have taken the proportions of a mounted skeleton of this species from Scott and Jepsen (1936) to give the body weight of *H. horridus* as entered in Table 11.1, 56 kg.

RELATIVE BRAIN SIZE

In the descriptions of the morphology of the brains of archaic Tertiary mammals, I review data on the absolute and relative size of the endocasts and of certain of their parts and of the bodies of our fossil sample. There are many problems in the estimation of these sizes, and the approach to be followed was discussed in Chapter 2.

The new data presented in this chapter are on 18 specimens from 17 species as tabulated in Table 11.1. In addition to several measurements of length and estimates of weight and volume that are used in the graphic analysis of relative brain size, I have presented my judgment of the habitus of each specimen ("average" or "heavy"), the model used in estimating total head and body length from the length of fragments of the skeleton, such as the skull, femur, and so on, and references used for determining data on body size. The references for the endocasts are indicated in the text and are usually the original reports in which the endocasts are described. Some of the endocasts are illustrated in Fig. 11.4. The most inclusive general source of descriptions of the endocasts, which includes numerous drawings and photographs (but is occasionally unreliable in its data on scale factors), is the "Traité de Paléontologie" (Piveteau, 1958; 1961).

Endocasts

It has been possible to make direct measurements by water displacement of the total volume of several of the endocasts. These measurements may be misleading, however, because artifacts in the endocasts, such as extensive representation of the cranial nerves or of casts of sinuses in the cranial bones, if they occur, are measured as part of the endocast. In the tabulated data on brain size (Table 11.1) I have, therefore, used indirect measures derived by the method of graphic double integration described earlier (Chapter 2); I will describe it again to indicate its special application to the present data.

Outline drawings of lateral and dorsal projections were prepared, their areas measured with a planimeter, and computations were performed as indicated in Fig. 11.3. Because the brains of archaic Tertiary mammals were all unflexured, we can estimate the volumes of several parts of the brain with this method: olfactory bulbs, forebrain, and hindbrain. The

Table 11.1
Endocast and Body Sizes for Archaic Mammals

Species[a]	Endocast size (ml or cm)[b]								Body size (cm or kg)[c]		
					Olfac-tory bulb	Foramen magnum		Total length (cm)	Skull length (cm)	Weight	Habitus
	Total	Fore-brain	Hind-brain			Height	Width				
Condylarths											
1. *Arctocyon primaevus* MHNP CR700	38	19	18		4.1	—	—	120	26.1	86	Heavy
2. *Arctocyonides arenae* MHNP CR733	8.3	5.7	1.5		0.7	—	—	61	13.8	11	Heavy
3. *Pleuraspidotherium aumonieri* MHNP CR252	6.0	3.2	2.2		0.4	—	—	52	12.0	3.3	Average
4. *Phenacodus primaevus* AMNH 4367	31	17	14		2.8	1.9	1.4	132	26.4	56	Average
5. *Meniscotherium robustum* USNM 23113	15	8.5	6.2		0.8	1.0	1.0	64	14.7	6.2	Average
6. *Hyopsodus miticulus* USNM 23745	3.2	1.5	1.7		0.14	0.88	0.62	30	7(?)	0.63	Average
Amblypods											
7. *Pantolamba bathmodon* AMNH 3957	19	15	9.4		1.0	—	—	85	—	30	Heavy
8. *Leptolambda schmidti* FMNH P26095	69	31	49		7.3	—	—	160	—	205	Heavy
9. *Barylambda* or *Haplolambda* FMNH P15573	102	28	74		—	3.3	3.8	230	—	620	Heavy
10. *Coryphodon hamatus* YPM 11330	93	45	49		6	2.6	3.9	175	—	270	Heavy

11. *Coryphodon elephantopus*	90	38	46	4	3.0	3.9	210	—	540	Heavy
12. *Uintatherium anceps* YPM 11036	300	150	140	41	4.0	5.4	300	—	1400	Heavy
13. *Uintatherium anceps* YPM 1537	250	130	120	11	5.6	5.3	300	—	1400	Heavy
14. *Tetheopsis ingens* YFM 1041	350	190	160	11	4.5	5.7	365	—	2500	Heavy
Creodonts										
15. *Thinocyon velox*	5.7	3.2	2.1	—	—	—	32.5	6.5	0.80	Average
16. *Cynohyaenodon cayluxi*	8.3	5.6	3.1	0.5	—	—	50.0	11.0	3.0	Average
17. *Pterodon dasyuroides*	62	31	31	—	—	—	120	30	42	Average
18. *Hyaenodon horridus*	85	58	30	7	—	—	132	39	56	Average

[a] Some of the species listed can be identified by museum numbers. MHNP, Muséum d'Histoire Naturelle, Paris; AMNH, American Museum of Natural History (New York); USNM, United States National Museum (Washington, D.C.); FMNH, Field Museum of Natural History (Chicago); YPM, Yale Peabody Museum.

[b] Endocast volumes by graphic double integration, two significant figures, of dorsal and lateral views as illustrated in Fig. 11.3 and in following references: Russell and Sigogneau (1965); Cope (1877, 1883, 1884); Gazin (1965a, 1968); Edinger (1956a, 1964b); Marsh (1876, 1886); Tilney (1931); Piveteau (1951); Scott and Jepsen (1936). Further discussion in text.

[c] Body weight from Eq. (2.2) or (2.3) depending on habitus. Notes and references for estimating body sizes, generally from figures of mounted skeletons, are as follows (numbered by species): 1, Modeled after *Mesonyx* (Scott, 1886); 2, modeled after *Mesonyx* and *Meniscotherium* (Gazin, 1965a); 3, modeled after *Meniscotherium* (Gazin, 1965a); 4, Osborn (1898c); 5, Gazin (1965a); 6, Gazin (1968); 7, Simons (1960); 8, Simons (1960); 9, modeled after *Titanoides* (Simons, 1960); 10, Marsh (1893); 11, modeled after *C. radians* (Osborn, 1898b); 12, Marsh (1886); 13, Marsh (1886); 14, Marsh (1886); 15, Denison (1938); 16, Piveteau (1961); 17, Piveteau (1961); 18, Scott and Jepsen (1936).

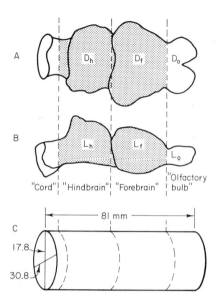

Fig. 11.3. The method of graphic double integration illustrated with the endocast of *Phenacodus primaevus* (cf. Figs. 2.7 and 4.4). A, Dorsal view; B, lateral view; C, equivalent elliptical cylinder: $V_{total} = abl\pi$ mm^3 = (17.8/2) (30.8/2) (81) $\pi \cong$ 35,000 mm^3 = 35 ml.

hindbrain was defined as ending caudally behind the exit of N.XII and anteriorly at the junction of cerebellum and cerebrum. The computations assume an elliptical cylinder as a model of the volume of the brain. I have illustrated the procedure (Fig. 11.3) with the well-known endocast of the Eocene condylarth *Phenacodus primaevus* (Cope, 1884; see Fig. 4.4). Some years ago I measured the total volume of this endocast (including olfactory bulbs) by water displacement and found it to be 35 ml. By the method of double integration "brain" volume was determined as 33.8 ml (Table 11.1). When the "cord" is included in the integration (Fig. 11.3), the total volume of the endocast is computed to be 35 ml.

The estimate of the total volume of the endocast was, surprisingly, exactly the same by the graphic method illustrated in Fig. 11.3 and in my earlier determination. In about a dozen instances in which the volume of endocasts could be determined by water displacement, the difference between that method and the method of graphic double integration did not exceed 10%, and in a number of instances (e.g., *Phenacodus* and *Tyrannosaurus*; cf. Fig. 2.7) it was less than 1%. If we keep in mind that in measures by water displacement there is no way of excluding the casts of nonbrain materials, such as cranial nerves or spinal cord, we may better appreciate the value of the graphic method used here.

Relative Brain Size

There is uncertainty in the measurements of endocasts with respect to the division into forebrain and hindbrain. My procedure has been to make no effort to estimate the size of the midbrain area, and, instead, I divided forebrain and hindbrain at the point of maximum inflection between those regions. In life, many of these archaic mammals undoubtedly had exposed midbrains, but a definite exposure was clear only in *Hyopsodus* (see below). For the others all that can be said is that the unflexured endocast surely signified an unflexured brain, and the cutting point chosen to separate forebrain from hindbrain actually cuts the midbrain. Little if any endocast of the region of rhombencephalon or medulla is incorrectly placed as "forebrain," and little if any exposed "brain stem" area of the hypothalamus and hypophysis is incorrectly added to the hindbrain. The error of the cut is probably all within the midbrain region, and that region is assigned in unknown proportions to forebrain and hindbrain. Dorsal, lateral, and ventral views of several of the endocasts are shown in Fig. 11.4. That of *Phenacodus primaevus* was shown in Fig. 4.4 and more sketchily in Fig. 11.3. (Since forebrain and hindbrain volumes were measured independently of total brain volume, the latter is not always exactly equal to the sum of the former measures, as is evident in Table 11.1.)

HABITUS AND BODY SIZE ESTIMATES

The estimation of body size in fossil mammals is an uncertain procedure at best, and the weight:length equations discussed previously (Chapter 2) were used for all the specimens of this chapter. In an unusual number of instances it was appropriate to take into account the generally bulkier proportions of many of these fossil species compared to living forms. I recall vividly my first impression of the femur and humerus of the Oligocene sabretooth *Hoplophoneus primaevus,* when I saw these next to the corresponding bones of the much larger living jaguar and the lynx to which it had been compared with respect to body proportions by Scott and Jepsen (1936). The long bones of this relatively small *Hoplophoneus* species were like those of a scaled-down jaguar in bulkiness rather than like those of the graceful lynx, although in body length *H. primaevus* was similar to a moderately large lynx. In order to take such valid impressions into account I have classified each specimen in the present sample with respect to habitus as either "average" or "heavy" according to a comparison of the general appearance of its long bones with those of living mammals of comparable size and niche. None of these fossils was unusually thin or graceful compared to living mammals of similar head and body length.

The estimates of head and body lengths for most of the species considered in this chapter were made from photographs or directly from the mounted skeletons. In some instances, mounted skeletons cited in the pale-

240   11. Archaic Tertiary Mammals and Their Brains

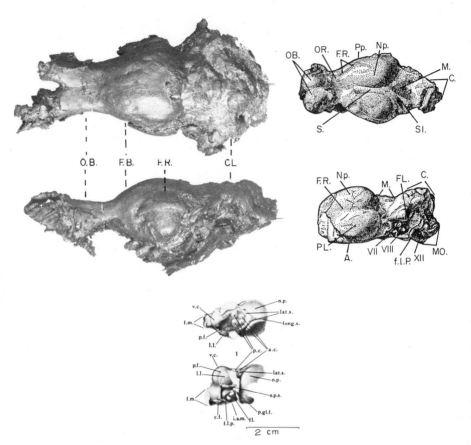

Fig. 11.4. Views of endocasts of condylarths (archaic ungulates). Top left, dorsal (upper) and lateral (lower) views of *Arctocyon primaevus* in Muséum d'Histoire Naturelle, Paris (MHNP CR 700), from Russell and Sigogneau (1965). Top right, dorsal (upper) view of *Meniscotherium robustum* in United States National Museum (USNM 19509) and lateral (lower) view of *M. chamense* in American Museum of Natural History (AMNH 19509), both from Gazin (1965a). Below, fragmentary dorsal (upper) and lateral (lower) view of *Hyopsodus miticulus* (USNM 23745), illustrating exposed midbrain, from Gazin (1968). Scale at bottom is for all specimens. Most abbreviations as in Fig. 4.4. Other relevant abbreviations: p.c., inferior (posterior) colliculi; a.c., superior (anterior) colliculi.

ontological literature as similar to those of our specimens were used as models for their body proportions (e.g., *Mesonyx* for *Arctocyon*). In a few instances, when only parts of the fossils were known and when several different body proportions were clearly possible, data on the length of the skull, scapula, and long bones of the legs were used to suggest an appropriate model.

Errors of Measurement

Before presenting the quantitative analysis it is appropriate to review some of the major sources of error. The greatest errors will certainly be in estimates of body size, even when complete skeletal reconstructions are available for this purpose. The decision to use weight:length relations to make the estimate of body weight was taken in order to reduce the subjectivity of the estimates and to have available an indication of the likely errors involved in the estimation, as discussed in Chapter 2.

There are several types of errors possible in estimates of brain size, none of which is very serious. Errors of one type are associated with the measurement of the volume of the endocast. These may arise either from unnoted crushing of the skull, which distorts the endocast itself, or from misapplication of the method of graphic double integration. The latter method, based on planimetric measurement of the areas of the lateral and dorsal projections of the cast, has generally given results within 5% or less of those obtained by water displacement or loss of weight of a cast in water. A more serious possible measurement error associated with graphic double integration occurs in those endocasts represented by dorsal projections only. These were *Arctocyonides, Pleuraspidotherium, Hyaenodon, Pantolambda,* and *Leptolambda*. In the first two specimens the lateral projections were estimated from the reconstructions by Russell and Sigogneaux (1965). In the other three specimens these were estimated as the same fraction as for the more completely known endocasts of the other specimens in the same order of mammals.

Another type of error in the estimation of brain size occurs in the assessment of the endocast as a model of the brain. In lower vertebrates the brain usually fails to fill the endocranial cavity, and, as described in previous chapters, I have made adjustments of the estimates of brain volume in these vertebrates to approximate their brain size in life as closely as possible. In living mammals the endocast is generally a remarkably good model of the brain, in some respects more accurate than the living brain removed from the skull, because the endocast retains its shape whereas the brain tends to become deformed when its cranial supporting mold is absent. When comparisons were made between the volume of an endocast and fresh brain weight, the two were found to be almost identical (Bauchot and Stephan, 1967). Contrary reports are usually attributable to using preserved brains; these are generally shrunken, although occasionally they expand, depending on the preserving fluid and the length of time they have been preserved (Hrdlička, 1906; Stephan, 1960).

In assessing body weight, brain weight, or brain volume, haphazard sampling in the collection of fossils can raise further problems for the analysis. I believe that these are less serious for the quantitative analysis of brain evolution than for other analyses. A sampling error could occur if a recovered specimen were not fully grown. In most instances this is not too serious because various criteria (e.g., status of sutures or wear on the teeth) are available for deciding whether a specimen was mature and whether to use its data in the analysis. If endocasts are taken from an animal of unknown status, particularly if all that is found is a natural endocast with minimal identifying material (as in the case of the United States National Museum's *Thinocyon velox*), one may be uncertain about the specimen's age. Even this is unlikely to produce serious sampling errors because adult brain size in living mammals tends to be achieved relatively early in life. For example, in the living horse, adult brain size is achieved before the end of the first 9 months of life, about 4% of the total life. Body size, on the other hand, may not reach the fully adult state until about 10–20% of a life span is passed (Count, 1947). A recovered impression of a brain in an endocast is, therefore, likely to represent the adult condition, even if the animal was not fully mature. We can note, finally, that intraspecific variability may also be a relatively minor problem. In most mammals brain size is more consistent than body size within a species. This point was well made and was documented by Radinsky (1967a) and is also discussed in Chapter 16.

All these classic sources of error are potentially important and I would not dismiss them casually. But it is also necessary to keep their potential effects in perspective. They are, cumulatively, unlikely to be critical; they are probably uncorrelated and hence likely to counterbalance one another.

Our quantitative analysis, based on between-species differences, will also tend to expose errors because large errors will probably produce inconsistencies in the data. In fact, the quantitative analysis may be judged on the basis of the internal consistency of the results. To the extent that we can develop a coherent picture of the evolution of the brain and to the extent that the picture is both simple and consistent with the biology of living animals, we may find the analysis generally acceptable. The fact of self-consistency and coherence is perhaps the best safeguard against major errors in individual measurements, and I have taken great care to make those measurements independently of, and prior to, undertaking the integrative analysis.

BRAIN:BODY RELATIONSHIPS

The data for all the specimens are presented in Table 11.1, and the relationship between brain size and body size is presented in the usual way on double logarithmic coordinates in Fig. 11.5. In addition I included data from Chapter 10 on mammals in "Mesozoic" niches: the Paleocene

Relative Brain Size 243

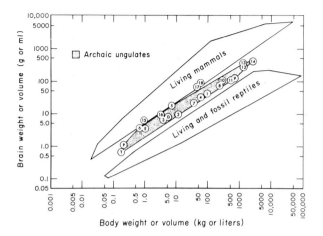

Fig. 11.5. Brain:body map of archaic ungulate data shown in relation to reptilian and living mammalian maps as discussed in Chapter 2. Note that these "ungulates" included carnivorous as well as herbivorous species and represent the early Tertiary archaic mammalian radiation with respect to body size. Numbers identify species as in Table 11.1. Creodonts (15–18) were excluded from the archaic polygon because they represent later (Middle Eocene to Lower Oligocene) species.

multituberculate *Ptilodus montanus,* the Jurassic triconodont *Triconodon mordax,* the living European hedgehog *Erinaceus europaeus,* and the living Virginia opossum *Didelphis marsupialis.* The specimens are numbered as in Fig. 11.1 and Table 11.1 in this chapter, and those described in the earlier chapter are identified by the initials of the generic names (P, T, E, and D) of the "Mesozoic niche" specimens.

Three minimum convex polygons are drawn in Fig. 11.5 to indicate the underlying brain:body relationships among these mammals. These include the two polygons that envelop the data on living mammals and living and fossil reptiles as discussed in Chapters 2 and 7, which are the background against which the other polygons are compared, and the minimum convex polygon, enclosing the data of 14 of the 18 archaic Tertiary mammals presented in this chapter as well as 4 of the mammals from "Mesozoic" niches presented in Chapter 10. The brain:body data of the 4 creodonts listed in Table 11.1 are not enclosed in a polygon; they fall at the bottom of the living-mammal polygon. The middle polygon maps the correlation of the first adaptive radiation of Tertiary mammals with the evolution of their brains.

Without further analysis Fig. 11.5 indicates that the broad adaptive radiation of the archaic mammals at the beginning of the Tertiary period took place with little or no gain in information-processing capacity (measured by relative brain size, see Chapters 1 and 3) beyond that achieved by the earliest Mesozoic mammals. A single polygon, similar in shape to that of living reptiles and mammals and oriented similarly in the coordinate plane, maps the level of brain development achieved in the early Tertiary period by most of the archaic mammals discussed in this chapter. That level is at or below the lower limits of the mammalian data of Crile and Quiring (1940), and when the creodonts are excluded, there is no overlap at all between the polygon for archaic mammals and that for living mammals in the sample described by Crile and Quiring. The conclusion is, essentially, that in the early radiation of the mammals the archaic forms invaded many niches by evolving larger and specialized bodies, but their brains increased in bulk only to the extent required to process the additional information to control the larger body. It is as if a population of opossums and small-brained insectivores expanded their niches by evolving new species that increased in body size, but the resulting animals remained magnified didelphids and insectivores with respect to their brains. The degree of magnification is established by the slopes of the polygons, that is, their orientation in the coordinate plane, which states the necessary minimal change in brain size that must accompany changes in body size.

Elsewhere in this book (e.g., Chapters 10 and 13) the data of Table 11.1 have been recast in the form of cumulative frequency distributions of the encephalization quotient *EQ*. This was done when an analysis, using minimum convex polygons as maps of the evolution of brain size, resulted in overlapping polygons that were difficult to "read" directly. For present purposes, however, this is not necessary because Fig. 11.5 presents an unusually clear picture of the advance in relative brain size that occurred with the evolution and the first adaptive radiation of the mammals. That the clarity due to the failure to overlap the living mammalian polygon is fortuitous should be obvious because the polygon for living mammals did not include either of the small-brain living forms, didelphids and insectivores, that fell within the archaic Tertiary ungulate polygon. This resulted simply from the fact that Crile and Quiring (1940) had not included data on a full-grown specimen of either *Didelphis marsupialis* or *Erinaceus europaeus* in their sample, and their specimens of didelphids and insectivores happened to be of larger-brained species. If a polygon on living mammals were developed on the basis of data on all living species, then the polygon would clearly overlap the upper half of the archaic Tertiary polygon; its lower limits would have been points D and E. This is done in the previous chapter in Fig. 10.5.

The full interpretation of the quantitative analysis thus far is, therefore, that early in their evolution the mammals achieved the status of large-brained animals beyond the level of their ancestors among the reptiles

(Chapter 10). The level that they achieved was sufficient for the behavioral adaptations associated with their first adaptive radiation of the Tertiary period, and that radiation was accomplished without significant further relative enlargement of the brain. Since that radiation entailed an invasion of niches with selection pressures for enlarged bodies, there was an evolutionary increase in absolute brain size that occurred in those archaic species of 65–40 m.y. ago. But the increase in brain size was no more than that necessary to handle the enlarged body.

FOREBRAIN, HINDBRAIN, AND OLFACTORY BULBS

It is sometimes asserted that an analysis based on the whole brain and its size is irrelevant for morphological and physiological analyses of the role of the brain in behavior (e.g., Holloway, 1966). One prefers, of course, to use the fine anatomy of the brain and to consider the subsystems that are effective in the brain's work. And despite the well-justified use of the gross brain size as a natural biological statistic for estimating the number of neurons, connectivity, and other relevant parameters, it is still important to determine the evolution of the parts of the brain, if that evolution is manifested in the fossil record.

In the illustrations of the brains of the archaic Tertiary mammals discussed in this chapter (Figs. 11.3 and 11.4), we could see that these early mammals were unusual in that their brains were unflexured and that one could estimate the volumes of at least three subdivisions—forebrain, hindbrain, and olfactory bulbs. Only the first two have been considered part of the brain, proper, in the gross analysis. The evolution of the relative sizes of all three parts can now be considered. In order to do this we must determine the outcome of the evolutionary process. That is, we must find a later group of mammals with which to compare the archaic Tertiary forms. We will do this by estimating the same brain fractions for living mammals in the case of forebrain and hindbrain and by comparing the relative size of the olfactory bulbs in the archaic Tertiary fauna with those of late Pleistocene and Recent samples.

Relative Size of Parts of the Brain

Although there are scattered data in the anatomical literature on the sizes of the parts of the brain in mammals (e.g., Latimer, 1956; Stephan et al., 1970[1]) most of the literature is restricted to relatively few species

---

[1] This reference was received during the final revision of the text along with Sacher (1970) in which its data are analyzed. Sacher's approach is discussed in Chapter 3 (pp. 72–74). The data of Stephan et al., limited to insectivores and

and to relatively small animals such as rabbits and insectivores. As background for our analysis we need broadly based data on the expected sizes of parts of the brain in living mammals of different body sizes. A major study by Wirz (1950) does cover the broad range that would be most useful for present purposes. She organized her analysis in terms somewhat similar to the present division into olfactory bulbs, forebrain, and hindbrain, which I chose because it was easy to estimate these volumes separately in the linearly organized, relatively unflexured, brains of early Tertiary mammals.

Wirz dissected fresh brains and weighed the total brain and the parts, defined as neopallium (cerebral hemispheres dorsal to the rhinal fissure), cerebellum, and stem complex (the medulla, midbrain, pons, and at least part of the thalamus, but excluding the internal capsule). For our purpose we will consider the neopallium identified by Wirz as comparable to, and certainly no larger than, the forebrain as we measured it, and the sum of the stem complex plus the cerebellum as comparable to the "hindbrain" in the endocasts. Details on her procedure and results are presented in Appendix I. By using her method and indices, I have estimated the weights of these parts of the brain in living mammals comparable in body size and niche to the archaic Tertiary group. These weights are summarized in Table 11.2, the data of which may be compared directly with estimates of the size of the forebrain and hindbrain in the archaic forms presented earlier in Table 11.1.

To validate our use of Wirz's indices, let us consider how well they estimate human data in the literature. Under the assumption that a typical human male weighs 70 kg, Wirz's analysis predicts a total brain weight of 1314 g, a forebrain weight of 1045 g, and a hindbrain weight of 158 g. The total weight agrees well with most sources as a typical human male brain weight. The most directly relevant data, by Marshall (1892), reported that in a group of normal men (167–173-cm tall and 50–60 years old) the average total brain weight was 1327 g, cerebral hemispheres ($\cong$ forebrain) were 1157 g, and cerebellum ($\cong$ hindbrain) was 145 g. It is clear that Wirz's indices led to reasonable estimates for man. It may be assumed that they also are reasonable for the species in Table 11.2.

The forebrain sizes of the archaic Tertiary sample (Table 11.1) are compared with those of the living mammals in Fig. 11.6. As in the case of the analysis of total brain size, these data are given in the form of minimum convex polygons enclosing the data points in log–log brain:body space. As in the case of the brain as a whole, the forebrain:body map of the archaic

---

primates, are less representative of the mammals than Wirz's sample. Within the limits of their sample the data of Stephan and his colleagues appear to be consistent with the present analysis, which is based on Wirz's monograph.

Table 11.2
Estimation of Relative Sizes of Parts of the Brain in Living Mammals from Wirz's (1950) Indices

Specimen[a]	E	P	Wirz's indices[b]				Basal[c] St.I.	Computed weights[d]		
			T.I.	N.I.	C.I.	St.I.		E_F	E_H	E_T
a. Hyrax (*Procavia capensis*)	19.2	3,500	15.8	8.5	1.74	2.82	1.45	12.3	6.6	22.9
b. Raccoon (*Procyon lotor*)	41	4,380	23.6	13.2	2.82	3.63	1.63	21.5	10.5	38.5
c. Brown bear (*Ursus arctos*)	407	197,000	40.3	23.3	5.10	4.65	10.10	235.3	98.4	407.0
d. Hyena (*Crocuta crocuta*)	168	43,500	25.4	16.9	2.82	3.56	4.89	82.6	31.2	124.2
e. Tapir (*Tapirus indicus*)	265	201,000	22.7	12.6	2.92	3.20	10.20	128.5	62.4	231.5
f. Giraffe (*Giraffa camelopardis*)	680	529,000	47.0	29.5	3.89	4.57	16.23	478.8	137.3	762.8
g. Elephant (*Elephas indicus*)	5,430	3,048,000	150.0	70.0	30.20	9.95	37.62	2,633.4	1,510.4	5,643.0

[a] Body (P) and brain (E) weights for specific animals from Weber (1896); species named following Walker (1964). All weights in grams.
[b] T.I., Total Index; N.I., Neopallial Index; C.I., Cerebellar Index; St.I., Stem Index.
[c] Data for basal stem index computed from Eq. (I.1 b), Appendix I, using values in body weight (P) column.
[d] Computed weights determined as follows: E_F = (Basal St.I.) (N.I.) = forebrain weight. E_H = (Basal St.I.) [(C.I.) + (St.I.)] = hindbrain weight. E_T = (Basal St.I.) (T.I.) = total brain weight.

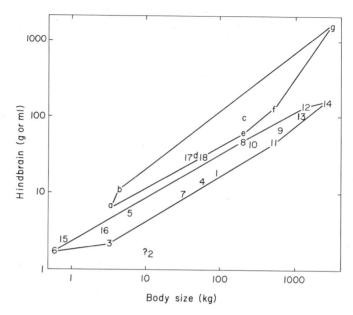

Fig. 11.6. Forebrain:body maps for archaic ungulates and for representative living mammals. Species identified as in Tables 11.1 and 11.2.

group, excluding the creodonts, is described by the minimum convex polygon similar in orientation to, but below, that of living mammals. The forebrain of living mammals is about 4 to 5 times as large as that of the condylarths and amblypods, according to the samples that have been compared here. The forebrain of creodonts, like the brain as a whole, is of the order of 50% or more larger than that of the archaic ungulates and overlaps the living species.

It is significant that these polygons, like those for the brain as a whole, maintain their orientation with a slope of about ⅔ on the log–log coordinate system. This suggests that the body size factor, which may be thought of as the requirement that an extended body "surface" be mapped on the brain, holds true for the forebrain as well as for the brain as a whole. Particularly significant for the analyses of gross brain size as an approach to the general problem of brain evolution is the fact that the results with this somewhat finer analysis are consistent with the more gross earlier results based on the brain as a whole. From the study of the volume of the functionally "highest" regions of a fossil brain, we reached results essentially the same as those achieved earlier with the more gross data on the whole brain, which are normally all that we have.

The portions of the endocasts described as the hindbrain in Table 11.1,

Forebrain, Hindbrain, and Olfactory Bulbs

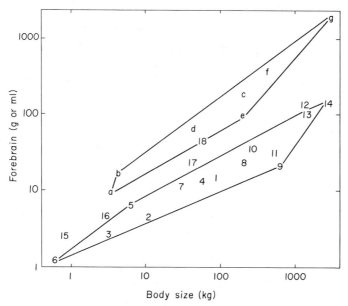

Fig. 11.7. Hindbrain:body maps for archaic ungulates and for representative living mammals. Species identified as in Tables 11.1 and 11.2.

compared with data derived from Wirz, are presented in Fig. 11.7. It is apparent, on inspection, that these data are almost as orderly as those of the previous graphs of brain:body relations in the archaic groups. Except for the creodonts, the archaic mammals have consistently smaller "hindbrains" than living mammals. The creodonts appear to be at the lower part of the range of the mammals from the living sample. (The hindbrain of *Arctocyonides*, specimen No. 2, was probably shown as too small in the reconstructions used for its measurement.) The general orientation of the hindbrain polygons is also at a slope of about $\frac{2}{3}$. There was some increase in relative size of the hindbrain in creodonts and perhaps again between creodonts and living mammals. The increase was comparable to that for the forebrain and for the brain as a whole.

It should, perhaps, be emphasized that the result in Fig. 11.7 is more surprising than the previous results. It indicates that the hindbrain, in particular the cerebellum, increased in size in about the same way and proportion as did the forebrain. The evolution of brain size from the Eocene epoch to the present was not merely "corticalization." The hindbrain was apparently involved to about the same extent as the forebrain.

In viewing archaic endocasts, such as that of *Arctocyon primaevus*, one often comments on the relatively large olfactory bulbs, and we noted in the previous chapter (pp. 221–224) that "enlarged" olfactory areas are frequently considered to be diagnostic of primitiveness in a brain. The meaning of this observation must be examined carefully. It arises from the fact that one typically looks at a brain or endocast without regard to its absolute dimensions. In the case of the olfactory bulbs, which contain mainly peripheral "receivers" in the neural apparatus that processes olfactory information (see Chapter 12), a certain amount of information can be or has to be received, and this should be independent of the size of the animal or the rest of its brain. Olfaction is not analogous to sensorimotor coordination in which a larger animal with more muscle fibers must have more neural elements to control the fibers and must, necessarily, have a larger nervous system to contain those neural elements. In the case of the olfactory bulbs we should, therefore, be concerned with the absolute rather than relative size. This structure in the endocast of *A. primaevus* is compared with that of the living timber wolf (*Canis lupus*) in Fig. 11.8.

The olfactory bulb region of *Arctocyon* as preserved in the endocast (Russell and Sigogneau, 1965) includes some of the turbinates, and the region as a whole is somewhat enlarged in the endocast. However, the limits of the bulbs are fairly clear, and their volumes can also be estimated by the method of graphic double integration. For *A. primaevus* the total volume of the bulbs was 4.1 ml. Whether this is a relatively small, normal, or large structure cannot be determined unless one knows the expected

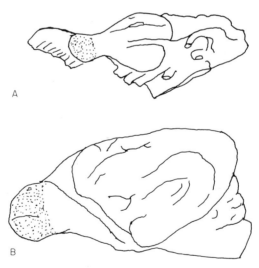

Fig. 11.8. Lateral views of endocasts. A, The condylarth *Arctocyon primaevus* (Table 11.1); B, the timberwolf *Canis lupus* (Tables 11.3 and 15.1), illustrating the absolute amount of olfactory bulb (stippled).

volume. The timber wolf is a living carnivorous mammal of about the same body size as *Arctocyon,* and it is apparent from Fig. 11.8 that the olfactory bulbs in the archaic form were, if anything, smaller than in the living carnivore. The primitive *Arctocyon* did not have a larger olfactory system.

OLFACTION AND THE EVOLUTION OF THE OLFACTORY BULBS

We cannot perform the same kind of analysis for the olfactory bulbs that we can for the rest of the brain because the impression of the bulbs in an endocast is normally considerably larger than in the brain. In the coyote *Canis latrans,* for example, the volume of the olfactory bulbs in the brain was 1.5 ml, whereas in the endocast the bulbs displaced 4.2 ml. It is likely that there is a consistent relationship between these two measures, however, and that mammals with truly large neural olfactory bulbs will also have large impressions of the olfactory bulbs on their endocasts. Our comparison is, therefore, between the volumes of the olfactory bulbs taken from endocasts as presented in Table 11.1 for the archaic Tertiary sample and the volumes in a sample of Quarternary mammals prepared for this purpose. The latter sample consists of fossils from the La Brea tar pits and a few living mammals and is presented in Table 11.3. The body weights in this table are not from the specimens that provided data on the olfactory bulbs; they are average data from Walker (1964) for the living forms and estimates using the weight:length equations for the fossils.

Table 11.3
Volume of Endocast Olfactory Bulb Region and Body Size in Quaternary Mammals[a]

Species[b]	Olfactory bulb (ml)	Body weight (kg)
a. Fox (*Urocyon* sp.)	1.4	5.0
b. Coyote (*Canis latrans*)	4.2[c]	12
c. "Lion" (*Panthera atrox*), r.l.b.	6.5	250
d. Black bear (*Ursus optimus*), r.l.b.	5.5	135
e. Draft horse (*Equus caballus*)	27.5	1,800
f. Badger (*Taxidea taxis*)	1.4	7.5
g. Timberwolf (*Canis lupus*), r.l.b.	5.2	60

[a] All endocasts from the Los Angeles County Museum, courtesy of Dr. J. R. Macdonald.

[b] Species marked r.l.b. are from fossils recovered from the La Brea tar pits.

[c] The measured volume of the olfactory bulbs in the fixed brain of a coyote (courtesy of Dr. W. I. Welker, University of Wisconsin, Laboratory of Neurophysiology) was 1.5 ml, indicating the difference between brain and endocast.

For the more complete quantitative analysis I have used the same procedure as in the analysis of the size of the brain. Data on the size of the olfactory bulbs were related to body size and graphed in brain:body space. Minimum convex polygons could then be drawn about the data of the archaic Tertiary forms and the progressive Quarternary species. This comparison is illustrated in Fig. 11.9.

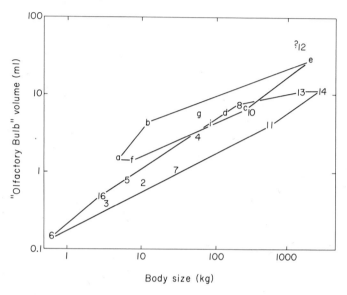

Fig. 11.9. Olfactory bulb:body map. Letters refer to specimens described in Table 11.3; numbers are archaic mammals of Table 11.1. Note that unlike previous maps, there is some overlap in these distributions, although living forms generally had relatively larger olfactory bulbs (cf. Fig. 10.7).

The results are unequivocal. The archaic mammals did not have relatively larger olfactory bulbs than recent mammals have. If there is a difference related to the 50 m.y. that separates these groups, it is in favor of the later group as having larger olfactory bulbs.

One individual of the early Tertiary species, *Uintatherium anceps* (No. 12), is excluded from the archaic polygon; the second individual of that species (No. 13), which is represented, had much smaller olfactory bulbs. A careful inspection of the endocast of specimen No. 12 convinced me that the olfactory bulbs had been incorrectly reproduced and that the anterior portion of the endocast was a result of an excavation of the cribiform area in which the ethmoidal bones had been removed. The anterior portion of the olfactory bulbs in that specimen was an artifact, an error in preparing the endocast.

Conclusions

I was somewhat surprised to find that the orientation of the olfactory bulb:body polygons still indicated an exponent of $\frac{2}{3}$ for the functional relationship. This means that the size of the olfactory bulbs is a function of body size and that body size has to be taken into account in estimating the significance of the size of the olfactory bulbs. Sacher's (1970) analysis, which appeared after the present analysis was completed, indicated that this might have been anticipated but that the contribution of a body size factor to the size of the olfactory bulbs might really involve causal variables other than those that contribute to the total brain size or the size of the forebrain and hindbrain.

To return to a functional and behavioral approach, if we think of the olfactory bulbs as receiving and transmitting information rather than as involved in integrative activity, then their absolute size may be an important concern only if they are unusually small. The archaic Tertiary mammals for which we have data may have, indeed, been macrosmatic, that is, dependent on olfaction for their distance information. But, in seeking to define their sensory dependence in behavioral terms, we need look no further than living large mammals such as the carnivores and ungulates of Table 11.5, rather than at less familiar insectivores. The present evidence is that the archaic Tertiary fauna used olfaction in a similar way or less effectively than do the most familiar living mammals.

CONCLUSIONS

Perhaps the most surprising result of this analysis of the early evolution of the mammalian brain is the orderliness of the brain:body function represented by the minimum convex polygon enclosing the data of archaic Tertiary mammals. This polygon is parallel to and slightly beneath the one determined for living mammals. It encloses not only the archaic Tertiary mammals but living mammals such as insectivores and didelphids and the lone Mesozoic mammal with known endocast, *Triconodon mordax*. It is definitely above the level of the "lower vertebrate" polygon and indicates a persistence of small-brained species.

The first conclusion, therefore, is that the early mammalian radiation was a major diversification of adaptations to which brain evolution did not contribute beyond the contribution required to maintain allometric relationships among the organ systems of the body. One may think of the mammalian species involved in this radiation as overgrown primitive mammals with respect to their neural control apparatus; their speciation and invasion of new niches was apparently accomplished within the limitations of a relative brain size characteristic of the most primitive of mammals.

In that respect they are analogous to the "lower" vertebrates in achieving a considerable radiation at a constant brain size level, although the archaic mammalian level already involved 4 or 5 times as much brain tissue as among reptiles and amphibians of comparable body size. The speciation in archaic mammals was accomplished by evolutionary changes in skeletal and other (unknown) structural or functional dimensions but did not involve an absolute increment in brain "power" beyond the level reached perhaps 100 m.y. earlier by Jurassic and possibly also Triassic mammals.

A second conclusion is based on the parallel orientation of the minimum convex polygons for each level of brain evolution. This must reflect some underlying body size factor that operates by fixing a brain size for a particular body size within a particular adaptive level of information-processing capacity. The recurring exponent of about $\frac{2}{3}$ suggests that there is a mapping of the body surface on the brain, that is, a surface:volume relationship (Gould, 1966). That surface would be a peculiar one, of course; for example, the sensory sheath of tissue making up the retina would cover a large area and the skin of the back might be relatively small in area. Yet given the expected distortions of the body surface in the creation of a body map, the idea that a mapping does occur is appealing because it is consistent with what we know about the living brain as an information-processing system. Beyond the information we have on the nature of representation of sensory and motor systems in the brain (Woolsey, 1958), which supports the idea of a mapping function, however, we are really too ignorant to say much more.

The third conclusion is tentative because of the limited material on which it is based and the relatively fine judgment of relative brain size and configuration that it requires. It is that the creodonts were somewhat more advanced than the other orders described as "archaic" in this chapter. Our criterion for considering an order archaic was, after all, arbitrary and not necessarily associated with brain evolution. It was simply that the order was entirely replaced within its niches by species of another order or orders. The creodonts were smaller brained than their contemporaries from the order Carnivora, as we shall see in Chapter 13. Yet they were larger brained than the amblypod and condylarth species that might have been among their prey. It is tempting to suggest a leapfrogging of selection pressures in which the larger-brained creodonts helped determine the successful replacement of amblypods and condylarths by larger-brained ungulates, and this put the creodonts at a selective disadvantage, compared to the Carnivora, as predators of the more progressive ungulates. However, the situation was probably more complex and involved competition between creodont and carnivore species in other predacious niches in which fish, rodents, rabbits, or birds may have been the characteristic prey. In

Conclusions

this case the adaptations of the prey species probably did not include significant enlargement of the brain. We consider this theme further in later chapters (Chapters 13 and 14).

The fourth conclusion is that there is no evidence from the organization of the brain and the olfactory bulbs that the archaic Tertiary mammals were peculiarly macrosmatic. Their reliance on smell was probably quite similar to that of many familiar living mammals. There is no evidence of excessive enlargement of the olfactory bulbs in the fossil forms. On the contrary, it is likely that living mammals such as the land carnivores and ungulates have relatively larger olfactory bulbs than early Tertiary mammals in similar niches. There may, thus, have been some perfection of the olfactory apparatus in the evolution of the mammals during the last 50 m.y., as well as a general increase in brain size in many groups.

In general it is approximately correct to identify the primary mammalian radiation of the Mesozoic era and early Tertiary period with a four- to fivefold increase in brain size above the level of the mammallike reptiles of the late Paleozoic and early Mesozoic eras. The shift from the primitive mammalian level to that of the living mammalian fauna involved a second four- to fivefold increase in brain size. Some of the details of that increase may be seen in the comparison of the creodonts to the other archaic forms described in this chapter, but additional details are our primary concern in later chapters.

It is well to keep in mind, finally, that an increase in brain size was not a necessary consequence or cause of the mammalian radiations of the Tertiary period (Andrew, 1962). The basal level achieved by the earliest mammals was sufficient for a variety of adaptive niches, including those occupied by the living opossums and some insectivores. In general, a level of brain evolution reflected by brain:body relationships is associated with a particular niche, and we can, in part, define niches by that level. Certain niches "explored" by the archaic radiation could be better occupied by larger-brained species than by those with the genetic adaptive capacities of condylarths, creodonts, and amblypods. But other niches apparently did not require such brain adaptations and little or no selective advantage accrued to enlargement of the brain. This is post hoc reasoning; yet it is a good guideline for thinking about the brain as a factor determining success or failure of species within their adaptive zones.

Chapter 12

Basic Selection Pressures for Enlarged Brains

Why should the brains of the earliest birds and mammals have been larger than those of their reptilian progenitors? I have suggested possible answers to this fundamental question at various points in previous chapters and elsewhere (Jerison, 1971a), and will devote this chapter to a more complete analysis of that question. We require two kinds of answers. First, we must specify the kinds of selection pressures that evolving "reptiles" suffered in entering the adaptive zones of birds and mammals and indicate the general adaptations required to survive in these zones. Second, we must indicate the kinds of neural adaptations that would have been necessary to control behavior in such zones, and we must show why these neural adaptations resulted in enlarged brains.

The basic facts for this general discussion have been presented in previous chapters. We have seen the evidence that the lower vertebrates have probably been at the same level of evolution of relative brain size throughout the 500 m.y. of their adaptive radiation, their entire evolutionary history, from the Ordovician period to the present. There are limits to that conclusion. There would have been no evolution had there not been varied behavioral adaptations among the lower vertebrates and varied neural apparatus associated with these adaptations. The evidence is, more precisely, that the varied adaptations were accomplished with approximately the same total amount of brain tissue if body size is taken into account, although different amounts of tissue were devoted to different functions in different species. Bottom fish or blind cave fish, for example, have had enlarged vagal lobes and forebrains, associated with the importance of chemoreception in their niches, whereas surface fish have had relatively enlarged optic lobes (midbrain), associated with their reliance on visual information (Poulson, 1963). But the total amount of brain tissue is about the same in all these fish (Chapter 5).

There were certainly some significant differences in brain size, both relative and absolute, among the lower vertebrates. As a group, however, their brain:body data could always be confined within the lower minimum convex polygon in graphs such as Fig. 2.4. The various specializations of the lower vertebrates did not result in sufficiently great increments in brain tissue in any of the species on which we have had data to raise them above that polygon. That is one sense in which I have considered brain size as a conservative evolutionary character: a single brain:body map covers the entire 500 m.y. evolutionary span of the lower vertebrates.

The second basic fact is that when selection pressures toward enlargement of the brain could first be identified they could be recognized among the earliest mammals, which actually evolved early in the Mesozoic "Age of Reptiles," about 200 m.y. ago. The enlargement of their brains seemed to provide essentially no competitive advantage to the mammals as opposed to the reptiles of that era. Both classes survived, but the Mesozoic mammals remained physically small and "insignificant" while the reptiles evolved into an amazing variety of forms and dominated a wide range of adaptive zones. The enlarged brain of Mesozoic mammals seems to have been a stable adaptation to an unusual adaptive zone because there was apparently no progressive increase in relative brain size between the time of their origin and the early Cenozoic era. The mammalian brain remained at the Mesozoic level in relative size for over 100 m.y., and the increase in relative size to the present level was a phenomenon of the last 50 m.y. or so.

The birds, as the other group of relatively large-brained vertebrates that descended from the reptiles, appeared in the fossil record some 30 m.y. after the earliest mammals. They were descended from a branch of ruling reptiles, and a third important fact that we should recognize is how completely different, that is, evolutionarily independent, the birds have been from the mammals. We should not be surprised by major differences in the brains and the behaviors between these advanced vertebrate classes because the comparable increase in relative brain size in these two major vetebrate classes must have been an instance of parallel evolution. We should expect, and seek to identify, very different adaptive zones that were open 150 and 200 m.y. ago for birds and mammals, respectively, both of which happened to be characterized by selection pressures toward the enlargement of the brain. But we should not be surprised to find that those selection pressures were different enough to result in significant differences between birds and mammals in their solutions to the problem of the neural control of behavior.

Evolutionary Trends

Selection pressures are often considered in terms of a better adapted species replacing a less well adapted species within a niche. Alternatively, environmental changes may modify a niche to make some genetic variants more successful than the basic stock that had occupied the niche. Such views are inappropriate for the beginnings of the evolution of birds and mammals because birds and mammals probably did not replace reptiles in Mesozoic niches. The mammals were contemporaries of the dominant reptiles during most of the Age of Reptiles. Only in the Tertiary period, after the massive extinctions of reptilian life, can one clearly identify a broad radiation of the mammals. Only in the Tertiary period did mammals evolve into gigantic land herbivores and carnivores and enter the ocean and air on the same scale as, and with the diversity of, the Mesozoic reptiles.

Although the birds probably appeared later in the history of life than did the mammals, their adaptive radiation may have begun somewhat earlier, as witnessed by the skeletal modernity of Cretaceous birds (Chapter 9). Some species of birds may have competed with and eventually replaced some pterosaurs in similar niches, for example, as flying and gliding water birds similar to loons and petrels. But not all the evolutionary challenges to reptiles had such a result. One major group of tailed flying reptiles, the long-tailed rhamphorhynchids, did not persist into the Cretaceous period and were replaced by other flying vertebrates. But they were contemporaries, not predecessors, of *Archaeopteryx,* the earliest known bird. It is an instance of the complexity of the problem that, on the evidence, the rhamphorhynchids were probably replaced by the tailless pterodactyls rather than by birds, although birds were evolving at that time.

The early adaptive radiation of birds is less well known than that of mammals and must have involved an adaptive zone in which fossilization was rare. It was apparently a phenomenon of the late Cretaceous period, as well as the Tertiary period, and overlapped rather than followed the great extinction of reptilian orders at the close of the Cretaceous. But there is no evidence that birds, as a group, should have been more successful than pterosaurs in any aerial niche that the pterosaurs occupied.

To identify the selection pressures for enlarged brains in Mesozoic mammals and birds we must identify the special characteristics of mammalian and avian adaptive zones in the Mesozoic era. As mentioned earlier, these must both have been relatively empty when first invaded; yet they potentially contained sets of stable niches that were not open to unmodified reptiles. The Mesozoic mammalian zone must have been particularly unusual because it was stable for at least 100 m.y., as evidenced by the general stability and undifferentiated status of most of the Mesozoic mammalian orders from the late Triassic period to the end of the Mesozoic era (Simpson, 1953, 1961; Olson, 1970).

Homoiothermy as an Adaptive Response

Speculations about mammalian and avian niches generally begin with the recognition of the importance of homoiothermy—the reflex maintenance of relatively constant body temperature by intrinsic mechanisms. Temperature regulation and maintenance in vertebrates is a broad topic with many implications for the analysis of adaptation (Heath, 1968; Nyberg, 1971; Prosser and Brown, 1961, Chapter 9). For the analysis of the special Mesozoic niches of mammals and birds as opposed to the reptiles that spawned them, two points should be emphasized. First, living reptiles are not cold blooded. When they are active, their body temperature is not normally significantly lower than that of the environment. In fact, one might properly describe them as warm blooded by day. Some lizards achieve reasonably stable body temperatures of 38° C by sunning themselves, even in an ambient temperature as low as 13° C (Prosser and Brown, 1961, p. 256), and they are normally active only at the higher body temperatures. This was presumably as true for Triassic reptiles as for living reptiles, and some Mesozoic reptilian groups may have been homoiothermic, in particular, the mammallike reptiles (Van Valen, 1960) and the pterosaurs (Bramwell and Whitfield, 1970). One of the selection pressures toward increased body size in dinosaurs may have been the slower rate of cooling of large bodies (Rensch, 1959; see also Chapter 2), and the large dinosaurs may also have been essentially warm-blooded animals.

The second point is that steady and raised body temperatures permit efficient cellular metabolism. For example, the chemical reactions of muscles and nerves in land vertebrates progress at proper high rates only at temperatures in the range of 30°–40° C (Fig. 12.1a). All animals become sluggish if their internal tissues are permitted to cool, and animals respond to the challenge of a cooler environment in a variety of ways. The basic advantage of homoiothermy is in providing more intrinsic control to avoid the cooling of tissues. The mechanisms may be shivering; developing layers of fat, hair, or feathers; migrating; moving into the sun or shade; control of peripheral circulation; and so on. These permit the muscles and nerves to function optimally. Many land vertebrates become torpid or hibernate when the ambient temperature is steadily low, as in the winter. But circadian and seasonal variations do not normally affect mammals and birds as much as they do reptiles (Fig. 12.1b).

The achievement of autonomic and other reflex physiological controls of body temperature, beyond the behavioral control typical of reptiles, such as moving into the sun or shade, was probably an important feature

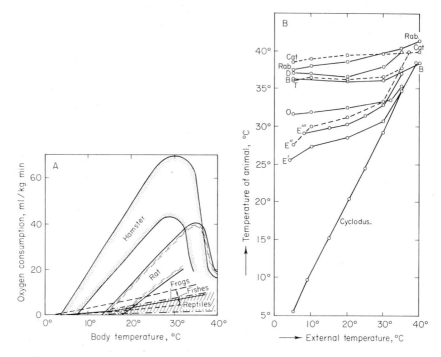

Fig. 12.1. The adaptive value and effect of homoiothermy. A, Body temperature in several vertebrate species as a function of ambient temperature (Martin, 1903); B, metabolic efficiency of muscle as a function of body temperature (Adolph, 1951). Abbreviations are rab., rabbit; D, *Dasyurus;* B, *Bettongia;* T, *Trichosurus* (dashes); E′, Echidna, No. 1; E″, Echidna, No. 2 (dashes); E‴, Echidna, No. 3; O, *Ornithorhynchus*.

in the evolution of the first birds and mammals. Among the developments that either accompanied or followed that achievement must have been the evolution of more efficient muscular metabolic systems and other changes in the body's chemistry. Such changes apparently were not enough to give birds and mammals a sufficient selective advantage over reptiles to replace them in the characteristic Mesozoic reptilian adaptive zones, however, because the reptiles continued to survive and evolve in those zones as contemporaries of their new relatives. Cretaceous birds and mammals are only retrospectively differentiated as classes; if their later history were unknown, they would undoubtedly have been classified as specialized orders of reptiles.

Mammals and birds may have failed to replace their reptilian relatives because of the climate and geography of the Mesozoic era. This was characteristically a dry, warm period (Schwarzbach, 1963). The oceans may have been 10° C warmer than they are now, there were no polar ice caps, and seasonal variations were certainly less severe than they are today. Most of the globe was probably semitropical or tropical, and the present temperate zones were probably warm and arid. This geological peace, with neither mountain building nor glaciation, lasted for the 150 m.y. of the Mesozoic era, with few breaks, ending about 65 m.y. ago. There was, therefore, little selective advantage for any "reptiles" of those times that evolved reflex temperature control mechanisms, especially if they were to sleep or be torpid at night and were to avoid persistently shady areas.

There is some evidence that the light–dark cycle characteristic of circadian activity in the major life zones of the present time may have been radically different when the mammals first appeared. The axis of rotation of the earth relative to the major land masses may have been closer to what are now temperate zones, and the magnetic poles of the earth have certainly moved radically during the last 200 m.y. (Tarling, 1971). Furthermore, polar regions of that period were quite probably considerably milder as life zones than they are today, and species of reptiles with some form of temperature control, whether extrinsic or intrinsic, could probably survive and evolve. In polar regions one particularly important feature would be the occurrence of the summer-long day and winter-long night. If the winter-long night was also relatively mild climatically, animals might adapt to it and survive in such an adaptive zone.[1]

There was, therefore, almost certainly a major adaptive zone open to "reptiles" that could live and be active in the shade or dark, whether darkness occurred on a circadian or annual cycle. That zone could have been filled during the Mesozoic era by homoiothermic and nocturnal "reptiles," and I suggest that the species of "reptiles" that filled it successfully were or became mammals.

A different story has to be reconstructed for the birds, and we can defer that to a later section. Let us turn now to the analysis of the evolution of the mammalian brain.

MAMMALIAN TRENDS AND NOCTURNAL ADAPTIVE ZONES

It is easiest to understand the early evolution of mammals if it occurred under selection pressures associated with their radiation within an adaptive

---

[1] The gist of this argument was developed in discussion with Malcolm McKenna, who pointed out the likelihood that a nonfrigid, arctic, winter-long night might have been characteristic of the ecosystems within which some Mesozoic land animals, including mammals, were evolving.

zone of nocturnal land vertebrates. Such a zone was available and relatively open in early Mesozoic times. The fact that "primitive" living mammals are usually nocturnal has frequently been recognized (e.g., Polyak, 1957), and the likelihood that the earliest mammals were nocturnal in their adaptations has also been proposed before. It has been insufficiently appreciated, however, that the adaptations of reptilelike animals to a life at night would make extraordinary demands not only on their temperature-regulating capacities but also on their sensory systems, and that this would have important implications for the direction of the evolution of their brains.

The earliest mammals could really not have been much different from reptiles, and the reptiles as a class were already complex organisms at the beginning of the Mesozoic era and were well adapted to lives in a variety of diurnal niches. It is likely that the earliest mammals retained many reptilian adaptations, even if they were evolving for life within the nocturnal adaptive zone, and a central assumption in this analysis is that special sensory adaptations were necessary if the early mammals, as nocturnal "reptiles," were to respond appropriately to events at a distance. We must imagine their new sensory adaptations as doing the same kind of information processing as done by diurnal reptiles, except that new sense modalities were brought into play. The first issue that should be faced is how reptiles normally handle distance information.

Reptilian Adaptations for Distance Information

Like most living reptiles, the late Paleozoic species were certainly visual animals, using vision for information about events at a distance. Too little is known about how such information is processed by reptiles, but it is likely that, as in amphibians (Lettvin et al., 1959), much of the work of defining distant stimuli, as associated with predators, prey, or other significant events, is handled by the elaborate neural networks of the retina. The reptilian visual systems of the brain itself, beyond the level of the retina, may have relatively little additional visual work to do. The major role of the visual systems of the brain in reptiles is probably to correlate the visual information with the activity of the neural control center for the appropriate motor response to a particular stimulus. It enables living frogs and lizards to strike at small (fly-sized) moving objects, but it does not enable these animals to distinguish between a fly and a black spot moved by an experimentalist (Hailman, 1970). In nature, however, it enables living reptiles and amphibians to capture their normal prey and, by analogous reflex mechanisms, to escape from predators by responding in appropriate ways to large moving objects. In general, it enables these animals to perform complex, though stereotyped, sensorimotor coordinations in response to distant events by using visual information.

If we differentiate the processing of visual stimulus information from

Mammalian Trends and Nocturnal Adaptive Zones

the coordination of stimulus and response, we should recognize that only the latter has to depend on brain structures rather than on the retina alone. The complete neural representation for the oriented or forced movements, which may be categorized as taxes (Fraenkel and Gunn, 1961) or as instincts (Tinbergen, 1951), involves the peripheral retinal system, the central, mainly midbrain, system, and central and spinal motor systems. Our concern here is primarily with the first of these in the consideration of the visual organization of behavior in living and fossil diurnal reptiles, and with the way information about events at a distance can be encoded by a sensory system.

With the evolution of the optics of the eye and the spatially organized retina it was possible to encode distance information directly as events at different points or regions of the retina. It is this code that had to be reproduced or approximated by any other sense modality that was to evolve to provide distance information for "reptiles" to which visual information was denied, if they were to live out their lives as otherwise normal "reptiles" in darkness or underground. To the extent that they could reproduce that code with another sense modality, they would require less extensive modification of the central midbrain systems and the central and peripheral motor systems that enabled them to use distance information as part of their adaptive behavior. In other words, they could continue to be "reptilian" in behavior, with the difference that they used senses other than typical reptilian vision as their modalities for processing information from events at a distance.

I suggest that audition and olfaction played such visionlike roles for the earliest mammals. Both of these sense modalities were available, and the example of living amphibians and reptiles suggests that they were already involved in relatively nonspecific, and in nonlocalizing, ways in the control of behavior. Hearing has served to trigger alerting or startle reactions to unusual sounds in lizards (Wever, 1965) and has permitted the analysis of auditory "sign stimuli" in mating behavior in living frogs (Aronson and Noble, 1945). Olfaction has been a relatively difficult modality to study, but it is clear from naturalistic observations and from the anatomical analysis of olfactory systems that this sense was not first used for locating objects, but rather for testing the olfactory characteristics of water and food (Carter, 1967). In the anatomical changes necessary to have information from these sensory systems serve as analogues to vision by giving accurate information about events at a distance, we have the basic clues to the mystery of the enlargement of the brain in the earliest Mesozoic mammals.

HEARING MODELED AFTER VISION AS A DISTANCE SENSE

To act in the place of vision as a distance sense hearing had to carry accurate information about where a sound came from, how far away its source was, whether the source was moving or stationary, if moving, the direction of motion, and even information about the size and shape of the source. How would the auditory system have to be organized to provide such information, which would be equivalent to that from a "normal" reptilian visual distance-sensing system? And how would auditory and visual systems subserving such a function differ from one another? To answer, let us first compare the two systems quantitatively (Table 17.1) to determine how they are organized to do their work. This can be done best in mammals and somewhat less adequately in reptiles.

In living mammals the flow of information through the visual system is differentiated peripherally (retinally) and centrally (at midbrain and forebrain levels) as diagrammed in Fig. 12.2A. The peripheral analysis by the neural retina is really quite elaborate. As a model nervous system, the neural retina is large enough and complex enough in its structure and function to be considered as a true peripheral brain (Sherrington, 1950; Granit, 1968). For example, the rat may have as many as 8 million neurons (bipolar, amacrine, and ganglion cells) in the retinas of its two eyes, which are involved in the processing of visual information (Lashley, 1950; there is some question about the count—rods and cones may have been included). It has only a fraction of that number in the strictly visual system of its brain: 70,000 neurons in the lateral geniculate bodies and 1.3 million neurons in the cortical visual centers. One may estimate from other sources that the total number of neurons in the rat's central nervous system is probably of the order of 25 million (Donaldson, 1924); the retinas of the eyes may thus approach the brain in rats with respect to information-processing capacity, as measured by the number of available neurons.

Information-processing capacity is reflected in the size of organs such as the eye and brain because the units (neurons) that do the processing take up space (Chapter 3; also, Young, 1964, Chapter 4). Visual species can process a good deal of information without enlargement of the brain because a significant part of the information can be processed at the peripheral sensory organ, the retina of the eye. But if a species evolves new adaptations in which other sensory systems, such as audition and olfaction, are used to obtain distance information, relatively little information processing can be localized peripherally.

In the auditory system of living mammals only a fraction of the elements that process information are in the periphery; most are in the brain (Fig. 12.2B). Peripherally, at the spiral ganglion in the mammalian (e.g., human) cochlea, there is a total of no more than 70,000 neurons in both ears. There is a tenfold increase in the number of neurons at the thalamic level and, at least in primates, a several hundredfold increase in central neurons at the cortical level (Chow, 1951; see also Table 17.1). In short,

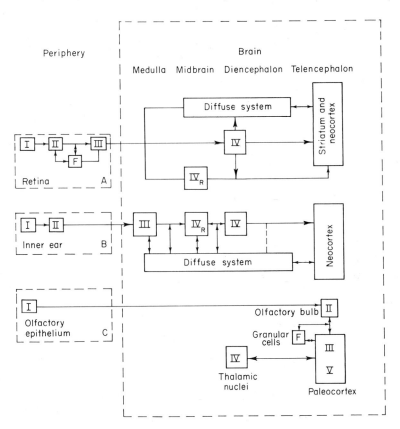

Fig. 12.2. Simplified schematic diagrams of visual (A), auditory (B), and olfactory (C) sensory systems of living mammals, illustrating fraction of system (and its information-processing capacity) within braincase and peripheral fraction external to brain near or at sense organ. Roman numerals give order of neurons—the minimal path. Direction of information flow indicated by arrowhead; recurrent, or feedback, loops identified by doubleheaded arrows and by F-boxes in a loop. Fourth order neurons in the diencephalon are in lateral geniculates for vision and medial geniculates for hearing. Stages labeled IV_R are in superior colliculi for vision and inferior colliculi for hearing and are part of reflex control systems for motor activities, such as eye movement and ear movement, as well as of reflex coordination of information from eyes and ears. Note that only the visual system has significant feedback loops within the (peripheral) retina, which are represented by the internuncial system and amacrine cells (cf. Table 17.1).

the auditory system is elaborated centrally at various levels of the brain. It can process only a fraction of available auditory information outside the brain. If animals are to use their auditory systems to process information about events at a distance, they have to develop enlarged brains.

The fact that reptiles have not developed auditory systems that contribute significantly to elaborate information processing is suggested by the small number of their primary receptor (hair) cells, as few as 50 and rarely more than a few hundred, in living lizards (Wever, 1965). The central projections of the hair cells to the medulla (acoustic tubercle and trapezoid body) and the midbrain (inferior colliculus) are not well known in reptiles, quantitatively, but they are probably not extensive and are subordinate to the vestibular system and the optic lobes (Papez, 1929).

If we exclude the living nocturnal reptiles, which we may regard as latecomers on the evolutionary scene, the degree to which living reptiles are visual rather than auditory animals is apparent from a comparison of the total number of receptor cells in the two systems in reptiles. From Polyak's (1957, p. 841) data one may estimate that the retinal surface of each eye of the American chameleon, *Anolis,* is about 40 mm^2. The cones of this lizard have a smaller cross-sectional area than in the human retina: central foveal cones are less than 1 μ in diameter, and peripheral cones are about 3 μ in diameter. There should be roughly 1 million cones in such a retina, or about the same number as in the human fovea, alone. The chameleon has an elaborate system of bipolar cells and ganglion cells, which could process information peripherally. The ganglion cells should equal the number of optic nerve fibers, which is of the order of 100,000 for each eye. This number may be contrasted with the few hundred fibers that can be assumed for its auditory nerve, and both figures may be related to the very small body size of this animal. The chameleon's eye is an order of magnitude smaller than the human eye. The central region of its fovea, however, is comparable to the human central fovea in the number of cones (about 100,000 in both species), and its entire retina corresponds, functionally, to the human fovea. In summing up the visual capacities of various vertebrates as he deduced them from the structures of their retinas, Polyak (1957) wrote, "the sharp-sighted chameleon probably has the most differentiated fovea of all the vertebrates, a veritable 'living microscope' " (p. 285). It was probably such a system that had to be approximated by auditory and olfactory information in the last of the mammallike reptiles and the earliest mammals, about 200 m.y. ago.

To be so successful an analogue for a functioning visual system, an auditory system must use some equivalent of the spatial code inherent in the structure of the retina. This could be accomplished only by encoding the information temporally: successive stimuli to the ear would have to be recorded, labeled, and compared with one another to construct the equivalent of visual space by analyzing the status of successively stored events. There is evidence that such a direct analysis is, in fact, performed by the central auditory system of echolocating mammals (Grinnell, 1970). An equally elaborate central auditory analytic system has been identified for encoding and comparing information from the two ears, which would be important for localizing sound in space (Goldberg and Greenwood, 1966; Masterton et al., 1968).

The use of hearing by early mammals as an analogue of vision, according to the sketch just presented, must have had two important effects. First, at a structural level it had to result in an enlarged brain because the many additional neural units to process the more elaborately organized information had to be somewhere, and there was no room for them at the periphery, near the sensory cells. They had to be in the brain. Second, at a functional and behavioral level the use of a temporal code in a way even approximating the spatial code evolved for visual information can be thought of as introducing "time" as a major element in the life of animals.

Olfaction as an Analogue of Vision

It is more difficult for us, as members of a species peculiarly unresponsive to olfactory cues, to imagine a finely discriminative olfactory world than a finely discriminative auditory world. It is, nevertheless, likely that olfaction, even more than hearing, was a centrally significant sensory dimension that evolved and became refined in the late mammallike reptiles and early mammals and that it was a particularly important sense modality for their adaptations to the nocturnal adaptive zone. At the structural level the olfactory system is directly represented in the forebrain, and the expansion of the forebrain that so obviously characterized the earliest mammals is most easily appreciated as an expansion of the significance of olfactory information.

The olfactory system, like the auditory system, does only a fraction of its information processing peripherally. The peripheral activity (Fig. 12.2C) is the spatial summation of messages from millions of receptor cells distributed in the olfactory epithelium. Although the receptor cells are neurons, their transmission is a one-way affair and involves no genuine integrative activity. Their role is analogous to that of the rods in night vision (see pp. 271–273). The picture of olfactory information processing that emerges from the anatomy of the system, supported by electrophysiological and behavioral studies (Adey, 1959; Heimer, 1968; Moulton and Beidler, 1967), is of responses by many receptor units to very small concentrations of some chemicals (wafted in by air but effective only in solution) in the olfactory epithelium. The responses are summed as many fibrils converge upon a limited number of bipolar (mitral) neurons in the olfactory bulbs. The axons from the mitral cells (60,000 in the rabbit and presumably similar numbers in other mammals) then transmit information without further processing to the primary receptor area, which is entirely in the forebrain: parts of the paleocortex of mammals and most of the forebrain in living reptiles. If smell is to be the basis of spatially localized information about distant events, the brain would have to be enlarged to make such use of the normally less specific information processed by reptilian olfactory systems.

The extent to which olfactory cues are used by vertebrates for the fine analysis of events at a distance is more difficult to visualize than such use of auditory cues. Experiments on object location and obstacle avoidance by sightless people are well enough known to make echolocation (the cue used by such men) an easily imagined ability. Odor, on the other hand, is notoriously difficult to localize, and the implied perceptual world is more exotic than the world of echolocating bats, which can avoid hanging wires or other minute obstacles (Griffin, 1958), or that of whales, which can discriminate among shapes of fish by the same method (Kellogg, 1961). Marler and Hamilton (1966) include a good brief discussion of spatial localization on the basis of olfactory cues, using examples of essentially reflex behavior or orienting behavior guided by olfactory cues.

An example of orienting by turkey vultures using olfactory cues is described by Stager (1964). Forcing air from a fixed bait, Stager was able to produce a dramatic effect on the search pattern of this bird (Fig. 12.3),

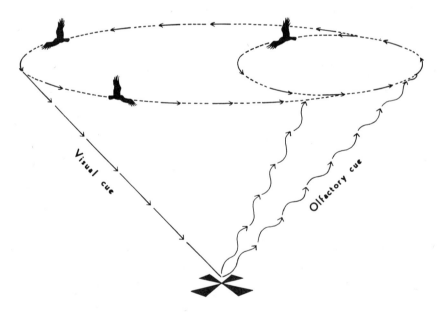

Fig. 12.3. Flight pattern of the turkey vulture (*Cathartes aura*) under exclusively visual control, narrowed as a result of the addition of an olfactory cue. From Stager (1964).

clearly illustrating the role of odor in ordering its activities. He also illustrated the correlation between such behavior capacities and the enlargement of the olfactory bulbs by comparing endocasts of various Pleistocene and Recent vultures. The living turkey vulture *Cathartes aura* had the largest olfactory bulbs in the group despite the fact that it was the smallest species and had the smallest brain. The orientation by the turkey vulture is not, however, precise in the sense that visually guided behavior can be; it does not have the kind of precision that is required of a visionlike distance sense.

The way olfaction is used by sniffing species, including relatively primitive hedgehogs and "progressive" carnivores, suggests that it may act as a fine distance sense only if the neural capacity for memory is extremely well developed. Among the more progressive species territories may be marked out by odor; by sniffing at a particular spot or succession of spots an animal could identify its own or a neighboring territory (Ewer, 1968). In order to construct a map a sniffing species would have to encode successively received information, with time intervals of minutes or more between discriminated cues.

It is clear that once the map had been developed, an olfactory cue received from a distance could be identified precisely by the remembered position of its source, which had been established by closer olfactory inspection on the previous occasion when the area involved was being mapped. The elaboration of olfaction as an accurate distance sense would, therefore, be analogous to audition in its reliance on temporal integration of input information, with even longer time periods involved in the integrative activity.

Peripheral Sensory Adaptations

Hearing and smell could become precise distance senses only by the evolution of central sensory-processing systems in the brain itself. That evolution helps one to understand the significance of the modifications of the peripheral sense organs of these systems at the transition from reptiles to mammals. These peripheral modifications are attested to by a fossil record and deserve a brief review.

The potential importance of hearing as opposed to vision for nocturnal land vertebrates suggests a set of selection pressures favoring species that could receive and process a wide range of auditory signals. Such species should be able to respond whether a signal were feeble or strong, whether it were a rustle of leaves by a potential predator or the cue to the presence of an unsuspecting prey. Invasion of a nocturnal niche would be more successfully achieved if the invader were equipped with a sound-sensing system in which the weakest possible signal was effective. Such a system would have to be very sensitive, but the penalty of sensitivity could easily be oversensitivity to stronger sounds. The listener would, therefore, also need protection from very intense sound pressures such as thunder.

The basic sound transmission system in the earliest land vertebrates persists in living reptiles (Fig. 12.4A) and in modified form in birds. It involves a large stapes acting via the oval window to produce standing

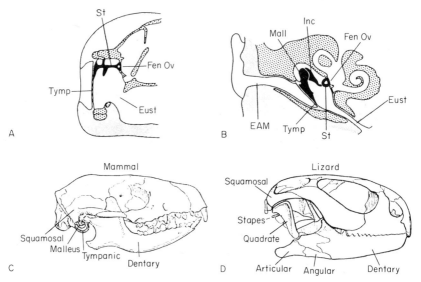

Fig. 12.4. Schematic cross section through the ear region of a lizard (A) and a mammal (B). Abbreviations are EAM, external auditory meatus in mammal; Eust, eustachian tube; Fen Ov, oval window; Inc, incus; Mall, malleus; St, stapes; Tymp, tympanic membrane. From Hopson (1967). Skull of a mammal (C) and a lizard (D), showing how middle ear bones of mammals evolved from jaw joint of reptiles, with expansion of the dentary as a single mandibular bone. From Watson (1951).

waves in the lymphatic fluid of the inner ear. Other bones in that region enter, via "bone conduction" of sound, into the sound transmission system. The mammallike reptiles were already reducing some of their skull and jaw bones under selection pressure to improve the activity of the masticatory apparatus (Hotton, 1959, 1960). If the selection pressures toward improving the dynamic range of the sound transmission system occurred before the "extra" maxillary and mandibular bones completely disappeared, these extra bones could have joined with the stapes to form a chain of ossicles to improve the transmission of sound from the air to the sense organ.

As it exists in mammals, the system of ossicles in the middle ear (Fig.

12.4B) effectively amplifies the sound pressures of almost inconceivably small displacements of the ear drum (of the order of the diameter of a hydrogen molecule) by reducing the area of the moving "piston" from that of the ear drum to that of the oval window. The resulting pressure amplification is from ten- to thirtyfold, depending on the frequency (von Békésy and Rosenblith, 1951). Of course, such amplification could be effected by a single appropriately tapered stapes (e.g., the columella of birds) without the intervention of the other two ossicles, the malleus and incus. The ossicular chain is superior because it has a dual function: amplification of weak sounds and, also, protection from intense sounds in a system that is supersensitive to weak sounds. As shown in various experiments by von Békésy and others the ossicular chain (and its associated neuromuscular reflex system) can act as a damping system, to reduce the displacement of the ear drum in the face of high energy vibrations of the column of air in the auditory meatus. The evolution of this system of bones in the mammals can be understood best in terms of unique selection pressures upon the mammals to use auditory information for a distance sense rivaling the visual system. Masterton et al. (1969) reviewed comparative studies of hearing in mammals, and Manley (1971) has published a similar review for all vertebrates.

The peripheral portion of the olfactory system, represented skeletally by the ethmoturbinal bones of the nasal cavity, also showed important evolutionary modifications correlated with the proposed nocturnal adaptive zone of the earliest mammals. The posterior turbinals provide the surface for attachment of olfactory epithelium, whereas the anterior portions are associated with the cleansing and warming of the air breathed in through the nostrils by land animals. These bones are known in mammallike reptiles, and their presence has been cited as evidence for homoiothermy in these reptiles (Van Valen, 1960). They are also evidence for the elaboration of the peripheral olfactory system with its unusual characteristics as noted in the previous section. Their presence, therefore, supports the notion that the invasion of nocturnal niches was begun by mammallike reptiles and that as part of their adaptation to these niches they evolved into at least partial homoiotherms and evolved more sensitive olfactory apparatus. Surprising additional evidence in favor of the invasion of nocturnal niches by the mammallike reptiles has been suggested by the unusual position of one fossil, *Thrinaxodon*, illustrated earlier (Fig. 7.6C), curled up like the living hedgehog in a position of a sleeping burrowing animal (Brink, 1958).

VISION IN NOCTURNAL MAMMALS

Our difficulty in imagining selection pressures on the auditory and olfactory systems is certainly due to our membership in an unusual order of mammals, the primates, in which the visual sense has probably been secondarily reemphasized (Chapter 16). Vision dominates our three-dimensional perceptual world and is the major source of sensory information for building that world. But fine vision is unusual in living mammals; most species have nocturnally adapted eyes that probably serve no better and often less well than their ears for localizing objects in space.

To achieve some insight into the question one should imagine oneself as adapted to move about and live in the dark. With a distribution of rods and cones in our eyes like that of cats or dogs (Walls, 1942), our central vision at night would be as good as our peripheral vision. We would not be forced to see things on dark nights by directing our gaze away from them and detecting their presence (more often their movement) out of the corner of our eyes. This kind of peripheral vision is readily trained, but those who have had to rely on it can testify to its inadequacy for accurate perceptual work. There is, in short, little precise visual spatial information normally available in a nocturnal environment. If we lived in underbrush in woodland areas, we would not even have the moon and stars to aid us, and we would be truly in the dark. For that purpose, the auditory sense, enhanced by functioning pinnae to gather sound, could be the modality of choice to localize objects and their movements.

The visual sense as a useful modality for nocturnal niches is differently organized from its role in diurnal niches. At night the problem is to use minimal differences among photic stimuli from various objects, when the energy reflected from those objects approaches zero. The solution of that problem by primates and by vertebrates active both during the day and at night has been to develop two separate sensory systems both served by the same optical system, with all the sensory elements located in the same anatomical locus—the retina. The cone system, common to most lower vertebrates, is a system that responds rapidly and with good resolution to relatively high photic energies, at least 1000 times as strong as the minimum energies for the rod system. A cone system is the characteristic visual system of nonmammalian vertebrate species. Rods, though present in most vertebrates, are almost peculiarly the source of information for vision in mammals; they are almost a mammalian specialization (Walls, 1942; Polyak, 1957). The development of an important rod system is, thus, probably a secondary specialization for the construction of a vaguely localizing visual perceptual space, and it is to be understood as an adaptation of the basic vertebrate visual system by the earliest Mesozoic mammals to provide information about events taking place in their nocturnal niches.

Although human visual abilities by day involve a cone system, as indicated earlier, it has been argued (e.g., Walls, 1942) that mammalian cone vision evolved from rod vision and that our cones are only analogous and not homologous to those of lower vertebrates. This is entirely consistent with my position because from my argument one would expect the mammalian retina to have responded to selection pressures to evolve a night vision system, and neither reptilian nor mammalian cones are sensitive enough for such vision. During the very long (100–150 m.y.) history of Mesozoic mammals, one would expect their retinas to become essentially pure rod retinas as they are in many living mammals. It was only when mammals could reinvade diurnal niches left empty by the great extinction of reptiles at the end of the Mesozoic era that they were, once again, under selection pressures to develop daylight vision. From their persistently small brains, I suspect that the archaic Tertiary species discussed in the previous chapter had not yet developed such vision and that the newly reevolved cone system of mammals required some further expansion of the brain. The relatively enlarged brains of the earliest primates (Chapter 16), and of some of the living squirrels that provided data for Fig. 10.6, may reflect the demands of a mammalian diurnal life, which I believe was an evolutionary novelty with novel central neural as well as peripheral adaptations.

Conclusions on Mammalian Nocturnal Adaptations

This chapter was intended as an integrative summary of points made previously, and new points that should be made, in presenting the detailed argument about the various selection pressures toward enlarged brains that were probably effective in the early evolution of mammals and birds. In the presentation of the case of the nocturnal adaptive zone many different approaches had to be integrated, and I wish now to summarize the basic argument and pick up some threads that may have been lost in the detailed presentation.

One underlying assumption was that the evolutionary process was conservative; the adaptations to the nocturnal adaptative zone by the earliest mammals would be made by minimal changes from their reptilian progenitors. With respect to the evolution of sensory systems this led to the proposition that distance sensing, which was part of the visual behavior repertoire of the reptiles, would remain part of the repertoire of the first mammals. But to achieve that end in darkness other sensory systems had to be used to provide the distance information.

In land vertebrates the only nonvisual sensory systems that are responsive to distant events are hearing and olfaction, and I, therefore, assumed that these systems became adapted in the earliest mammals to receive and analyze information from events at a distance. The assumption was supported by anatomical and paleontological evidence about the evolution of peripheral parts of these systems: the auditory ossicles and the ethmoturbinal bones of the olfactory system.

An important structural difference among these distance-sensing systems lies in the neural structures that are used for the analysis of sensory information, and this provided the basic clue to the enlargement of the brain in the earliest mammals. The analysis of visual information in reptiles can take place at the retinal level because the retina is a kind of peripheral brain equipped with numerous neural circuits that appear to be adapted for such analysis. Hearing and smell, on the other hand, are essentially central nervous system senses with respect to where information in these modalities would have to be analyzed to provide accurate information about the location of events at a distance. If these senses were to be used by the earliest mammals as distance senses, various structures in the brain itself, the inferior colliculi for hearing and the forebrain for both olfaction and hearing, had to become enlarged to permit the additional neurons to be packaged somewhere in the body.

The argument led to a second important conclusion that was functional rather than structural. Regardless of where the auditory and olfactory information was to be processed, the very fact that these senses were to be used to provide accurate spatial information about events at a distance implied important and novel developments in behavioral capacities. The auditory and olfactory systems had to produce information equivalent to that from the normal reptilian visual distance-sensing system. But the basic organization of hearing and smell is not spatial; the primary information is not encoded spatially (as it is by the retina for visual information). A spatial representation could be constructed only by taking successive events that are heard or smelled and integrating them into something comparable to a spatially organized pattern. That integration is over time, and the use of the new sense modalities could only be effective if time were comparable to space as a dimension of analysis.

In other words, there was a selection pressure toward taking a series of stimuli and organizing them into a unitary pattern that was analogous to the pattern of a spatial array of simultaneously seen visual stimuli. Although our own experience may be a treacherous source of analogies, since we process information via elaborate imagery and rich perceptual worlds (Chapter 1), one can probably consider the patterning of melodies and bird songs into recognizable and distinctive structures as comparable or analogous to the patterning of a visual design to be distinctive from other visual designs.

The selection pressures referred to the outcome of neural integrative analysis, not to the process of analysis. It is, of course, true that neural analysis as a process involves temporal as well as spatial organization of patterns of excitation and inhibition. But the outcome of the analysis in man (and other "higher" animals) is to create a projected real world, and it is the nature of that kind of projection that concerns me here. It is possible to imagine a simple perceptual world of primitive animals in which the representation of external stimuli is essentially by an abstract code referring to their location, without imagery, consciousness, or awareness of a perceptual world. I am pointing out now that even in such simple perceptual worlds one may have a code referring to when as well as where a stimulus occurred relative to other stimuli. It was at least such an advance that was implied in the use of hearing and smell as accurate distance senses. These can add up to a time sense.

AVIAN TRENDS AND ADAPTIVE ZONES

Evolutionary, structural, and functional arguments all lead to the conclusion that the enlargement of the brain in birds could not have resulted from the same selection pressures as in mammals. With respect to evolution we know that birds arose from an entirely different reptilian stock than the mammals and probably many millions of years after the first mammals appeared. The structural and functional arguments are that most living birds are diurnal animals, with eyes beautifully constructed as refractive devices to project visual images on elaborately organized retinas. Their brains are enlarged, relative to those of reptiles, and recent anatomical evidence (Karten, 1969) would have them performing significant auditory as well as visual analysis at the level of the brain, although they also have elaborately organized central neural visual systems that contribute to the size of their brains.

The endocast of *Archaeopteryx,* described in Chapter 9, was clearly avian and had the typically avian enlargement of the optic lobes (superior colliculi). The size of this first bird's eye was also impressive, as it is in modern birds. Like most living birds, it must have been a diurnal, visual animal. What adaptive zone had it invaded to provide selection pressures for its evolution of an enlarged brain?

The answer for *Archaeopteryx* and later birds is necessarily more speculative than for mammals. There is much less useful information about the reptilian progenitors of birds, certainly nothing like the almost continuous gradation that has been identified between certain lines of mammallike reptiles and the earliest mammals. The fossil evidence is also much less adequate on the adaptive radiation of Mesozoic birds than it is on the mammals; the 50 m.y. step from *Archaeopteryx* to the Cretaceous birds was probably as great, morphologically, as the 150 m.y. step from *Sinoconodon* to the Oligocene mammals. We are not, of course, restricted to endocasts when we try to speculate knowledgeably about behavioral adaptations. There are many other fossil remains, such as teeth, skull, long bones, sternum, and so on that enable us to reconstruct probable behavior patterns. *Archaeopteryx* seems in every way a missing link between reptiles and birds (de Beer, 1954); yet the Cretaceous birds seem almost modern with respect to the information they offer about adaptive zones. This means that the selection pressures leading to modern birds were present during, and must be identified as part of the world of, the Age of Reptiles.

In our previous discussion of selection pressures on birds (Chapter 9), the possibility of adaptations to flight as a reason for the enlargement of the brain was discounted on the grounds that such adaptations should be possible without major enlargement of central neural coordinating and motor systems. We can be reasonably certain, furthermore, that the adaptive zone of water birds, such as sea gulls and albatrosses, was on the whole preempted by flying reptiles during the Cretaceous period, until their extinction at the end of the Mesozoic era, although the Cretaceous birds *Ichthyornis* and *Hesperornis* are considered to have been water birds of that general type. We are, therefore, left with the unresolved problem of identifying possible avian niches that pterosaurs were poorly adapted to and that would be characterized, among other ways, by selection pressures toward the enlargement of the brain.

If the major early evolution and adaptive radiation of birds occurred in woodland niches of the sort hypothesized for *Archaeopteryx* and if the Cretaceous fossil birds were not typical of the birds of their time and have been discovered only because their niches made fossilization and preservation more likely than for birds adapted to woodland niches, then we may consider the problems of visual adaptation as a potential selection pressure for an enlarged brain. These problems may have been severe enough in woodland niches to require more neural information-processing tissue than is available in the retina, giving an adaptive advantage to species evolving central neural representation of vision in addition to the retinal representation.

The available niches of any time should be comparable with niches of other times, and we may consider the early radiation of birds in comparison with the Cenozoic adaptive radiation of mammals into diurnal niches vacated by the extinction of many species of Mesozoic reptiles. The early part of that mammalian radiation was, as we have seen, accomplished with little expansion of the brain. But we will see in Part IV that a considerable expansion of the mammalian brain occurred later in the

Cenozoic era. Furthermore, even at the beginning of the Cenozoic era, to the extent that evidence is available, species from relatively progressive orders of mammals, such as the primates, were evolving with relatively larger brains than their relatives from the archaic orders. Let us consider the possibility that the earliest birds had invaded niches comparable to those successfully invaded by tree-dwelling diurnal mammals with well-developed visual senses. We may also consider the possibility of some evolution of the auditory system in early birds to permit a more varied use of auditory information than occurred in reptiles.

THE CHALLENGE TO VISION IN WOODLAND NICHES

The woodlands are the homes of many birds, tree squirrels, and small primates, the diurnal animals par excellence of the present. The selective advantage to birds like *Archaeopteryx* was that they were probably reasonably good flyers, limited more by endurance than by aerodynamics (Chapter 9). They could probably fly well and gracefully from branch to branch, forage on fruits and berries, and, contrary to the reconstruction of their life habits by Heilmann (pp. 198–199), they could readily catch insect prey on the wing. They could also pounce on small ground and tree-dwelling reptiles by flying (instead of leaping as the living primates might) to a branch or to the ground and could escape from their predators in similar fashion. In a world without diurnal mammals one may think of the earliest birds as invading a primatelike niche. This would have placed them under unusually strong selection pressures to develop central neural representation of the visual system.

The "eye-as-brain" sufficient for many living amphibians and reptiles as well as for their early Mesozoic ancestors may not have been sufficient for a "reptile" in a "primate" adaptive zone. The special adaptation to that zone, according to this suggestion, was effected in the speciation of the birds as specialized "reptiles." When a similar zone was later entered by the mammalian descendants of another branch of the reptiles in the early Tertiary period, these mammals redeveloped the cone system of vision that had been lost in the evolution of night vision by nocturnal mammals. The primates in particular provide evidence on the adaptations that are appropriate for life in that adaptive zone.

By analogy with the primates it is clear that such a zone puts an unusually strong selection pressure on the development of the central neural representation of the visual system. The difficulties are of the sort that make it necessary to use more elaborate information about the external world than can normally be provided by sensory systems acting without elaborate systems for analysis. Visual information from the environment is in the form of the mottled background of leaves, bark, and changing lights and shadows as the organism moves rapidly through that environment. It is clearly a situation in which perceptual constancy, as discussed in the opening chapter, would be adaptive. It would be more useful to code elements in the environment as "figure" (or "object") and "ground" and to have the figure maintain its "constancy" in the face of changing representations at the retinal level.

I have suggested that the earliest birds had invaded a set of niches most nearly like those of living tree-dwelling primates, and we can appreciate the selection pressures involved in the successful occupation of such niches when we consider the behavior repertoire of living tree-dwelling primates. There was a major difference between the primate adaptation and that of birds, however, which is discussed in later chapters. I argue that the primates, as mammals, began their adaptive radiation with an auditory system encephalized to provide temporal information, and their encephalized visual system was modeled after that auditory system. This could not be the case in birds, and we may assume that there is nothing in birds homologous to the visual world of primates, although objects existing in space and time of the sort that one accepts unquestioningly as the "real" world could be part of the avian world by convergent evolution. Such parallel evolution would have been necessary to maintain a steady state of existence in a world of mottled textures, changing whenever there is motion, and the birds would have required additional encephalic representation of the visual system in order to maintain some constancy of visual information. As a result the brain in birds had to become enlarged as a visual information-processing center supplementing the processing taking place at a retinal level. That, at least, is the conclusion that would follow from this speculative discussion.

OTHER SELECTION PRESSURES ON EARLY BIRDS

In the lives of living birds a number of behavioral dimensions are central, and many kinds of behavior are radically disrupted by lesions of the forebrain or the optic lobes (Stettner and Matyniak, 1968). In the previous section I emphasized the perceptual problems faced by birds if they were to organize "perceptual worlds" from the sensory information received from trees, leaves, the mottled sky that may be visible, and so forth. But more dimensions of bird behavior than the purely perceptual are involved in the neural adaptations to particular life zones.

A major dimension of bird behavior, as it is known in living birds, has to do with the social organization of birds in groups, in mating, in nesting, and in the care of the young. Establishing and maintaining such social organization depends on many coordinated perceptual and motor activities. Visual cues are organized in ways surprising to the human

Avian Trends and Adaptive Zones

observer, as "sign stimuli," in which no object quality (Chapter 1) seems to be assigned to a pattern of stimulation. Responses may be to configurations that seem very abstract to human eyes. In gulls, for example, there is a well-known feeding response of the young described by Tinbergen (1961) to the general shape of a beak with a spot placed appropriately; the response is "released," in nature, when the young bird sees its parent's beak with normal markings. It can be released even more easily by artificial models that exaggerate certain aspects of the stimulus configuration. Such perceptual organizations that are tied to the social behavior of animals are known among many living lower vertebrates, including fish, amphibians, and reptiles, and many examples are reviewed in ethological texts (e.g., Marler and Hamilton, 1966; Hinde, 1969a). The special feature of these phenomena in birds is that they are often organized by combining "prewired" stimulus–response relationships with the capacity to learn new relationships.

I have suggested no models of the selection pressures on early birds that would require advances beyond reptilian fixed action patterns. It seems to me that such advances might more easily be considered as results rather than causes of additional neural tissue in the bird brain. I think it likely that secondary effects of enlarged brains include the capacity for increased plasticity in the development of control of a behavior pattern but that the basic behaviors originally evolved as rigid patterns, much as they remain in lower vertebrates. This point is made again later in this chapter and is developed in a separate chapter (Chapter 17).

Another as yet unconsidered dimension of bird behavior, which is very important in many living species but of unknown significance for the earliest birds, is the role of sound and sound communications in effecting social control. In living birds, songs and chirping are used to establish territories, attract sexual partners, attack invaders of a nesting area, and probably in many other ways. Although songbirds are relatively late arrivals on the evolutionary scene, known as fossils from later Tertiary deposits, they represent an evolutionary achievement of some complexity. In other orders of birds the vocal apparatus is sufficiently developed to produce a variety of social calls: warning cries, danger signals, territorial signals, and so on, and this implies a level of sensorimotor integration, or perhaps more accurately motor-sensory integration, in which the motor (vocal) behavior of one individual is coordinated with sensorimotor behavior of another individual of a species. It is now known that, as in visually released fixed action patterns, some of this behavior is developed by a complex interaction between "prewired" and learned components. The learning is partially of the semiautomatic type termed "imprinting," which is, itself, a kind of "prewired" capacity to learn during certain critical periods of life (Hinde, 1969a,b), and partially of the type seen in the animal behavior laboratory or in the human classroom. In the latter type of learning almost any arbitrary response pattern (e.g., striking keys on a typewriter in an appropriate way) can be associated with an equally arbitrary stimulus pattern (e.g., marks on a manuscript page) by following a few elementary rules for training animals or people.

The evolution of the use of sound in the manner of birds was certainly related to the evolution of an enlarged forebrain, or generally enlarged brain, and forebrain damage is known to affect such behavior (Hinde, 1969a). But as in the case of the evolution by birds of elaborate visually organized fixed action patterns that transcend those of the lower vertebrates, it does not seem likely that this evolution was the source of the selection pressures toward brain enlargement of the earliest birds. Their environmental and social adaptive zone was probably invaded and successfully occupied first, and then, in the course of the adaptive radiation within that zone, the further evolution of social mechanisms already present in reptiles, amphibians, and fish could be extended to new levels of complexity. That further evolution would have produced adaptations during life of the sort implied by imprinting and learning as part of the establishment of the successful neurobehavioral adaptations of the adult.

Conclusions on Early Avian Adaptations

There has been a mass of research contributed by ethologists and comparative psychologists during the past two decades on the behavior of birds and (still more recently) the neural control of this behavior. As a source of clues to the initial evolution and differentiation of birds from reptiles, and the enlargement of the brain correlated with that evolution, that effort in the study of animal behavior has been less relevant than the speculations, presented earlier, about the visual demands of woodland niches. The recent research is of great evolutionary importance in describing the extent and dimensions of the adaptive radiation of birds, but the clues to their reaching a stage from which that radiation could take place must arise from speculations about the original niches that they first invaded.

The difficulty is best recognized by reference to my analysis of the early mammals. My hypothesis was that the earliest mammals could only be slightly different from reptiles with respect to brain and behavior, and the only way to keep the behavioral difference minimal in the nocturnal niches that I assumed for the first mammals was for them to have somewhat larger brains. I have emphasized the point that throughout the mammalian radiation of the Mesozoic era and early Tertiary period their brains remained at approximately the same level of relative size. The evidence of the next few chapters points to the Mesozoic and early Tertiary level achieved by the early mammals as providing an appropriate take-off point for the further evolution of the brain, which then advanced clearly beyond the reptilian ancestors in all respects, including behavior mechanisms. The progressive mammals include diurnal forms that live very different lives from those of diurnal reptiles.

I can suggest no comparable story for the birds. There is no clear evidence of an adaptive zone within which they achieved a stable semi-reptilian life for a period of time with somewhat larger brains that were necessary for them as modified "reptiles" in their niches. There was no equally clear transition demanded by shifts from reptilian adaptations to avian adaptations. This failure may, actually, be no failure at all. It may simply indicate a greater similarity in the adaptations of birds and reptiles as compared to mammals and reptiles. The course of avian evolution may have involved the gradual expansion of "reptilian" capacities in niches that taxed those capacities only slightly. It is possible that in the process of expansion the brain reached a "critical mass" in which plasticity of behavior became a genuinely useful behavior mechanism that could be integrated with fixed action patterns—the more typically reptilian mechanisms.

Part IV

Progressive Evolution of the Brain

There have been only a few truly revolutionary changes in the otherwise stable or slowly progressive evolution of the vertebrate brain. We are familiar with one of these "revolutions," the evolution of the human brain. That topic is a scientific paradox: either much is known or little is known depending on the perspective from which it is viewed. The paleontological perspective, which is most direct and which I have emphasized, is usually thought of as providing one of the more opaque lenses for studying brain evolution in man (von Bonin, 1963), but in our context it provides a reasonable amount of evidence even at the relatively low taxonomic level of genera and species. This analysis is clearest when applied to changes among classes, orders, and occasionally families of vertebrates, but only rarely is there sufficient evidence to consider the evolution within a limited line of descent such as the hominid line.

There was probably another equally dramatic "revolution" in the evolution of the vertebrate brain, which occurred about 50 m.y. ago. It extended almost throughout the mammals, as a class, and may have been the beginning of the evolution of intelligence. The morphological change was unusually simple: the appearance of a flexured rather than a linearly organized brain and the touching or overlapping in the endocast of the anterior cerebellum and posterior cerebrum. This contiguity has been remarked on as signaling the end of "midbrain exposure" in mammals, although there are, in fact, many living mammals in which the midbrain is clearly exposed in the endocast (e.g., most bats). In discussing this topic, Edinger (1964a) pointed out that midbrain exposure was probably accidental and reflected the fact that few living mammals have highly specialized behaviors in which the midbrain has a major role. She pointed out that "specialized brains reflect specialized behaviors," and invoked what I have termed the principle of proper mass in explaining differences in brain morphology in different species.

The most important "revolution" in the brain of about 50 m.y. ago, in my judgment, was the beginning of a major trend toward increase in relative brain size in mammals. Both the development of flexured brains and the frequent disappearance of midbrain exposure may be recognized as correlated with that effect. These were secondary results, I believe, of a solution of a packaging problem: how to fit the amount of brain evolving to handle more advanced behavior into a skull that sits comfortably on the neck and body and contains, anteriorly, a mouth, teeth, snout, eyes, and so on, which are of the general type that had evolved during the mammalian radiation.

Material is packaged most efficiently into a sphere, less so into a cylinder. The flexured brain with cerebrum and cerebellum associated in the familiar ways of birds and progressive mammals is approximately spherical or at least ellipsoidal. The linearly organized brain of lower vertebrates and of "primitive" living mammals, such as opossums, is more or less cylindrical. If we refer back to Fig. 11.3, we will note that the length of the brain and olfactory bulb of *Phenacodus primaevus* totaled 7.4 cm, and the volume of these structures was about 35 ml. The total skull length in that archaic ungulate (Table 11.1) was 26.4 cm, about the same as in a small deer. If the fraction of the length of the skull of *Phenacodus* devoted to the brain case were exactly spherical and if the olfactory bulbs lay somewhat under the forebrain, as they do in living progressive mammals, the volume of the brain could have been over 200 ml, that is, a volume comparable to or approaching that of living deer. This may help us to appreciate the advantage of a flexured over an unflexured brain, although the specific example is somewhat exaggerated because the flexured brain is never perfectly spherical.

The main point is that some time during the Eocene epoch the relative size of the mammalian brain increased beyond the archaic mammalian level in many groups of mammals, and the facts and reasons associated with that increase are the subjects of the next four chapters. That increase necessarily entailed morphological changes of the sort just noted, as well as the development of fissurization. It has been argued, I believe correctly, that even the process of fissurization, which results in convoluted rather than smooth brains, should not be considered evidence of an advanced or progressive brain. It is very likely, also, a secondary effect of brain enlargement. It is presumably a consequence of the mechanical difficulty of packing a particular amount of cortical surface, which is, effectively, where most of the cortical neurons in mammals lie (in a thin sheet). Since the area is generally considerably greater than the smooth surface of even a spherical cerebrum, fissurization is one solution to the problem of packing the actual surface into the available volume (LeGros Clark, 1945; Chow and Leiman, 1971).

In Chapter 13 I develop the discussion in greater depth with the best and most complete of the available fossil data on progressive mammals, those on the ungulates, or hoofed mammals, of the orders Perissodactyla and Artiodactyla, and the progressive carnivores of the order Carnivora. The next chapter (Chapter 14) is devoted to the South American ungulate radiation of the Cenozoic era, a radiation of ungulate orders that were, according to our definition, also archaic, in that they were entirely replaced in their niches by the progressive ungulate orders just considered. The circumstances of both the radiation and the replacement of the South American fauna are unusual, however, because they provide the data of a genuine, if retrospective, evolutionary experiment on the effect of the prey–predator relationship upon the progressive evolution of the brain.

Much of the material on the mammals that did not fit easily elsewhere is considered in a chapter in this section (Chapter 15) on "Special Topics." In that chapter I treat aspects of the evolution of the brain in aquatic mammals such as whales and pinniped carnivores, the evolution of the brain in elephants, and endocasts of ancient mammals of known geological age but of uncertain affinities with other mammals. I also discuss the Quaternary mammals, in particular the well-known Pleistocene fauna from the La Brea tar pits of California.

The last substantive chapter of this section (Chapter 16) is devoted to the evolution of the brain in primates, including man. A final chapter is devoted to the possible causes for the progressive evolution of the brain, with emphasis on the human brain and the evolution of human cognitive capacities.

Chapter 13

Progressive Tertiary Evolution: Ungulates and Carnivores

This is the first chapter in which we study an evolutionary expansion of the brain that went beyond the enlargement correlated with the evolution of new sensory capacities. (No such capacities are known to have appeared in the progressive Tertiary mammals, although it is likely that an old vertebrate adaptation, the use of vision as a distance sense, took on new meaning in several orders of mammals.) A straightforward "explanation" of that expansion as the source of the "greater behavioral capacity" mentioned by Lashley was a theme of the first chapter (pp. 18–25), but it is really not adequate as a scientific explanation. The primary purpose in the present chapter is to describe the major events in the history of the brain in those groups of Tertiary mammals that would normally be thought of as average or typical in the layman's sense and that are also "average" in the sense that their brains are presently at or near the typical relative size level of living mammals. The possible reasons for the expansion to that level are discussed more fully in later chapters (especially Chapter 17). Our subjects are the ungulate orders Perissodactyla and Artiodactyla and the fissipedes of the order Carnivora.

The ungulates (hoofed mammals) and carnivores are easily thought of as ecological units of prey and predator species, and it is natural to review them together. The combination is also heuristically appropriate because we know more of the fossil history of their brains and bodies than we do in other groups. Many endocasts from these orders have been described, and many more can be examined in museums around the world. These descriptions are reviewed only briefly because they have been published and illustrated in easily accessible texts and monographs. The bodies of many of these mammals can also be reconstructed from mounted skeletons because of efforts, now relatively rare but once among the major activities of museums and universities, in which fossil remains were assembled and skeletal reconstructions were attempted. As a result of this activity by museums and collectors enough material is available for a fairly detailed quantitative analysis of patterns of brain evolution within the order Carnivora and two major ungulate orders, the odd-toed order Perissodactyla and the even-toed order Artiodactyla.

The basic issue is whether the brain increased in relative size during the Tertiary period. Can we support the generally affirmative answer as offered by Lartet (1868), Marsh (1874), and their successors, or should we join Edinger (1949, 1962) in her belief that, except for the hominids, "there has been no perceptible progress of the brain after the initial early Tertiary differentiation" (1949, p. 14)? The quantitative analysis will force us to reject Edinger's conclusion, although it will not support a simple deterministic version of Marsh's "law" (pp. 14–16). We will be enabled to make significantly finer and more detailed judgments about the pattern of the evolution of brain size than is implied by either of these two extreme positions.

In this chapter numerical methods first introduced in Chapter 10 replace the graphic methods of analysis by convex polygons that have been used as much as possible so far. This is inevitable because, as one approaches the condition of living mammals, differences among groups that can be assigned to various levels of brain evolution should be expected to be smaller and smaller, and more powerful analytic methods have to be used to determine whether particular observed differences are likely to be anything more than sampling artifacts. In order to perform a valid quantitative analysis relatively large samples have to be assembled, and it is partly because our samples of ungulates and carnivores are sufficiently large that the numerical analysis could be undertaken.

The numerical methods used are based on the index of cephalization in the version described by von Bonin (1937). In particular, the encephalization quotient EQ (Chapter 3) is computed for each of the mammals in the sample, and that measure is analyzed to determine changes in relative brain size within and among the several orders of mammals under consideration. Changes within an order are determined over time, between the earlier Tertiary or Paleogene (Paleocene, Eocene, and Oligocene) epochs, dating from about 65 to about 22 m.y. ago, and the later Tertiary or Neogene (Miocene and Pliocene)[1] epochs, dating from about 22 to

---

[1] This definition of the Neogene is the one preferred by many vertebrate paleontologists. The term is also frequently used by geologists (and in Europe) to include the Pleistocene or even the Recent epochs. For the purposes of this book it is a period in geological history following the Paleogene in which most of the mammalian fauna was still to be replaced by species now living. In Chapter 14 several Pleistocene South American ungulates are included in a "Neogene" sample.

2 m.y. ago (Berggren, 1969). These are related to the present situation in (living) carnivores and ungulates.

EVOLUTIONARY HISTORY

As the Eocene epoch drew to a close about 40 m.y. ago, all the presently surviving orders of mammals had appeared, and their brains were either similar to or were evolving toward the external shape of their living descendants. Oligocene endocasts (25–35 m.y. ago), such as those of the three-toed horse *Mesohippus,* the ancestral camel *Poëbrotherium,* and the early sabertooth "cat" *Hoplophoneus,* are all appropriately and recognizably like those of living ungulates and carnivores in appearance. The carefully studied record of the horse brain (Edinger, 1948) shows that in the equids some major changes in shape may have occurred earlier, between the level of the "dawn horse" *Hyracotherium* ("eohippus") of the Lower Eocene and its descendant *Orohippus* of the Middle Eocene. The transition to *Mesohippus* was from a smooth linearly arranged brain as in the archaic mammals to the typically flexured and convoluted "modern" brain, and it was completed about 30 m.y. ago (Fig. 13.1).

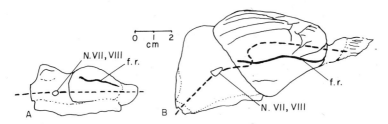

Fig. 13.1. Side view of endocasts of *Hyracotherium* (A) and *Mesohippus* (B) to illustrate the evolution of a flexured brain in the evolution of the equids (cf. Figs. 13.3 and 13.6). f.r., Rhinal fissure differentiating neocortex (dorsal) from paleocortex (ventral). Flexure shown by dashed line.

Despite the evident morphological modernity of the brains of Lower to Middle Oligocene mammals, my own earlier allometric analysis of relative size (Jerison, 1961) indicated that in that middle Tertiary epoch of 30–35 m.y. ago, as represented by the fauna of the South Dakota Badlands, brain size was still at about half its modern level. Information on the history of the camel brain (Jerison, 1971b) suggests that the modern level of brain size evolution may not have been reached until late Pliocene or even Pleistocene times and that camelids of 10–20 m.y. ago, such as

Procamelus and *Protolabis,* had brains that were between ½ and ⅔ the size of those of their living descendants if body size is taken into account.

The two ungulate orders and the order of living carnivores considered in this chapter are not the only hoofed and carnivorous orders that are known. In addition to the living elephants, the last representatives of the once flourishing Proboscidae, several extinct orders of hoofed animals went through essentially all their adaptive radiation in South America. Flesh-eating orders other than the Carnivora are also known and include Primates, as we now know from the ethology of baboons and chimpanzees (Howell, 1967) and from our own human experience, and marsupial carnivores that flourished in South America (Simpson, 1965). Other ungulate or carnivorous species are considered in Chapters 14 and 15.

Phylogenetic Lines

Although it is the most ancient of the three orders considered in this chapter according to the age of its oldest skeletal fossils (at least Paleocene), the order Carnivora is known endocranially only from the late Eocene or early Oligocene. It was also at about that time that the true carnivores were probably replacing creodonts in carnivorous niches. We will see that the endocasts of the true carnivores were relatively larger than those of nearly contemporaneous (actually, slightly more ancient) creodonts. A few species genuinely overlapped in time, *Pterodon* and *Hyaenodon* among the creodonts and *Eusmilus, Hoplophoneus, Dinictis,* and *Daphoenus* among the true carnivores; in this set, the true carnivores were, in fact, larger brained.

It would be both instructive and very important to have appropriate data to compare earlier forms, in particular Lower Eocene or even Paleocene species of creodonts and miacid carnivores. We will see in the section on brain morphology that there were evidently significant morphological changes in the external configuration of the brains in perissodactyls sometime between the early Eocene and either late Eocene or Lower Oligocene times, and there is a suggestion of similar transformations among the artiodactyls. Data on the carnivores on this point, in particular a comparison between the more archaic and more progressive carnivorous orders, Creodonta and Carnivora, would obviously be useful.

The first group of true carnivores, the miacids, is essentially unknown endocranially. Those with known endocasts, which we emphasize, are late Eocene and Oligocene or later specimens, by which time the living families of carnivores had become differentiated. The relationships are indicated in Fig. 13.2.

The two ungulate orders are believed to have been derived from the archaic order Condylarthra, and impressive evidence of skeletal similarity

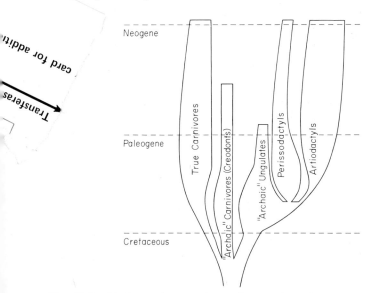

Fig. 13.2. Phylogenetic tree of ungulates and carnivores.

between the Lower Eocene *Hyracotherium* and an early phenacodontid *Tetraclaenodon* of the Middle Paleocene has been presented by Radinsky (1966). The relationship between the condylarths and the artiodactyls, though not as well understood, is also generally considered to be direct (Romer, 1968) through another condylarth family, the hyopsodontids rather than the phenacodontids. In Chapter 11 we discussed relative brain size in genera from both of these condylarth families, although these (*Phenacodus* and *Hyopsodus*) were too late on the scene to be considered, themselves, as possible ancestors. Our sample of archaic ungulates was sufficiently representative of the range of adaptations and eventual developments of ungulate evolution to be an appropriate comparison group for the ungulates that are considered in this chapter.

The most complete record is for the order Perissodactyla, the odd-toed ungulates, which presently includes horses (zebras), tapirs, and rhinoceroses but which was a more varied order at one time. Our sample includes forms such as *Hyracotherium* ("eohippus") and its close relative, the ancestral tapir *Heptodon,* both from the Lower Eocene about 50–55 m.y. ago. Within the lineage of the horse we have all the specimens reviewed by Edinger (1948), which provide a completely representative group from each successive epoch of the Tertiary (six species). We have less adequate but also fairly extensive data on the several other perissodactyl lineages mentioned above, including the rhinoceroses (five species) and the titanotheres (three species). The total number of perissodactyl species is actually smaller than that for the other groups, but, because of the less diverse evolution of the perissodactyls, our picture is representative of their adaptive radiation.

If we keep in mind that the history of the order Carnivora begins in the Paleocene and that our earliest specimen is Upper Eocene, we realize that 20 m.y. of evolution of that order is missing from our data. The perissodactyls and artiodactyls are first known from the Lower Eocene. We have two perissodactyls from that time, *Hyracotherium* and *Heptodon;* hence, that order is reasonably completely represented (compared to other samples available to us). Our earliest artiodactyls are Upper Eocene, and we, therefore, lack the first 10–15 m. y. of evolution of that order. Despite these limitations we will be able to develop a consistent picture of the evolution of brain size in these groups of progressive mammals. One important point is left as a hypothesis, which has also been raised by Dechaseaux (1959). This concerns the evolution of the mammalian brain during a second great Tertiary radiation. There is the possibility that a fundamental change occurred in many groups of mammals at about the same time during the Eocene, resulting in an enlarged brain with all the concomitants of enlargements discussed in the introduction to Part IV. The best evidence is from our perissodactyls, although similar evidence can also be seen in our data on Upper Eocene artiodactyls, compared with their descendants and relatives of the Upper Oligocene, the Miocene, and the Pliocene.

Evolutionary Rates and Related Phenomena

The evolutionary facts in the discussion that follows are the raw data for inferences that have been described under the headings "ecological incompatibility" and "relay" by Simpson (1953, Fig. 19; 1959, Table 12). They show when and at what rate the Carnivora (in particular the fissipedes) replaced the Creodonta in carnivore niches by the end of the Paleogene; they offer comparable information on how the two ungulate orders of this chapter had earlier (late Eocene) replaced the archaic ungulate orders discussed in Chapter 11, and they show how the artiodactyls have tended to replace the perissodactyls during the middle and late Tertiary.

Within the two ungulate orders there are several families or superfamilies that have been considered archaic in some respects, in that they evolved and radiated relatively rapidly and eventually became extinct as families. Among the perissodactyls, the titanotheres (superfamily Brontotheriodea) are such a group, arising in the Lower Eocene to become the largest land mammals of the Lower Oligocene and then disappearing rapidly. One measure of evolutionary rate for this group can be derived from gross body size. This changed dramatically as the titanotheres evolved (Osborn, 1929; Hersh, 1934; Robb, 1935a,b), and is evident in our data on three species. According to my computations (Table 13.1), there was a six- or sevenfold increase in body size from *Mesatirhinus* (350 kg) to *Menodus* (2300 kg).

Several groups of artiodactyls are also archaic in this sense. These are from what Romer (1966) has referred to as the "tylopod assemblage," of which only the camelids have survived as a family to the present time. The extinct families from that assemblage include the oreodonts and entelodonts in the New World and the anoplotheres and cainotheres of the Old World; all these groups are represented in our sample.

An important aspect of evolutionary rates refers not so much to the rate of change of species as to the rate of change of diversification of species. Thus, if during an appropriate span of time species of one order become more diverse and those of another order less diverse and if the two orders are relatively similar with respect to their adaptive zone, one may reasonably think of the diversifying group as replacing the undiversified group. This was the case of the fissipedes, which displaced and replaced the creodonts. It has also been true of the artiodactyls, which since the late Eocene have been replacing the perissodactyls.

The apex of evolution of the perissodactyls, with respect to the size of their largest individuals, was achieved by the titanotheres in the Lower Oligocene in North America with the species *Menodus giganteus* as the largest in our sample (the largest genus was, probably, *Brontops*). In the Upper Oligocene or Lower Miocene there appeared the "rhinoceros," *Baluchitherium*, a specimen for which an endocast is known but has been unavailable to me. It weighed, perhaps, 15–25 metric tons and was probably the largest land mammal that ever lived. In numbers and diversity of species, however, the artiodactyls have clearly outpaced the perissodactyls. Walker (1964) listed only three living families (6 genera, 16 species) of perissodactyls, whereas he could list nine living families (82 genera, about 200 species) of artiodactyls.

BRAIN MORPHOLOGY

The characteristic appearance of the brains of the older specimens in our sample was either similar to or only slightly "in advance" of those of the archaic Tertiary ungulates discussed in Chapter 11. The endocasts of the Lower Eocene perissodactyls, *Hyracotherium* and *Heptodon,* are essentially indistinguishable from those of Lower and Middle Eocene condylarths (Fig. 13.3), and it is only in the Middle and Upper Eocene spec-

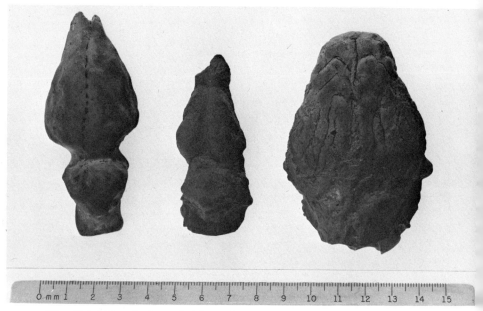

Fig. 13.3. Dorsal views of, left to right, endocasts of *Heptodon* (MCZ 17670), *Hyracotherium* (YPM 11694), and *Mesohippus* (USNM 22539). Midline of *Heptodon* and some of the sulci of *Mesohippus* emphasized by pencil markings. See Table 13.1 for explanation of museum numbers; scale numbered in centimeters.

imens that one can identify a clear advance above that level (Edinger, 1948, 1956a). The advance seems to have been independent of family or order, and it is this point that is badly in need of further analysis by the examination of more endocasts.

The least equivocal evidence of the advance in brain morphology associated with its evolution is in the perissodactyl lineages. We have data on specimens from Lower, Middle, and Upper Eocene strata, as well as later evidence. The different families cannot be differentiated on a primitive–progressive dimension. They include the "progressive" equid lineage beginning with *Hyracotherium* and the "primitive" tapiroids, such as the Lower Eocene *Heptodon,* all of which had condylarthlike brains. But they also include the Middle Eocene rhinoceroses *Hyrachyus* and *Colonoceras*, with brains that were intermediate between the archaic and living mammalian level. If we were to arrange the endocasts of all the Eocene mammals in the sample, it would be most natural to group together the Lower Eocene specimens on the one hand and most of the Middle and Upper Eocene specimens on the other. The first group would have condylarthlike endocasts, displacing only 10–40 ml of water, which makes them relatively small even for animals of their body size (Table 13.1).

Brain Morphology

Living mammalian brains of that size are generally at least somewhat convoluted (with exceptions among the large rodents), but these Lower Eocene endocasts suggest smooth brains.

The Middle and Upper Eocene endocasts, on the other hand, are uniformly convoluted in their forebrain areas. There are clear, detailed distinctions among the orders Carnivora, Perissodactyla, and Artiodactyla in the patterns of the convolutions, which we will consider presently, but the main point is that gyri and convolutions have begun to appear. These occur even when the absolute brain size was about the same as in the Lower Eocene forms. Specifically, the Upper Eocene camelid *Protylopus* had a clearly convoluted brain, which was about the same size as that of *Hyracotherium* and considerably smaller than that of *Heptodon* (Table 13.1).

I have indicated earlier (p. 284) my judgment that convolutions were not necessarily indicative of advanced brain development. Nevertheless, when two brains are viewed and when both have the same volume but one is smooth and the other convoluted, it is clear that the second would support more brain surface, and, hence, a greater cortical extent, regardless of gross size.

The most active recent workers in this field, Edinger (1948, 1966), Dechaseaux (1969), Piveteau (1951, 1961), and Radinsky (1968b, 1969), have attempted to use data on fissural patterns to analyze the evolution of the brain in various lineages of ungulates and carnivores. Although, as Radinsky has pointed out, "endocranial casts of most mammals reproduce almost all of the detail seen on the surface of the brain" (1968, p. 495), it is occasionally possible to be badly misled by an endocast. I have noted in Chapter 11, that the olfactory bulbs can be exaggerated in size in the endocast as compared to the brain. In primates even the most notable of fissures, the Sylvian, which separates the temporal from the frontoparietal lobes, may be displaced in the endocast compared to the brain (Connolly, 1950; von Bonin, 1963). (There is, of course, a real question on which is the correct and which the erroneous configuration in the latter case. The brain is quite malleable compared to the skull and may become distorted in dissection and after fixation, and one may not be certain that the impressions of fissures in the endocast do not provide the more accurate representation.) Let us review the specifics of the various endocasts considered as brains, assuming that distortions were relatively unimportant in the analysis of the brain, proper.

CARNIVORES

An extensive amount of work on carnivores by Radinsky (1968b, 1969) is beginning to appear on the endocasts of this group, and earlier work by Piveteau has been summarized in the "Traité de Paléontologie" (1961) edited under his supervision. Piveteau (1951) has also published a theoretical account of the evolution of the fissural pattern in various carnivores. I have indicated my uncertainty about the validity of the analysis of fissural patterns for information about the evolution of the brain, but the rationale for the approach is simple and direct. In living carnivores (and mammals generally) there is enough localization of function to enable one to judge the significance of a particular class of behaviors in an animal's life by the superficial extent of the brain structures subserving these functions, as represented by identifiable gyri in the brain. This has been demonstrated in many anatomical and physiological studies by Welker and his colleagues (Welker and Seidenstein, 1959; Welker and Campos, 1963), for example, and it is a familiar tenet of comparative neurophysiology (Adrian, 1947). I have termed it the principle of proper mass. Much of this work and that of other investigators has been done on carnivores, and some of the most brainlike endocasts are from living and fossil species of this order (Figs. 2.2, 13.4, 13.5).

Radinsky has used these facts to good advantage in his analysis of the evolution of the brain in otters, felids, and canids (Radinsky, 1968b, 1969). His procedure was to compare the fissural patterns of a (more or less) phylogenetic series in an attempt to infer the development of behavior patterns from the development of characteristic fissures in the brain as mirrored in the endocast. In the otter, for example, he reasoned from his observation of an "enlarged coronal gyrus" in *Potamotherium,* an Oligocene "otter" (Savage, 1957), and from the presence of such a gyrus in the living otter, in which this region is a projection area for the vibrissae, that the sensitive use of vibrissae was an early adaptation of the otters (Fig. 13.4). He emphasized other lutrine adaptations including tactile sensitivity, in which other parts of the brain were involved, in other species that he reviewed (Radinsky, 1968b).

I hesitate to be critical about such reasoning because it is making the best of minimal information by inferring habits from the details of brain structure. I must confess, however, that in viewing endocasts of the two "otters" I was more impressed by the differences than by the similarities. The Oligocene form was much more typically a carnivore endocast, like those of wolves and cats, whereas the living otter's endocast seemed peculiarly "busy" in its convoluted pattern, with narrow, intricate convolutions apparent in both brain and endocast. Lacking detailed modern studies of the localization of function in living otters, I am hesitant to make quite as much of the comparison of the Recent with the Oligocene. It would, of course, be surprising if the Oligocene form, so similar in body configuration

Brain Morphology

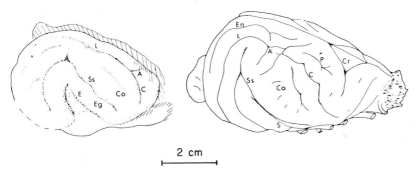

Fig. 13.4. Endocasts of Oligocene and Recent otters as interpreted by Radinsky. Left, *Potamotherium valentoni* (AMNH 22520); right, *Lutra canadensis*. Abbreviations: A, ansate sulcus; C, coronal sulcus; Co, coronal gyrus; Cr, cruciate sulcus; E, ectosylvian sulcus; Eg, anterior ectosylvian gyrus; En, entolateral sulcus; L, lateral gyrus; P, postcruciate sulcus; S, sylvian sulcus; Ss, suprasylvian sulcus. From Radinsky (1968b).

to the living form, had not achieved reasonably similar behavioral adaptations and the appropriate neural structures to handle them.

Studies like Radinsky's will generally tend to confirm suspicions about brain–behavior relations, and it would probably be necessary to have very unusual brain adaptations in fossil forms to suggest that one reject these suspicions. In short, the difficulty with studies such as Radinsky's is that they are necessarily more subjective than one would like and more influenced by the research worker's expectations. The problem is one of the philosophy and sociology of science, attempting to derive objective conclusions in the face of strong expectations by the scientist and limited data that could redirect his expectations and temper his conclusions. Yet this risk must be accepted if we are to make full use of the fossil evidence on the evolution of the brain. In his studies of canid and felid evolution Radinsky (1969) follows the same method, with comparably uncertain results. His analysis was accompanied by some of the most attractive drawings of endocasts that I have seen, and several of them are reproduced here (Fig. 13.5). In viewing these, I have difficulties even more serious than those with the analysis of the otter brain.

In the case of the canids and felids there was an increase in both brain and body size during the evolutionary sequence, and it is difficult to ascribe changes in fissural patterns entirely to changes in neural organization (as Radinsky recognized). Enlargement of the brain as a concomitant of enlargement of the body almost inevitably results in increased fissurization, as I have mentioned elsewhere. (It is sometimes referred to as "Bail-

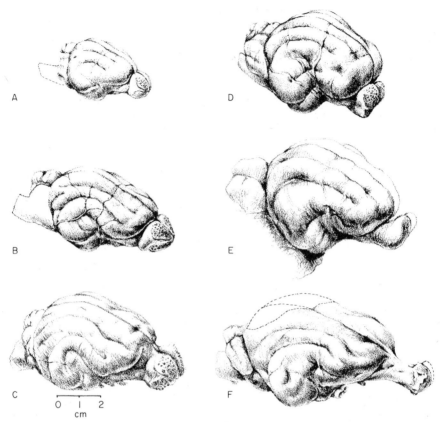

Fig. 13.5. Fossil canid (A–C) and felid (D–F) endocasts. A, *Hesperocyon* ("*Pseudocynodictis*") *gregarius* (AMNH 39475); B, *Mesocyon* sp. (AMNH 6946); C, *Tomarctus* cf. *euthos* (F.AMNH 61074); D, *Hoplophoneus primaevus* (AMNH 490); E, *Nimraevus brachyodus* (YPM 14835); F, *Pseudaelurus* sp. (F.AMNH 61835). All but *Nimraevus* (on which it was impossible to estimate body size) are included in Table 13.1. From Radinsky (1969).

larger's law" that large brains tend to be fissured and small brains smooth.) Radinsky has concluded, first, that there may have been more frontal and prefrontal cortex in later species, and he correlated this with the evolution of inhibition of primitive behavior and the "enabling" of social behavior. Second, he considered the evolution of the cruciate sulcus as independent (analogous rather than homologous) in canids and felids, a secondary effect of expansion of primary somatic motor and sensory cortex in the adjacent sigmoid gyri, which led to the development of a cruciate sulcus.

There is insufficient space to present Piveteau's equally abstract and speculative analysis of sulcus formation. He was concerned (Piveteau, 1951) with the development of the specialized convoluted pattern of gyri

Brain Morphology

and sulci from what he considered to be a more primitive pattern present in creodonts and some Eocene artiodactyls. The "primitive" pattern had more or less parallel sulci, and he related the appearance of "sigmoidal" convolutions to the flexure of the brain and related phenomena. The argument was similar to Elliot Smith's (1903, 1908).

PERISSODACTYLS

It is possible to present a detailed analysis of the endocasts of many perissodactyl species, and Radinsky is presently in the midst of the necessary analysis. A detailed exposition of the evolution of the horse brain from eohippus (*Hyracotherium*) to the living *Equus caballus* was published by Edinger (1948) and is a classic in its field. Her point of view included many statements about quantitative relations that are often inadequately grounded (Jerison, 1971b), but as a morphological analysis it stands alone in its completeness and in the extent to which its information can be interpreted independently of her conceptions. Simpson (1951) has presented a useful popular review of the evolution of horses.

Some of the material with which Edinger worked is included in the sketches of the dorsal views of equid endocasts in Fig. 13.6. This figure also presents information on the body size of each species. Edinger gives very complete descriptions for endocasts of *Hyracotherium, Pliohippus,* and *Equus*. She includes excellent photographs of dorsal and lateral views of the endocast of *E. occidentalis* (LACM-17 C-3), from the Pleistocene La Brea tar pits of California. She also discusses other specimens of all genera shown in Fig. 13.6. Quantitative data on the Tertiary equids of Fig. 13.6 are included in Table 13.1. Data on *Equus occidentalis* are $E = 870$ ml, $P \cong 640$ kg, and $EQ = 0.98$.

In her long monograph Edinger presented many detailed statements that are difficult to summarize in the available space. The main conclusion was that the brain and its parts did not evolve at consistent or constant rates during the evolution of the horses and certainly not consistently with the evolution of the rest of the skeleton. She considered the most important changes to have been in the increased relative amount and convolutedness of the cerebral cortex, in particular of the neocortex, although she also emphasized the increasing complexity of the cerebellum. The greatest step, according to Edinger, was between eohippus and *Mesohippus*, and this agrees with my earlier remarks on the probable major evolutionary changes in the brain that occurred sometime between the Lower Eocene and Lower Oligocene epochs. She believed that with the appearance of *Mesohippus* in the Lower Oligocene the horse brain had taken on an essentially modern configuration, although there was further advance during the Miocene and Pliocene to "less straight" fissures and additional fissurization.

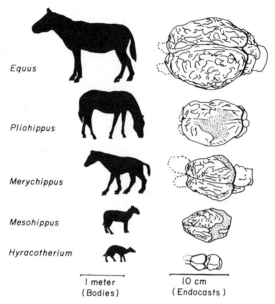

Fig. 13.6. The evolution of brain and body size in horses. Sources of endocasts given in Table 13.1 and in the text. Body sizes estimated from various sources, especially Edinger (1948) and Stock (1956). See Edinger (1948) for additional views of endocasts. Stippled areas are portions of the endocasts in which the fissural pattern was masked in the author's copy. Olfactory bulb areas shown as dashed lines.

The pattern of evolution of the brain as revealed in endocasts of titanotheres and rhinoceroses is generally of the sort described by Edinger for the horse brain. There was a consistent increase in size, accompanied by increased fissurization apparent in most progressions. There are a few exceptions to the latter statement, probably due to the failure of the endocast to preserve the fissural patterns of the brain. This was particularly true of the very large specimens such as *Menodus* and *Dicroceras* (Table 13.1).

Edinger noted the fact of flexure in the brain; it was apparent in *Mesohippus* as opposed to eohippus. I would make little of this fact, beyond the statements presented earlier (p. 284) and Fig. 13.1, in which flexure was related to efficient packing: a flexured brain is brain is more nearly spherical than is an unflexured one and, therefore, requires less cranium to surround and protect it.

ARTIODACTYLS

In the analysis of this final group of endocasts several recent references should be considered: Dechaseaux (1969), on the early evolution of the brain in the group as a whole; Sigogneau (1968), on one particular set of early deerlike forms; Repérant's (1970, 1971) papers on fossil camelids; Edinger's (1966) last substantive contribution to paleoneurology, on the evolution of the camelid brain; and a quantitative reassessment of Edinger's paper (Jerison, 1971b).

The outstanding morphological feature of the evolution of the brain in artiodactyls was certainly the transition from the parallel orientation of the major gyri in the Upper Eocene forms of 40 m.y. ago, such as *Dacrytherium* or *Anoplotherium,* and from the linear alignment of forebrain and hindbrain in these forms to much more convoluted and complex gyri and flexured brains, in such Upper Oligocene forms as *Aletomeryx* (about 25 m.y. ago). The brains of those late Oligocene "deer" are indistinguishable, except in detail, from typical neocortical expressions in living deer and antelope (Fig. 13.7).

The evolutionary picture of the camelids is different. The small-brained and small-bodied *Protylopus* of the Upper Eocene was much more modern with respect to its endocast than were the other known Upper Eocene artiodactyls. Viewing its endocast, we have no reason to suspect that it was from a fossil mammal—it could as easily have been the endocast of many living forms, and many living species of other mammalian orders of comparable body size (Table 13.1) have significantly more "primitive" brains as far as external appearances go. The argument on the evolution of relative brain size is more complex and is presented in the next section. To anticipate, however, it is clear that despite its relatively modern appearance *Protylopus* had a brain that was no more than half the size to be expected of a living mammal, a living camelid, of its body size.

Conclusions on Brain Morphology

I have already indicated my reservations about elaborate analyses based on the details of fissural patterns, even for specimens in which the endocast obviously mirrors the brain very well. The basic difficulty is with assumed similarities of the functional organization of living brains to brains of fossil animals as known from endocasts. This difficulty should not be exaggerated, and efforts such as Radinsky's are of the greatest importance in any attempt to reconstruct the evolution of the brain as a functional as well as structural system. My preference for a quantitative analysis based on gross brain size can be subjected to equally severe criticisms, since I emphasize the importance of the quantity of available analytic neural tissue, and purposely disregard even "known" regional specializations or functional localization.

The morphological review of the endocasts considered in this chapter reveals a second feature, which is also evident in the quantitative analysis

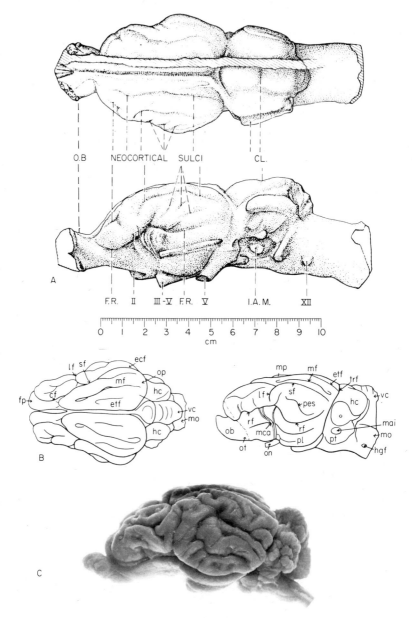

Fig. 13.7. Endocasts and braincast of artiodactyls (see Table 13.1). A, *Anoplotherium*, from Dechaseaux (1969); B, *Protoceras*, from Lull (1920); C, living white-tailed deer (*Odocoileus virginianus*), braincast courtesy of W. I. Welker. Note similarity of B and C.

of the next section. There was an increasing diversity in the morphology of the brain. Our oldest endocasts, those of *Hyracotherium* and *Heptodon*, could have come from the same sample as most of the archaic ungulates of Chapter 11. Many of the Middle and Upper Eocene ungulate endocasts, in particular from the artiodactyls, have a certain common form with elongated linearly arranged brains that lack the flexure typical of later mammals. The sulci of the artiodactyls, in particular, are arranged as longitudinal bulges and are remarkably like the sulci of the creodonts described in Chapter 11. (See *Anoplotherium* in Fig. 13.7A.)

As the progression entered the Upper Eocene and Lower Oligocene strata, the endocasts were more characteristic of perissodactyls, artiodactyls, and carnivores, that is, they were flexured and with at least some familiar fissures (in particular in the carnivores). With the further passage of time beyond the Oligocene, we can see differentiation within each of these orders. This is apparent in the lineage of the horses and camelids as illustrated by Edinger (1948, 1966) and by the several carnivore lineages documented by Radinsky (1968b, 1969). In short, one of the major evolutionary trends that is documented by the morphological analysis is the increasing diversification of the external form of the brain. If other progressive orders, such as the elephants or primates, were under review in this chapter, that trend could be documented in even greater detail. Its discovery in the quantitative analysis of relative brain size is probably the most important result to be presented and documented in the next section.

RELATIVE BRAIN SIZE

At the risk of being repetitious, I will review the meaning of relative brain size (Chapters 2 and 3), and in the extensive analysis of the following pages I present the most thorough use of the encephalization quotient, EQ, as its measure (Chapter 3). In many applications one refers to brain:body ratios, and a careless reader may immediately think of a simple ratio, dividing the brain weight or volume by the body weight or volume. This is, of course, inappropriate, and, without belaboring the point, we may note that there does exist a more appropriate brain:body "ratio." It is not a simple ratio; rather, it relates gross brain size to the $\frac{2}{3}$ power of the body size. The use of such a modified ratio is justified by the appearance of graphs such as Fig. 2.3, in which one can see that brain size is an approximately linear function of body size when these are graphed on logarithmic coordinates. The slope of a "best-fitting" line to such an array of points is approximately $\frac{2}{3}$, and that is the reason for the $\frac{2}{3}$ power in the ratio.

We will see the slope of $\frac{2}{3}$ in Fig. 13.8, when we examine a graph like Fig. 2.3 for the species reviewed in this chapter and in Chapter 11. The slope appears again when we examine data of a sample of living ungulates and carnivores that represent the present status of relative brain size of mammals within comparable adaptive zones.

Measurement of Brain and Body Size

Brains of fossils are known from their endocasts, natural or artificial casts of the cranial cavity, of the kind presented earlier in this chapter. In most instances the brain volume (treated as equal to endocast volume) in the fossil samples could be measured directly from the endocast by water displacement. For some the method of graphic double integration was used to estimate brain volume, when the endocast was either not available or obviously crushed. In this method, the brain is modeled by an elliptical cylinder with the dimensions derived from illustrations of the lateral and dorsal projections of the endocast (Figs. 2.7 and 11.3).

Body sizes in the fossils were estimated by weight:length equations developed from the data on living carnivores and ungulates [Chapter 2; Eq. (2.2) and Eq. (2.3)] or from scale models of reconstructions. The head and body lengths of the fossils could be estimated either from mounted skeletons or from the skull and long bones as described in the paleontological literature (Chapter 2, Fig. 2.8).

Brain size and body size in living species were taken from published reports such as Count (1947), with the restriction that body weights had to be in the range reported by standard compendia such as Walker (1964). A detailed discussion of these methods applied to living and fossil camelids was recently published (Jerison, 1971b), and they were summarized in Chapters 2 and 3.

Cenozoic Ungulates and Carnivores

We can now analyze the history of brain size in Tertiary mammals and their living descendants in the Quaternary period. The Tertiary is presently dated from about 65 m.y. ago, the beginning of the Paleocene epoch, to about 2 or 3 m.y. ago, the end of the Pliocene (Berggren, 1969). Our comparisons involve the major holarctic (Northern Hemisphere) carnivorous and ungulate orders during that period. We include the "archaic" ungulates and carnivores (Chapter 11), orders that were entirely replaced in their niches, and the "progressive" orders that replaced them. The terms "archaic" and "progressive" refer mainly to the fact of extinction and replacement, but they also reflect a diagnosis of skeletal features in the known fossils as described, for example, in Romer (1966).

Relative Brain Size

A number of years ago I described a progression between an "archaic Eocene," an Oligocene, and a Recent sample of mammals (Jerison, 1961) with respect to mean brain size, and the mean EQ's for these samples were 0.25, 0.5, and 1.0, respectively. That is, the brains of the respective fossil groups were one-quarter and one-half the size expected of living mammals of the same body size and from similar ecological niches. The samples were small and heterogeneous, but the results are confirmed in this more complete analysis (cf. Jerison, 1970b), in which emphasis is also on the diversity of relative brain size. The four archaic and eight progressive fossil species of the earlier report may be contrasted with the sample that can now be presented: 17 archaic and 52 progressive fossil species. This larger sample is also more homogeneous, being limited to holarctic carnivores and ungulates. The fossil specimens are identified and data on their brains and bodies are presented in Table 13.1. Comparable data on living carnivores and ungulates are summarized in Fig. 13.9.

Table 13.1
Brain and Body Size Estimates and Encephalization Quotients (EQ) in Fossil Ungulates and Carnivores[a]

Genus and identification[b]	Endocast volume (ml)	Body weight (kg)	EQ
Archaic ungulates			
Arctocyon MH CR700	38	86	0.16
Arctocyonides MH CR733	8.3	11	0.14
Pleuraspidotherium MH CR252	6.0	3.3	0.23
Phenacodus AM 4367	31	56	0.18
Meniscotherium US 23113	15	6.2	0.37
Hyopsodus US 23745	3.2	0.63	0.36
Pantolambda AM 3957	19	30	0.16
Leptolambda FM P26095	69	210	0.16
?Barylambda FM P15537	102	620	0.12
Coryphodon I YPM 11330	93	270	0.19
Coryphodon II(a)	90	540	0.11
Uintatherium YPM 11036	300	1400	0.20
Tetheopsis YPM 11041	350	2500	0.16
Archaic carnivores			
Thinocyon US	5.7	0.80	0.55
Cynohyaenodon MH (809)	8.3	3.0	0.33
Pterodon MH (809)	62	42	0.43
Hyaenodon (b)	85	56	0.48
Paleogene ungulates			
Heptodon MCZ 17670	39	46	0.25
Hyracotherium YPM 11694	15	13	0.23
Mesohippus US 22539	78	20	0.88
Hyrachyus AM 11651	90	60	0.49
Colonoceras YPM 11082	71	47	0.45

Table 13.1—(cont.)

Genus and identification[b]	Endocast volume (ml)	Body weight (kg)	EQ
Subhyracodon US 22540	280	550	0.35
Amynodon YPM 11453	210	890	0.19
Palaeosyops (B) AM 1544	240	390	0.37
Mesatirhinus (B) PU 10041	190	350	0.32
Menodus (B) YPM 12010	630	2300	0.30
Dacrytherium (A) MH (1091)	35	12	0.56
Diplobune (A) MH (1092)	53	40	0.38
Anoplotherium (A) MH (1092)	80	82	0.35
Archaeotherium (E) PU 10908	190	230	0.42
Merycoidodon I (M) (c)	20	23	0.21
Merycoidodon II (M) (c)	50	46	0.32
Protylopus FM P12203	13	2.6	0.57
?Eotylopus UT 40504-1	22	10	0.40
Poëbrotherium I AM 1482	31	19	0.36
Poëbrotherium II AM 636	60	41	0.42
?Amphitragulus MH (d)	40	25	0.39
Protoceras YPM	58	12	0.92
Paleogene carnivores			
Plesictis MH (815)	11	1.3	0.77
Potamotherium NB SAU2280-1	50	9.7	0.92
Daphoenus I FM UM-1	49	26	0.46
Daphoenus II PU 12588	66	30	0.54
Hesperocyon AM 39475	15	2.0	0.79
Pachycynodon MH (812)	39	9	0.75
Hoplophoneus I PU 10156	47	20	0.53
Hoplophoneus II US 22538	52	49	0.32
Eusmilus MH (817)	38	21	0.42
Herpestes MH (816)	13	2.1	0.66
Cynelos (e)	110	49	0.64
Neogene ungulates			
Merychippus LA 368/3929	270	110	0.98
Neohipparion FM P15871	160	150	0.47
Pliohippus FM P15870	270	220	0.62
Teleoceras MCZ	1500	1500	0.95
Cainotherium (C) NB 2653	5.2	0.46	0.73
Samotherium (b)	420	630	0.48
Promerycochoerus (M) YPM 11002	160	210	0.38
Merycochoerus (M) (c)	77	120	0.26
Protolabis US 2128	140	97	0.55
Procamelus F.AM 40366	380	310	0.69
Dremotherium MH (1098)	53	19	0.62
Aletomeryx YPM 10765	45	16	0.59
Dicroceras MH (1099)	150	47	0.96

Relative Brain Size

Table 13.1—(cont.)

Genus and identification[b]	Endocast volume (ml)	Body weight (kg)	EQ
Neogene carnivores			
Plesiogulo GU 268	140	38	1.03
Mesocyon I AM 6920	52	10	0.93
Mesocyon II AM 6946	37	9.5	0.69
Cynodesmus (b)	36	13	0.54
Tomarctus F.AM 61074	58	15	0.80
Pseudaelurus F.AM 61835	89	43	0.60

[a] The data in this table are assembled from many sources. The most comprehensive secondary sources are Edinger (1929) and Piveteau (1958, 1961). Additional morphological descriptions and illustrations are in Black (1915, 1920) and Dechaseaux (1969), Edinger (1948, 1956a, 1964b), Gazin (1965a, 1968), Gervais (1872), Hürzeler (1936), Moodie (1922), Orlov (1948), Osborn (1898a), Radinsky (1967c, 1969), Russell and Sigogneau (1965), Savage (1957), Sigogneau (1968), and Tilney (1931). Methods used in estimating brain and body size are described in Chapter 2.

[b] Specimens identified by genus following Romer (1966) and by museum number or other reference as follows: AM, American Museum of Natural History (New York); F.AM, Frick Collection in American Museum of Natural History; FM, Field Museum of Natural History (Chicago); GU, Geological Museum of the Academy of Sciences of the Ukranian SSR (Kiev); MCZ, Museum of Comparative Zoology (Harvard); MH, Muséum d'Histoire Naturelle (Paris); NB, Naturhistorischen Museum (Basel); PU, Princeton University, Department of Geology; US, United States National Museum (Washington, D.C.); YPM, Yale Peabody Museum. Boldface letters are initials to identify the "nonprogressive" superfamilies of Paleogene and Neogene ungulates discussed in the text. Roman numerals identify species (or specimens) of a genus in which there were notable differences in size. Specimens identified by lowercase letter are described in the following references: (a) Cope (1877, 1884), (b) Edinger (1929), (c) Thorpe (1937), (d) Sigogneau (1968), (e) described as *Amphicyon* in Edinger (1929) and *Cephalogale* in Gervais (1872). Specimen numbers for MH are page numbers in Piveteau (1958, 1961).

The assemblages discussed in this report were differentiated according to the statistically "significant" comparisons of the type described in Appendix II. For example, it was impossible to differentiate among subjects from the Eocene and Oligocene progressive orders with respect to *EQ*, and these were, therefore, combined as the Paleogene. Similarly, it was impossible to distinguish the artiodactyls from the perissodactyls of any epoch, and these were, therefore, combined as progressive ungulates. The following are brief descriptions of the assemblages.

1. Archaic ungulates. The 13 species in this group are of the extinct orders Condylarthra and Amblypoda. Our sample is from the mid-Paleocene to the late Eocene and dates from about 60 to 40 m.y. ago.

2. Archaic carnivores. This is the smallest sample and includes four mid-Eocene to Lower Oligocene species from the extinct order Creodonta, 50–35 m.y. ago.

3. Paleogene ungulates. Although the Paleogene technically begins with the Paleocene, the 22 species in this group range from the Lower Eocene to the end of the Oligocene (55–22.5 m.y. ago). Included are 11 Eocene and 11 Oligocene species; categorized by order, there are 10 perissodactyls and 12 artiodactyls. There are several extinct ("archaic," see p. 204) superfamilies, identified by their initials in Table 13.1. The remaining ungulate families still survive and include the ancestors of horses, rhinoceroses, tapirs, camels, and the deerlike groups (tragulids, cervids, bovids).

4. Paleogene carnivores. All 11 species of Carnivora in this group are from the Oligocene, 35–22.5 m.y. ago, and for statistical purposes they could be compared with Oligocene ungulates. The analysis was essentially the same whether the comparison groups covered all the Paleogene or were limited to the Oligocene. Our Paleogene carnivores include ancestors of the sabretooth as well as of living bears, dogs, cats, and mustelids.

5. Neogene ungulates. The Neogene (Miocene and Pliocene) sample of 13 species of perissodactyls and artiodactyls includes two species from the extinct artiodactyl superfamily, Merycoidodontoidea, labeled M in Table 13.1. Others were relatives of the familiar species. The period dates from 22.5–2.5 m.y. ago.

6. Neogene carnivores. With only six species represented, this group is unfortunately small, and more data could be assembled for these animals and for the archaic creodonts by developing museum materials more adequately. The sample includes ancestors of dogs, cats, and the glutton, 22.5–2.5 m.y. ago.

7. Recent ungulates. The 25 living species in this group were more or less matched in niche and body size to the fossil samples. The group is somewhat limited because only a few perissodactyls have survived to the present time (horses, rhinoceroses, and tapirs), and we have data on only five species. The remaining 20 species are artiodactyls.

8. Recent carnivores. Fifteen species of land carnivores (fissipedes) are represented, more or less matched in niche and body size to the fossils. The Recent samples of ungulates and carnivores were assembled from the literature, much of which has been summarized by Count (1947).

Brain:Body Relations

The data on brain size and body size presented in Table 13.1 are shown graphically in Fig. 13.8. Figure 13.8 presents the measurements of our 69 fossil species. Figure 13.9 presents the measurements of 40

Relative Brain Size

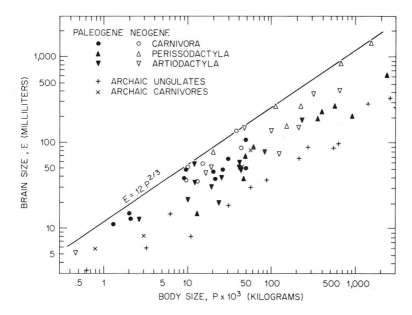

Fig. 13.8. Brain size (endocast volume) as a function of body size in 69 fossil ungulates and carnivores, logarithmic scale. The line is Eq. (3.6), the "average" for living mammals fitted to a large and diverse set of species. From Jerison (1970b).

living species. The diagonal in each figure is the allometric function giving the "expected" brain size for any given body size in an "average" living mammal (Chapter 2).

Two results are apparent. First, the similar orientations of the line and the arrays of points in both graphs are notable, especially in view of the independent origin of the lines and of all of the points in Fig. 13.8 and most of those in Fig. 13.9. The orientation, which is the graphic representation of the exponent of $\frac{2}{3}$ in Eq. (2.1), is found repeatedly in the analysis of brain:body data (von Bonin, 1937; Bauchot and Stephan, 1961; Jerison, 1969) and surely represents an important biological process at work. The line may be thought of as a "best fit" for the data in Fig. 13.9, although it is not a "least-squares fit" (Chapter 2, footnote 1, p. 45).

The second clear result apparent from Fig. 13.8 is that, with a single exception, the fossil data points fall below the line of Eq. (3.6). This is evidence for Lartet's (1868) principle, which was restated as one of Marsh's "laws": that the brains of Tertiary mammals were smaller than those of living mammals (Chapter 1, p. 15). But the "law" or "principle" has many exceptions (Edinger, 1962), which we can understand only by examining the variability or diversity of our samples.

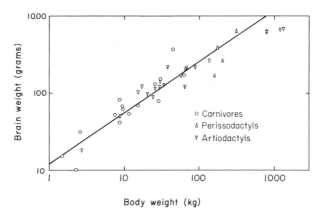

Fig. 13.9. Brain weight as a function of body weight in 40 living species of carnivores and ungulates, more or less matched in body size and niche to the fossil sample of Fig. 13.8. Data from Count (1947), Warnke (1908), and Weber (1896). Line is Eq. (3.6), $E = 0.12\ P^{2/3}$, and was derived independently of this sample.

Diversity of Encephalization Quotients, EQ

By casting the EQ data on relative brain size of each assemblage into a cumulative frequency distribution, we are able to analyze changing diversities of the assemblages. Let me review the method (see Chapter 3) with our smallest sample, the archaic carnivores. The four specimens had EQ's of 0.33, 0.43, 0.48, and 0.55, respectively. These were treated as representing midpoints of equal portions of the range which, in this case, involved successive quartiles. The specimens were, therefore, assigned percentile scores of 12.5, 37.5, 62.5, and 87.5. The graph of these percentiles as functions of EQ is the cumulative frequency curve for the archaic carnivores and is one of the cumulative frequency curves for our groups (Fig. 13.10).

A number of interesting comparisons can be made among the groups when the data are viewed in this way. The continuing increase in relative brain size as measured by EQ is clearly shown, and the geologically later groups appear to be more variable (diversified) than the earlier ones. There is also an interesting relationship suggested between the carnivores and ungulates. At any given time the carnivores were generally larger brained than their ungulate contemporaries, and this difference was more or less maintained as the two groups evolved. It would be reasonable to consider that the carnivores and the herbivorous ungulates were in a kind of feedback condition. At the beginning of this process, the selection pressures exerted by larger-brained archaic carnivores eventually gave an evolutionary advantage to the larger-brained, progressive ungulates over their

Relative Brain Size

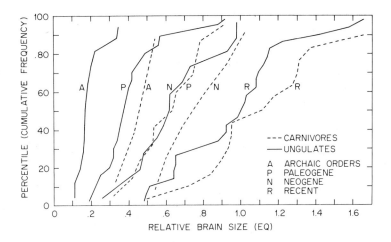

Fig. 13.10. Cumulative frequency distributions of encephalization quotients in fossil mammals of Table 13.1 and Recent samples of Fig. 13.9.

archaic contemporaries in their group of niches. With the emergence of the Paleogene progressive ungulates, the larger-brained, progressive carnivores had an advantage over their contemporary small-brained archaic carnivores (Andrew, 1962).

There has been a natural experiment testing the hypothesis that this kind of feedback occurred. This was in the evolution of South American ungulates free of predation by either creodonts or true carnivores during most of their adaptive radiation. We examine that experiment in detail in Chapter 14.

Our present purpose is to examine the diversity that is indicated in Fig. 13.10 and consider its implications. In order to learn more about the underlying distributions of relative brain size in the populations from which our samples were taken, the data of Fig. 13.10 were regraphed on normal probability paper as shown in Fig. 13.11. For clarity the data on ungulates and carnivores are presented separately, in order to show the individual data points. If the underlying distributions are normal, the arrays of points in each sample will fall on a straight line, and departures from normality will appear as departures from linearity.

Inspection of Fig. 13.11 shows that the underlying distributions are approximately normal for all the groups, although there are extended right-hand tails in the two older ungulate samples and in the Recent carnivores. The solid lines in Fig. 13.11 are least-squares fits to the data between the sixteenth and eighty-fourth percentiles ($\pm 1\ \sigma$). These lines were used to determine the means and standard deviations as shown in Table 13.2.

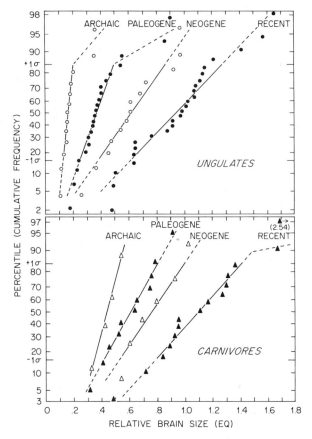

Fig. 13.11. Cumulative frequency distributions of relative brain size as measured by the encephalization quotient (EQ) in fossil and Recent ungulates and carnivores (plotted on probability paper). Lines were fitted within the range of ±1 standard deviation (σ), and the dashed extensions indicate representativeness of normal distributions for extreme cases. From Jerison (1970b).

STATISTICAL ANALYSIS

Since the data for the various assemblages overlap, and since there are sufficient data to enable us to undertake a statistical evaluation of the differences among the assemblages, it is appropriate to introduce statistical procedures at this point. (In the results considered previously the main issue was whether or not sets of data overlapped one another, and inferences could always be based on the presence or absence of overlap.) The details of the analysis are presented in Appendix II as part of the statistical analysis of all our quantitative data on mammals with the exception

Relative Brain Size

Table 13.2
Estimates of Relative Brain Size Parameters[a]

	Parameter	Ungulates	Carnivores
Archaic	$\overline{X}$	0.18	0.44
	σ	0.05	0.09
	n	13(9)	4(4)
Paleogene	$\overline{X}$	0.38	0.61
	σ	0.12	0.19
	n	26(16)	11(9)
Neogene	$\overline{X}$	0.63	0.76
	σ	0.24	0.23
	n	13(9)	6(4)
Recent	$\overline{X}$	0.95	1.10
	σ	0.33	0.31
	n	25(17)	15(11)

[a] n is sample size; number in parentheses is sample size used for fitting the lines in Fig. 13.11 from which mean $\overline{X}$ and standard deviation σ were estimated, cf. Appendix II, Table II.1.

of those described in Chapter 15. It is a simple analysis of variance of the *EQ* data, followed by a new modification of the Duncan Multiple Range Test (Dixon and Massey, 1969). I restrict my review, here, to the results of the statistical analysis, indicating when differences apparent among the assemblages should be thought of as reflecting real differences and when they should be considered artifacts of sampling.

A philosophical issue is involved in this interpretation. Most readers familiar with statistical tests and the use of statistical inference in a general way are likely to believe that a difference should be significant at "the 0.05 level" or some comparable level before it should be considered real. Such a belief can be misleading, however, especially in cases such as those of the present data. Involved is the issue of risks of errors of falsely rejecting a true difference (type II errors, in the statistician's jargon) as opposed to the risks of falsely accepting two sets of data as different ("rejecting the null hypothesis") when they are in fact not different (type I error). In many experiments one evaluates the risk of type I errors as being much greater than those of type II errors and seeks to minimize the former risks (for example, by using a probability criterion of 0.05 of making an error) without too much concern about the latter risk. In our case, however, we are equally concerned with either error and should seek "significance levels" that equate the probability of either error. Simple methods for setting such rational significance levels have recently been developed (Hodges and Lehmann, 1968).

The analysis in Appendix II supports the conclusion, first, that there was an increase in average relative brain size among the assemblages of the carnivores and ungulates as they evolved. Second, it supports the conclusion that until recent times carnivores had larger brains than ungulates, so much so that a particular "generation" of carnivores had brains that were typically equal in relative size to those of the next "generation" of ungulates. The analysis also shows that in recent times this differential has been overcome by the ungulates. (It supports the "null hypothesis.") This is the conclusion suggested by the appearance of Figs. 13.9 and 13.10. There was probably a feedback interaction between these two major assemblages during the first half of the Tertiary, but they apparently achieved equality by the end of that period.

An important substantive result with respect to diversification is implied by the statistical analysis presented in Appendix II. The inhomogeneity of variance that is obvious in Fig. 13.11 and Fig. 13.12 disappears if a logarithmic transformation of EQ is used. This is also true for the various groups considered in other chapters. This means that the amount of diversification of relative brain size was approximately proportional to the average relative brain size in each group. Figure 11.5 may be reviewed in order to obtain a visual picture of what happened in the evolution of brain size. The mammalian succession and diversification would be represented in brain:body space as a set of overlapping minimum convex polygons, each similar in orientation and shape to that drawn about the archaic Tertiary ungulates. These would all be contained within the larger "mammalian" polygon. Where there was an increase in relative brain size, as in the groups discussed in this chapter, the successive groups would be represented by a set of upwardly displaced polygons. Where specialization involved little increase in brain size, as in the insectivores, the polygon of a living group might be near the lower edge of the mammalian polygon. Because of the logarithmic metric of brain:body space, EQ's of groups in upwardly displaced but similarly shaped polygons would be more diverse than those in lower polygons.

As the final statement in this overview of the statistical analysis, it is appropriate to describe the populations represented by our assemblages. These are biologically real populations in the sense that they were made up of living individuals, but they were populations in both time and space. The individuals contributing to a given assemblage may have lived millions of years apart; they are a sample representing species that may have been spread over several continents. It is the species that is the element in each population rather than an individual. Each species is characterized by what has been treated as a typical, or mean, value of EQ, although it was

Relative Brain Size

obtained from a single individual. With the species as the unit, we have been studying evolution above the species level (Rensch, 1959).

EVOLUTION OF DIVERSITY IN BRAINS

The most intriguing substantive finding is the changing diversity shown by the slopes of the lines fitted to the data of Fig. 13.11. To appreciate the changing diversity in brain size in the temporal succession of assemblages that have been sampled, we should reconstruct the populations that they represent. Using the parameters determined from the lines in Fig. 13.11 and presented in Table 13.2, a frequency distribution of EQ in each population could be determined as shown in Fig. 13.12. The normal curves are equal in area, and they may be viewed as probability distributions for

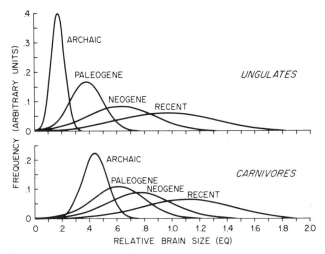

Fig. 13.12. Changing distributions of relative brain size as the brain evolved. Means and variances are based on fitted functions of Fig. 13.11. From Jerison (1970b).

EQ in the populations represented by the assemblages. These curves are our best guess about how the evolution of brain size in carnivores and ungulates actually occurred.

Two conclusions about diversity of brain size are inescapable. First, diversity evolved just as average size evolved. In the evolution of the mammal groups these evolutionary trends were correlated. Second, despite the evident general trend toward increase in average brain size, there is an interesting an important overlap in the region of low brain size which indicates that there were at least some small-brained species present at all times. The evolution of enlarged brains, though generally a route to success and survival of new species, was not universal even among progressive orders.

The evolution of diversity in relative brain size implies a diversity in behavioral adaptations. Some may have required little or no change in relative brain size, but many behaviors required changes. The extent to which relative brain size is different among species of an assemblage must be indicative of the diversity of niches occupied by the species of that assemblage. More specifically, it refers to diversity with respect to the presence or absence of selection pressures toward enlarged brains. The result shown in Fig. 13.12 is, therefore, evidence that as the carnivores and ungulates evolved they invaded more diverse niches, and correlated with this invasion was the evolution of more diverse brains.

The second conclusion from the analysis shown in Fig. 13.12 was that all the assemblages have included some small-brained species. By the logic just presented, this implies that there are and were niches in which selection pressures towards enlarged brains have been either minor or nonexistent. This does not mean that small-brained species are themselves survivors, or relict forms. Evolution in the niches they occupy could proceed at essentially the same rate as in other niches. It implies, merely, that survival and succession of species in those niches was largely determined by factors unrelated to relative brain size, and small-brained species tended to be replaced by other small-brained species. The data on living insectivores (Chapter 10), for example, suggest that they evolved in one such adaptive zone.

Issues on the Significance of Brain Size

Several important issues are resolved by the observation of diversification combined with some persistence of small-brained species. There is, first, the question of the uniformity of Lartet's principle or Marsh's "law" of increase in brain size in the Tertiary. Edinger (1962) discussed this problem and presented many examples of apparent contradictions, for example, in the reversal of relative brain size in the lineage of the rhinoceros. Her particular examples were *Teleoceras* of the Miocene (Table 13.1) and the living rhinoceros. The forms were similar in body size, but the latter have brains about half as large as in the known fossil specimen of *Teleoceras* (Scott and Osborn, 1890). The fossil had $EQ = 0.95$ and in the living rhinoceros, $EQ = 0.50$, approximately. The present analysis is consistent with this degree of diversity as shown by the overlapping curves of Fig. 13.12.

There remains a problem to guess at aspects of the niche of *Teleoceras* that would be associated with selection pressures toward brain enlargement. This is a speculative issue raised by our discussion of diversity of relative brain size in terms of the diversity of ecological niches. But, empirically, the datum on *Teleoceras,* like that on the living rhinoceros, is consistent with the distributions of Fig. 13.12. In other words, we expect some species of the Neogene to overlap other species of the Recent assemblages; specifically, we expect to have EQ's like those of *Teleoceras* and the living rhinoceros. If there were never such overlap, modern statistical approaches to evolution would be inappropriate.

Another issue is the apparent success of many small-brained species and small-brained orders of mammals. The rodents are usually thought of as comprising a progressive order in terms of the number of individuals, the number and diversity of species, and the rate of evolution of species, yet they are consistently small-brained today, with $EQ < 1.0$ in most species (Chapter 10). The present analysis merely acknowledges the fact as part of the empirical characterization of the set of niches occupied by most rodents. These niches, among other things, present little or no selection pressure toward increased brain size.

In contrast with my conclusion that there have always been niches for small-brained species, even in progressive orders such as ungulates and carnivores, Marsh (1886) had proposed in two "laws" that species destined for extinction had smaller brains than their contemporaries that were to survive (Chapter 1). In our samples there are data relevant for these propositions with respect to the Paleogene and Neogene ungulates because 12 of the species in these groups were members of "nonprogressive" superfamilies that became extinct in the Lower Pliocene or earlier. These may be compared with the remaining 23 species, which were from superfamilies that still survive.

The "nonprogressive" species within the progressive orders were labeled in Table 13.1 according to the initials of their superfamilies, A, B, C, E, and M (anoplotheres, brontotheres, cainotheres, entelodonts, and merycoidodonts). Inspection of the data permits us to see immediately that there is no basis for Marsh's position. There were small-brained and large-brained nonprogressive species within each group of ungulates, and their characterization as nonprogressive on the basis of survival is manifestly not correlated with the measure of relative brain size. In fact, among these particular families, the oreodonts (Merycoidodontoidea), which were lowest in relative brain size, also survived the longest as a family.

The oreodonts were actually unusually successful in many ways, being the most frequently found species in the great White River Oligocene fossil beds and relatively common in various Miocene beds. They were a successful group by almost all evolutionary criteria until their extinction.

Neither their success during the Oligocene or Miocene nor their eventual failure should be related to the size of their brains, which were characteristically small throughout their adaptive radiation. They were clearly a group that had found a set of niches in which relatively large brains were irrelevant for survival.

We must ask, finally, whether small-brained species also evolved with respect to brain size, that is, was there an increment beyond the very lowest mammalian level of EQ? The evidence, presented on insectivores in Chapter 10, is that there was some increase in relative brain size beyond the most basal level; no living species of mammals are as relatively small-brained as several of the archaic ungulate species described in Table 13.1 and in the figures. The smallest brained of living mammals is the Virginia opossum (*Didelphis marsupialis*). If one takes its typical brain size as 6.5 g and its body size as 3.5 kg (Weber, 1896), then $EQ = 0.24$. This is at the upper margin of the distribution of archaic ungulates as normalized in Fig. 13.11, and we note that it could fall within the fitted distribution in Fig. 13.12. It would be at the upper end, however, at about the eighty-fifth percentile. There are apparently no data on living mammals that would be near the lower bound of the range of EQ in the archaic ungulates.

CONCLUSIONS

The progressive evolution of the brain can almost be characterized by a single word, "diversification." The evolution of increased diversity from the minimum or basal mammalian level necessarily resulted in the evolution of increased average size and the appearance of many species with enlarged brains. At the same time other species were evolving skeletally, but their brains remained relatively small. The key factor is probably that the brain evolved in a way appropriate to behavior within a particular niche, and, as more diversified niches were invaded, more diversified brain adaptations were required. But the brain did not evolve in an exuberant way. It has been a "conservative" organ; even fissural patterns in the mammalian brain have been consistent enough over time within some major groups to permit a probably valid reconstruction of the evolution of major fissures (Dechaseaux, 1969; Edinger, 1948; Radinsky, 1968b, 1969, 1970; Sigogneau, 1968). The same is true of relative brain size, and in general one finds relative brain size among vertebrates to have been stable over long periods of time. Only occasionally have there been great changes beyond those required by the evolution of larger bodies, that is, changes reflected as increments in EQ.

Conclusions

Some apparently revolutionary changes occurred in the brain as it evolved, in particular during the Upper Eocene epoch and the surrounding times. These were the appearance of a flexured brain, of fissurization, and of neocorticalization. The importance of these changes should not be underestimated, but they can all be considered as correlates of a unitary set of selection pressures toward increased relative brain size. The net effect of these selection pressures was to displace upward a fundamental brain:body function.

Conservatism in brain size has characterized all the lower vertebrates. The basal (archaic) mammalian level was above the level of lower vertebrates, but, given that basal level as a minimum, many orders of mammals, including so-called primitive or conservative orders such as the insectivores, have advanced only slightly from that condition. Progressive orders such as carnivores and ungulates did evolve enlarged brains in many niches, and these also became more diversified with respect to the characteristic brain sizes of their species.

An unusual aspect of the conservatism of the brain as an evolving organ is the stability of its association with body size. Figures 13.8 and 13.9 are good examples of that stability because the graphed equation is derived from a broad range of species of living mammals and is completely independent of the points of our fossil and Recent samples plotted on the same coordinate system. Yet the parallel orientation of these two sets of data is obvious and impressive and is consistent with all of our other results.

We have yet to consider the evolution of the primate brain, however, and in particular the evolution of the human brain. Our judgment about the conservatism of brain evolution will have to be revised somewhat within that context, although even there we will find the principle of conservatism of brain evolution useful in understanding the changes that actually occurred within the various primate lineages, including the hominids.

Chapter 14

Neotropical Herbivores: An Evolutionary Experiment

Isolated during much of the Tertiary period, from the Paleocene to the Pliocene epochs, the South American fossil mammals participated in what Simpson (1965) has described as "an evolutionary experiment on a grand scale." The evolution of the brain in these animals was an aspect of that experiment, in which the Neotropical (South American) ungulates were "control" groups and the Holarctic (Northern Hemisphere) ungulates described in the preceding chapter were "experimental" groups. Among the selection pressures under which the "experimental" groups evolved were those resulting from their predation by progressive carnivores with enlarged brains. The Neotropical ungulates were "control" groups because their evolution never involved that selection pressure. Progressive carnivores that could prey on them did not reach South America until the Panamanian land bridge was reestablished at the end of the Tertiary, ending that continent's isolation. During the 50 m.y. of continental isolation the major Neotropical predators were marsupial carnivores whose brains, though unknown, must have been relatively small, like those of living marsupials. The experiment compares the effects of the presence and absence of progressive carnivore predators on the evolution of the brain in their ungulate prey.

Quantitative data are available on brain and body size in 20 Neotropical fossil ungulates. Nine specimens from the Oligocene and Eocene are grouped together as Paleogene species, which lived between 55 and 22.5 m.y. ago. The remaining 11 species include several Pleistocene specimens as well as others from the Pliocene and Miocene and are grouped together as a Neogene assemblage from the last 22.5 m.y. We will compare the distributions of relative brain size in these two groups and ask two questions: Did relative size increase and did the dispersions differ? These questions have already been answered for the Holarctic carnivores and ungulates. There was an increase in both relative size and in the diversity of relative size that continued throughout the Tertiary (Chapter 13).

EVOLUTIONARY BACKGROUND

The earliest Neotropical mammals evolved from the well-known, broadly ranging, Paleocene marsupials and archaic ungulates and from the spatially much more restricted and less well-known edentates. Of the marsupial radiation one need only note that living marsupials are all relatively small-brained, and there is no reason to suppose that ancestral forms differed in that respect. The archaic Holarctic ungulates are also known to have been relatively small-brained, with brain sizes generally between 10–25% of those in comparable living mammals (Chapter 11).

Species of the archaic ungulate orders, Condylarthra and Amblypoda, reached South America prior to the isolation of that continent, when there was a land bridge to North America, probably in the early Paleocene (Simpson, 1965). They radiated extensively, after the land bridge disappeared, to evolve into at least three additional ungulate orders: Notoungulata, Litopterna, and Astrapotheria. Furthermore, some condylarths persisted in South America long after they became extinct elsewhere. Most of our specimens are notoungulates, although we have data on one astrapothere and two litopterns. Other large herbivores that we consider, briefly, are one Miocene and four Pleistocene edentates. An evolutionary tree showing the probable relationships among most of the species discussed in this chapter is presented in Fig. 14.1.

The complete picture of the radiation of mammals in South America as summarized by Simpson (1965) included a minor invasion by "island-hopping," progressive carnivores, small relatives of the raccoon, in mid-Neogene times, but this probably involved no special selection pressures on the ungulate groups. At the close of the Pliocene and throughout the Pleistocene, following the reestablishment of a land bridge at the isthmus of Panama, diverse species of Holarctic carnivores and ungulates were able to interact significantly with their counterparts in South America, and in most instances the result was the extinction of the Neotropical fauna and its replacement by Holarctic invaders.

BODIES AND BRAINS

The extensive adaptive radiation of the Neotropical ungulates is known from the diversity of their skeletal adaptations. Elephantlike, horselike, rabbitlike, their forms as mirrored in their skeletons are reminiscent of the familiar shapes of various groups of living mammals, even though the South American fossils are entirely unrelated species. A few reconstructions were presented in Fig. 14.1, and more may be seen in Barnett (1960) and especially in Scott (1937). For our purposes the im-

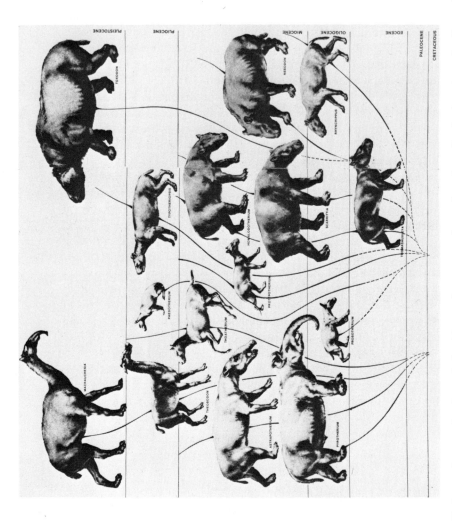

Fig. 14.1. Phylogenetic tree showing relationships and reconstructions of various South American fossil ungulates, including some of the species represented in Table 14.1. Drawings by Jean and Rudolph Zallinger, from Life Magazine (Jan. 26, 1959).

portant point is that there was considerable diversity in body size and shape, a situation that contrasts with that of the brain.

The 20 endocasts that are the basic ungulate data in this chapter are unusual and quite different from those of other mammals. If a random sample of a few of these were placed among a larger sample of Holarctic endocasts of comparable age, even a relatively untutored observer would sort the endocasts of most of the Neotropical forms into a different group from the Holarctic forms. In some instances they are somewhat like the endocasts of the archaic Tertiary mammals (Chapter 11), or of the more primitive or older specimens of the "progressive" orders (Simpson, 1933a). A few of these Neotropical, ungulate endocasts (Figs. 14.2 and 14.3) illustrate these facts. In most instances the endocasts have atypical patterns of fissurization or almost no fissurization. They were almost all relatively less neocorticalized than progressive late Eocene Holarctic ungulates, with dorsally displaced rhinal fissures and laterally expanded pyriform lobes. There is no clue about functional localization in the brains of these animals, beyond the recognition of the placement of the rhinal fissure, of the relatively large size of paleocortical structures, such as the pyriform lobes, and of the size of the pituitary in the larger species. The best approach to the evolution of the brains of these animals is through the quantitative study of relative brain size.

THE EXPERIMENT

The South American Tertiary ungulates are treated separately from the other ungulates that were their contemporaries elsewhere in the world in order to emphasize the fact that an evolutionary experiment was, in fact, in progress as these various ungulate groups evolved. The experiment can be described in classic form as a 2 × 2 design as follows.

	Role of Progressive Carnivores	
	Present	Absent
Paleogene	A	C
Neogene	B	D

Groups A and B have already been described in the previous chapter. These were the Paleogene and Neogene Holarctic ungulates. Groups C and D are described in this chapter as Paleogene and Neogene Neotropical ungulates. I stretch a point slightly by including a few Pleistocene ungulates in group D. This is actually not much of a concession; invertebrate paleontologists frequently define the Neogene as including the Pleistocene,

14. Neotropical Herbivores

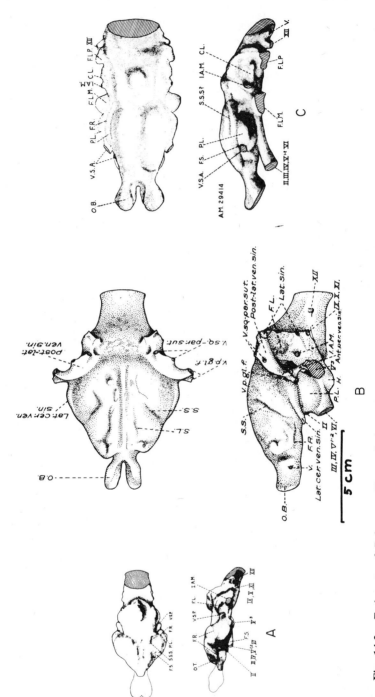

Fig. 14.2. Endocasts of Paleogene Neotropical ungulates. A, *Notostylops pendens*, AMNH 28614, from Simpson (1933a); B, *Rhynchippus equinus*, FMNH P13410, from Patterson (1937); C, *Rhyphodon* sp., AMNH 29414, from Simpson (1933a). Labels as in Figs. 4.4 and 14.3.

The Experiment

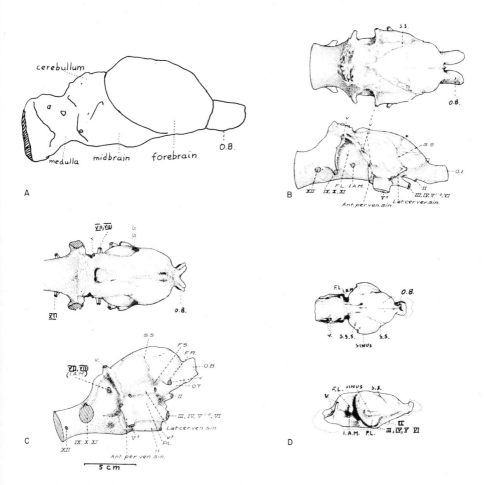

Fig. 14.3. Endocasts of Neogene (including Pleistocene) Neotropical ungulates. A, *Astrapotherium magnum,* FMNH P14259 (lateral view only); B, *Homalodotherium cunninghami,* FMNH P13092, from Patterson (1937); C, *Nesodon imbricatus,* FMNH P13076, from Patterson (1937); D, *Typotheriopsis internum,* FMNH P14420, from Patterson (1937). Labels: S.S. sylvian sulcus; Ant. per. ven. sin. and Lat. cer. ver. sin, sinuses. Others as in Fig. 4.4.

as do European vertebrate paleontologists. Furthermore, the forms that are included are *Mesotherium* (*Typotherium*) and *Toxodon,* both of which are also known from Pliocene strata and both of which suffered extinction, presumably following competition with Holarctic ungulates that invaded South America, or as a result of their inability to survive predation by progressive Holarctic predators, after the reestablishment of the Panamanian land bridge.

Methods of estimating brain and body size were the same used previously, and the method of analysis is the same. The statistical analysis on the present data is included in Appendix II, although the statistics applied to the "experiment" are discussed in this chapter.

Relative Brain Size

The basic data on Neotropical ungulates are summarized in Table 14.1, which presents brain and body size estimates, the encephalization quotients

Table 14.1
Brain:Body Relations in Extinct South American Ungulates

Genus[a]	Endocast volume (ml)	Body weight (kg)	EQ
Paleogene			
Leontinia FMNH P13285	390	450	0.55
Notostylops I MNBA 10506	17	5.9	0.43
Notostylops II AMNH 28614	9.7	3.5	0.35
Oldfieldthomasia AMNH 28780	13	3.1	0.51
Proadinotherium FMNH P13590	110	88	0.48
Rhynchippus FMNH P13410	110	32	0.91
Rhyphodon AMNH 29414	41	70	0.20
Trachytherus FMNH P13281	60	74	0.28
Protheosodon FMNH P13418	55	32	0.46
Neogene			
Astrapotherium FMNH P14259	510	610	0.59
Proterotherium AMNH 9245	57	14	0.82
Adinotherium FMNH P13110	90	88	0.38
Hegetotherium AMNH 9223	23	5.1	0.65
Homalodotherium FMNH P13092	210	400	0.32

The Experiment

Table 14.1—(cont.)

Genus[a]	Endocast volume (ml)	Body weight (kg)	EQ
Mesotherium MH (127)[b]	71	110	0.26
Miocochilius UC 39651	27	9.3	0.51
Nesodon FMNH P13076	220	290	0.42
Protypotherium AMNH 9246	12	3.0	0.48
Toxodon MH (124)[b]	570	1100	0.45
Typotheriopsis FMNH P14420	80	60	0.43

[a] Specimens are identified by museum number as follows: FMNH, Field Museum of Natural History (Chicago); MNBA, Museo Nacional de Historia Natural (Buenos Aires); AMNH, American Museum of Natural History (New York); MH, Muséum d'Histoire Naturelle (Paris); UC, University of California (Berkeley). Endocasts described by Dechaseaux (1958), Patterson (1937), Simpson (1932, 1933a,b), and Stirton (1953) (cf. Figs. 14.2 and 14.3). Bodies described by Riggs (1935, 1937) and other sources [Piveteau (1958)].

[b] Page number in Piveteau (1958), for identification at MH.

corresponding to those sizes, and other identifying data on all the specimens. These data are presented separately for the Paleogene and Neogene fossils, and the brain:body size relationships are illustrated in Fig. 14.4 on logarithmic coordinate axes, as part of an allometric analysis. Data on five Neogene edentates, to be discussed briefly later, are included in Fig. 14.4. The solid line is Eq. (3.6), which has been used consistently as a reference line to estimate "expected brain size" of living mammals and in the computation of *EQ*. The dashed line is a least-squares fitted line for the South American ungulates, Eq. (14.1).

Three issues that have been raised previously are nicely illustrated by Fig. 14.4. First, the array of points is obviously parallel to the line of living mammals. As in the Holarctic fauna discussed in the preceding chapter, there is evidence of a fundamental constant of brain:body relations in the Neotropical mammals, which is represented by the slope of $\frac{2}{3}$. Second, like the Holarctic assemblage these fossils were uniformly below the level of brain size of the living fauna, and this further confirms observations such as Lartet's (1868) and Marsh's (1874) that Tertiary mammals were relatively small-brained.

The third important issue is the goodness of fit of the brain:body equation, that is, how closely associated the points in Fig. 14.4 are to a

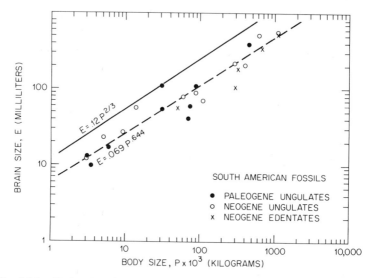

Fig. 14.4. Endocast:body relations in South American (Neotropical) fossil ungulates and edentates. In addition to Eq. (3.6), a "best-fitting" line for the fossil ungulates is drawn, illustrating the parallel orientation of the lines.

line that could be fitted to them. The relationship is conventionally evaluated by determining a product-moment correlation coefficient r for the logarithms of brain and body size, and for these data $r = 0.96$. This means that 92% of the variance in log brain size may be predicted from log body size. The high correlation is typical of brain:body data and is one reason for using Eq. (3.6) (Chapter 3) as an estimator of "expected" brain size in living mammals. It signifies that if one knows only the fact that a body size is of a mammal, one has enough information to make a very reasonable guess about the size of the brain of that mammal.

At various points in this text I have calculated the parameter of a line of best fit by the objective method of least-squares, and since this is a methodological exercise important for "experimental" analyses, I have done it again here, with errors assumed in both variables (Burrington and May, 1970). The equation of the least-squares line for the Neotropical assemblage is:

$$E_n = 0.068 \ P^{0.644} \tag{14.1}$$

E_n is the brain size in milliliters, and P is body size in grams or milliliters. It is apparent that the exponent is approximately $\frac{2}{3}$, and I prefer to write the equation with that exponent because it is appropriate for a surface:volume relationship. Adjusting the multiplier to place the best-fitting

The Experiment

line through the centroid of the points in Fig. 14.4 (the geometric means of E_n and P), the equation becomes:

$$E_n = 0.053 \; P^{2/3} \tag{14.2}$$

Equation (14.1) is shown as the dashed line in Fig. 14.4, to represent the South American forms, although we use Eq. (14.2) to determine a typical EQ for that assemblage. The two are almost indistinguishable graphically (as the reader can verify), and the graph of Eq. (14.2) is omitted to avoid cluttering Fig. 14.4 more than is necessary.

The definition of the encephalization quotient EQ, as presented in Chapter 3, Eq. (3.7) and applied to these Neotropical ungulates, would give an average EQ for them, which is designated EQ_n. Thus,

$$EQ_n = \frac{0.053 \; P^{2/3}}{0.12 \; P^{2/3}}$$

$$EQ_n = \frac{0.053}{0.12}$$

$$EQ_n = 0.44 \tag{14.3}$$

The assemblage, therefore, had brains that were about half the size of those of living mammals.

DIVERSITY IN RELATIVE BRAIN SIZE

Another interesting result on relative brain size in the Holarctic faunas was with respect to changing diversity as evolution proceeded. To evaluate diversity in the South American ungulates we, first, construct cumulative frequency curves for the EQ's in the Paleogene and Neogene assemblages given in Table 14.1. It is apparent from inspection of Fig. 14.5, which is a graph of these frequency distributions, that the groups were the same with respect to both means and variances. There was no increased diversification as this fauna evolved.

We may compare the Neotropical ungulate assemblage with the Holarctic ungulates and carnivores described in the preceding chapter by constructing a single cumulative frequency distribution for Neotropical fauna and graphing that on normal probability paper. If this fauna is like its Holarctic relatives, we may expect the points to be aligned reasonably linearly on such paper, which would be evidence for an underlying normality of the distribution of EQ. We will then be able to estimate the mean and standard deviation from a fitted line.

The data are graphed with a normal probability scale as the ordinate in Fig. 14.6, which also includes a summary of the results of the earlier

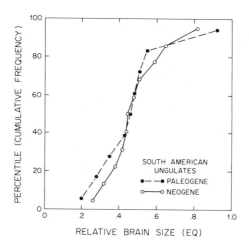

Fig. 14.5. Cumulative frequencies of relative brain size in the Paleogene and Neogene Neotropical fossil samples, illustrating absence of change between these groups (cf. Fig. 13.10, curves for Paleogene and Neogene Holarctic ungulates).

graphic analysis (Fig. 13.11) that had been performed with the Holarctic data. All the lines were fitted by least squares only in the range of $\pm 1\ \sigma$, in order to emphasize central effects and limit the role of outliers. The heavy line in Fig. 14.6 is fitted to the Neotropical data. It is evident, by

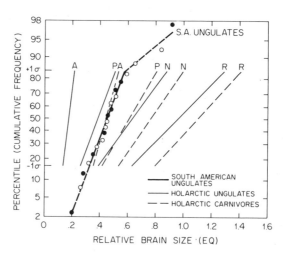

Fig. 14.6. Cumulative frequency curve for entire sample of Neotropical ungulates, compared with basic data on Holarctic ungulates and carnivores. A, Archaic; P, Paleogene; N, Neogene; R, Recent. Filled circles for Paleogene; open circles for Neogene. See Appendix II for statistical analysis.

The Experiment

inspection, that the Neotropical Paleogene and Neogene ungulates were comparable to the Holarctic Paleogene ungulate assemblage in diversity and were almost identical with the Holarctic archaic carnivores in that respect.

Several parameters can be estimated from Fig. 14.6. The median EQ determined from this graphic analysis was 0.46, essentially the same as the mean of 0.44 derived by Eq. (14.3). There are outliers at the upper extreme of the South American assemblage, just as there were in several of the Holarctic groups, suggesting the presence of species that were "experimenting" with enlarged brains. The standard deviation determined from the best-fitting line was 0.14, which is similar to that of the archaic carnivores and Paleogene ungulates of the Holarctic (Table 13.2).

Evaluation of the "Experiment": Statistical and Graphic

A conventional statistical analysis of the evolutionary "experiment" of this chapter is part of the overall analysis presented in Appendix II. That analysis was performed on the logarithms of the EQ data of the various groups described in this book, and mean values relevant to the "experiment" (antilogs of the means of log EQ) were almost the same as those derived from cumulative frequency distributions of the type shown in Fig. 14.6 and presented in Table 14.2. The data in Appendix II are from complete samples in each assemblage. The analysis in Table 14.2 is based only on points lying between ± 1 σ in a cumulative frequency graph drawn on probability paper with the "mean" read from the fiftieth percentile value

Table 14.2
Statistics on Relative Brain Size (EQ) in Neotropical and Holarctic Ungulates[a]

	Parameter[b]	Neotropical ungulates	Holarctic ungulates
Archaic	Mean	—	0.18
	σ	—	0.05
	n	—	13
Paleogene	Mean	0.44	0.38
	σ	0.14	0.12
	n	9	22
Neogene	Mean	0.47	0.63
	σ	0.13	0.24
	n	11	13

[a] Statistics based on cumulative probability graphs and lines fitted between sixteenth and eighty-fourth percentile (± 1 σ).

[b] σ, standard deviation; n, sample size (cf. Table 13.2 and Appendix II, Table II.1).

and the standard deviation taken from the sixteenth or eighty-fourth percentile value.

The multiple range test used with the analysis of variance presented in Appendix II, using the "significance level" of 0.05, indicated that the Paleogene and Neogene Holarctic ungulates were in different groups. It was possible to construct a group in which either of the Holarctic assemblages could be joined by the Neotropical fossil ungulates of South America. According to this analysis we would, therefore, accept a division of the groups in which the Neotropical fauna was intermediate between the two Holarctic groups. From the actual values of the mean EQ's in Table 14.2, we would consider the Neotropical fauna to be undifferentiated with respect to geological strata: the Paleogene and Neogene Neotropical assemblages can be considered as two samples from the same population. We would, further, consider the Neotropical group as slightly higher in EQ than the Paleogene Holarctic ungulates and considerably lower than the Neogene Holarctic ungulates.

If we wish to be equally wary of both types of errors inherent in statistical decision making, that is, of errors of commission and omission (type I and type II errors) as discussed in the previous chapter, we may specify confidence limits other than 95%. To repeat the argument, in this "experiment" we really wish to equate the risk of calling assemblages equal in EQ when they are in fact unequal (type II error) with the risk of calling them unequal when they are in fact equal (type I error). By using the 0.05 level we are stating, however, that we are willing to accept no more than a 5% risk of a type I error, that is, we want odds of 19 to 1 in favor of a difference between two groups before calling them different, and are biasing ourselves heavily in favor of a decision of "no difference."

Following the argument of Hodges and Lehmann (1968), we might do well to seek a significance level in which we equate the risks of type I and type II errors. To choose that level rationally we must state how large a difference there would have to be between two assemblages before we would be concerned with failures to detect it. If we consider the standard deviation of the more diverse of two samples being compared as a meaningful difference, then with our sample sizes we should take the 0.20 "level of significance" for statistical tests. The tests used are t-tests.

Following that procedure, we reach the decision that the two South American assemblages were not significantly different from one another and that they were intermediate between and significantly different from both the Paleogene and Neogene Holarctic ungulates. Their position in the evolution of relative brain size is, therefore, accurately represented by

The Experiment

the ungulate lines in Fig. 14.6, and these lines can be considered to represent true differences among the groups.

A final graphic analysis is presented in Fig. 14.7 to enable one to picture clearly the position of the South American ungulates relative to

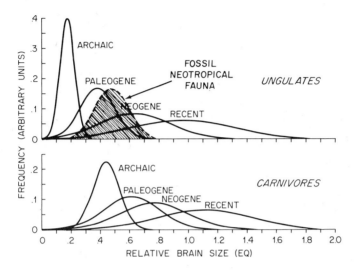

Fig. 14.7. Evolution of relative brain size in ungulates and carnivores, indicating the changing diversity with the advance of time. The isolated South American fauna is comparable to Archaic carnivores and Paleogene Holarctic ungulates in diversity, though somewhat superior in average brain size to the Holarctic ungulates.

the Holarctic fauna. In this figure the normal curves derived from the analysis of Fig. 14.6 are shown; Fig. 14.7 is essentially the first derivative of Fig. 14.6. The line that indicated a mean equal to 0.46 and σ equal to 0.14, the parameters for the South American ungulates derived from Fig. 14.6, was used to construct the shaded normal curve in Fig. 14.7. Had we used the parameters (including outliers) from Appendix II, we would have had a mean $= 0.45$; $\sigma = 0.16$.

It is apparent that the South American fossil ungulates formed an assemblage that was relatively uniform compared to Neogene and Recent ungulate groups and slightly more diverse and slightly in advance of the Paleogene Holarctic ungulates in mean brain size. The Neotropical ungulates had increased significantly in mean relative brain size beyond their immediate ancestors among the archaic ungulates, and they were also more diverse than that ancestral stock. Some diversification, therefore, did occur early in the adaptive radiation of the South American ungulates. But there was no further diversification comparable to that in the Holarctic faunas, and the South American assemblage remained at the Paleogene level of brain size and diversity through over 50 m.y. of their Tertiary radiation.

Fossil Edentates

If the South American radiation of mammals is to be used as a "control" for the effect of progressive carnivores on the evolution of the brain in their Holarctic prey, we must consider at least briefly the status of the edentates. The perissodactyls and artiodactyls considered in the preceding chapter were the major large herbivore prey of the Holarctic zone. (The Proboscidea could have been considered with them, although the only specimen available for such an analysis would have been the Upper Eocene *Moeritherium*. Other proboscid endocasts are all Pleistocene and essentially the same as in living elephants; see Chapter 15.) The large herbivorous mammals of South America included not only the descendants of the archaic ungulates but the entirely separate stock represented by the ground sloths and armadillos.

There are remains of these groups directly suitable for this analysis only from the Miocene ground sloth *Hapalops*. Later ground sloths known from the Upper Pliocene are represented in our assemblage by North American Pleistocene specimens considered to be the results of an invasion from South America (Simpson, 1965). We also have data on one Upper Pliocene armadillo, *Plaina*.

The data on these specimens are summarized in Table 14.3; the points giving the brain and body data were included in Fig. 14.4 but not in the derivation of Eq. (14.3) or the drawing of that line. Had they been included, there would have been almost no change in any of the parameters that were estimated. The correlation between log brain and log body size

Table 14.3
Endocast and Body Data in Fossil Edentates

Species[a]	Endocast volume (ml)	Body size (kg)	EQ
Hapalops	57	49	0.35
Megalonyx	340	370	0.37
Paramylodon harlani	550	1100	0.43
Nothrotherium shastense	280	320	0.50
Plaina FMNH P14424	110	290	0.21

[a] *Plaina* from Field Museum of Natural History. Endocasts of other specimens are described by Dechaseaux (1958), Edinger (1929), and Woodward (1898). Body size estimates from illustrations in Piveteau (1958), Scott (1937), and Stock (1956).

remains at 0.96. The slope of the "best-fitting" line as determined by least squares is still 0.64, but the centroid is somewhat lower. If a line is drawn with a slope of $\frac{2}{3}$ through the centroid, the allometric equation becomes

$$E_n = 0.051 \ P^{\frac{2}{3}}$$

It is apparent that the fossil edentates could have been grouped with the Neotropical fossil ungulates without changing the present analysis. This is also true for the analysis of diversity. If the edentates were added to the curve-fitting analysis of Fig. 14.6, the mean EQ derived from the fiftieth percentile of the redrawn line would be 0.44 instead of 0.46, and the slope would be slightly more steep: σ equal to 0.12 instead of 0.14. These are very small differences that have no effect on the interpretations that can be offered of these results. One may, therefore, consider the conclusions with respect to the Neotropical ungulates as extendable to all larger herbivores of that continent, although in the discussion that follows references are restricted to the ungulates because, after all, these are much better known with respect to relative brain size.

EXPERIMENTAL CONCLUSIONS

The most striking result of this analysis is the discovery that a major faunal assemblage in South America went through a significant adaptive radiation without any notable changes in relative brain size during the 50 m.y. span of that radiation. This is apparent in the data of Fig. 14.5, in which the distributions of relative brain size in the Paleogene and Neogene ungulate assemblages were practically indistinguishable.

Although an important finding, this should not completely overshadow the real increase in relative brain size that must have occurred early in the evolution of the South American forms. These ungulates were derived from Holarctic archaic ungulates (Simpson, 1948). The initial radiation in which archaic species evolved into something different was clearly accompanied by significant increments and diversification in relative brain size, which may be seen by comparing the archaic ungulates to the South American forms (Figs. 14.6 and 14.7). Once the South American orders were established, however, they remained remarkably stable with respect to relative brain size, though in other respects the radiation of species and their progression continued at normal rates (Simpson, 1965).

If the Neotropical assemblage is to serve as a genuine control group to test the effect of the presence of advanced predators on brain evolution in the experimental group of Holarctic ungulates, it is important that we have evidence that the South American group could evolve larger brains.

Thus, the difference between the archaic ungulates and the Neotropical ungulates is an important one which establishes the genetic potential of the latter group. At some point in their evolution they could and did evolve larger brains, and this resulted in their achieving approximately the same relative brain size as did the Paleogene Holarctic ungulates.

We must now consider two problems. First, why was there any enlargement in relative brain size in the Neotropical ungulates if they were simply descendants of small-brained condylarths? Second, given that there was an early expansion of the brain in the Eocene Neotropical ungulates, why did it not continue as it did in the Holarctic ungulates?

Early Enlargement of the Brain

Although the answer must be speculative, the following speculations seem reasonable, and, more importantly, they can be tested with a relatively small effort on the part of a few museums around the world. There would be at least two kinds of selection pressures on the descendants of the condylarths to make it advantageous for species to have enlarged brains. As a first possibility, there is the very fact of geographic dispersion, which suggests that the species invading the Neotropical zone before the isolation of the southern continent may have been those that could find alternate solutions to the problems of competition with successful ungulate species in a limited home range. They would tend to be wider ranging species with behavioral (and hence brain) adaptations suitable for a more varied set of environments. I do not regard this as a particularly strong argument, but it is a possible true history.

A second possibility is more attractive to me not because it is intrinsically superior, but because it makes specific predictions for future analysis. This is that the Neotropical ungulates, like their Holarctic relatives among the archaic ungulates, were under selection pressures from predacious carnivorous species. In the latter case I assumed that these species were the larger creodonts, which apparently did not reach the Neotropical zone. In the case of the Neotropical herbivores, in particular the ungulates, it does not seem too farfetched to assume that some predacious carnivorous marsupials appeared early enough to fill a creodontlike adaptive zone with respect to brain evolution. Their selection pressures on the larger Eocene Neotropical herbivores would have been analogous to those exerted by the creodonts on the archaic ungulates.

The Holarctic Carnivore–Herbivore Balance

Let us reconstruct a possible history of brain size in carnivores and ungulates, beginning with the archaic ungulate assemblage described in Chapter 11, the most homogeneous population (smallest σ) in Fig. 14.7.

Experimental Conclusions

This assemblage was made up mainly of large herbivores but included several arctocyonid species that are usually considered to have been at least partly carnivorous, although perhaps carrion feeders rather than active predators on their contemporaries (Russell, 1964). They had been considered by taxonomists as species of archaic carnivores, the Creodonta, and only in the past decade has it been recognized that they are more properly included among the archaic ungulates and are properly members of the order Condylarthra (Romer, 1966). From Fig. 14.7 it is apparent that all the archaic ungulates were similar in relative brain size, and, despite the fact that they included both carnivorous and herbivorous species, the probable differentiation in habits was not reflected by the specializations in relative brain size that characterized the assemblages that succeeded them.

It was this group that we considered to have been subjected to the unusual selection pressures of the larger creodonts, such as *Oxyaena*, in the late Paleocene and Lower Eocene, and it was partly for this reason that we considered the more progressive ungulates to have had something of a selective advantage, since they had the genetic capacity (however that may be defined) to evolve larger brains—presumably associated with more efficient behavior patterns. There is no information on brain size in the earliest creodonts; those of Middle Eocene times were much larger brained than the arctocyonids, though they were relatively small-brained "archaic carnivores" compared to species of the order Carnivora (Table 13.1). If the Paleocene creodonts were comparable to their Middle Eocene successors in brain size, one may guess that they would have entered rapidly into the perhaps unoccupied niche of active predators rather than scavengers on the archaic ungulates.

Body size was important. If their habits were appropriate, some of the arctocyonids were certainly large enough to be dangerous predators upon the large herbivorous archaic ungulates of their time. Could they have had the habit patterns appropriate to the predator's role, or would they have been strictly limited to the role of carrion feeders, relying on disease or old age to kill their "prey" and to provide them with a meat diet? Their small brains suggest the second role, and, if they were not active predators, they would not have put selection pressures on their herbivorous contemporaries to develop enlarged brains necessary to support appropriate defensive habits.

The earliest creodonts that were large enough in body to provide the selection pressures of a predacious species appeared in late Paleocene times and flourished in Lower and Middle Eocene times. These were probably intermediate in relative brain size between the archaic ungulate level and the Paleogene carnivore level indicated in Fig. 14.7, and it seems likely that they were in the range of the archaic carnivores indicated in that figure.

The true carnivores of those times, the miacids, were small mustelid-like forms that probably could not prey on the ungulates of the size of most of those included in our archaic ungulate assemblage. (They could have preyed on at least two of the smaller archaic ungulates, *Hyopsodus* and *Meniscotherium,* which were the size of squirrels or hares, and it is interesting that these two ungulates were the ones that were "experimenting" with enlarged brains, having $EQ = 0.36$ and 0.37, respectively.) The predacious niches were probably invaded by the creodonts, and this could have been part of the source of a selection pressure on the larger archaic ungulates to evolve appropriate defensive adaptations, including new habit patterns and appropriate increments in brain size associated with such patterns.

In the Holarctic zone the emergence of the progressive carnivores, indicated as the Paleogene carnivores in Fig. 14.7, could have been a crucial element in the further evolution of the Holarctic ungulate and carnivore brains. There is, thus, a rationale for the leapfrogging of brain size, in which the Paleogene carnivores were the source of the selection pressure on their contemporary ungulates, eventually forcing them to evolve as the large-brained Neogene ungulates. The Neogene ungulates would have, in turn, produced a selection pressure on the surviving Paleogene carnivores to achieve the Neogene level of carnivore brain size, in order to be successful predators on the more advanced ungulates. When the level of living carnivores and ungulates was reached, we have seen that the orders became approximately equal in relative brain size.

NEOTROPICAL CARNIVORE–HERBIVORE BALANCE

Since neither the creodonts nor the progressive carnivore orders are known to have reached South America early in the Tertiary, when the Neotropical ungulates achieved their basic level of relative brain size, some other selection pressure must have been acting on these early descendants of the condylarths. If, as I suggested earlier, there were selection pressures from predacious marsupial carnivores that had reached South America before the isolation of that continent in the Paleocene or early Eocene, there are some interesting predictions for future research. At least one carnivorous marsupial species is known that could have been large enough to prey on phenocodontid-sized ungulates; this was the "lion-sized" *Arminiheringia* of the Upper Eocene (Scott, 1937).

It is natural to suppose that selection pressures by the marsupial predacious carnivores of the Neotropical zone would have been approximately equivalent to those of the creodonts in the Holarctic. In fact, from Fig. 14.6, one might even consider that the selection pressures were slightly stronger because the Neotropical ungulates were slightly but significantly (according to statistical analysis) larger brained than the Holarctic Paleogene ungulates. One would, therefore, expect the predacious marsupials of South America to have relative brain sizes that were about as great as or slightly greater than those of the creodonts. The data to test this hypothesis are available in many specimens lying in museum drawers, awaiting the preparator's skill and patience and the scientist's initiative.

A corollary of this hypothesis is that the marsupials could not respond to the kind of feedback of selection pressures that produced the "relay" (Simpson, 1953) in relative brain size in the Holarctic fauna. Instead, they were limited to a maximum, at about $EQ = 0.50$, that is at about half the brain size of the "average" placental living mammal. This hypothesis is approximately correct, as evidence from living Australian and South American marsupials would suggest. No marsupial species has an $EQ > 1.0$, and the range of EQ's for the species summarized by Brummelkamp (1940) was 0.24–0.62.

The pattern of brain evolution in the Neotropical zone, therefore, represents good evidence in favor of a hypothesis, occasionally in disfavor, but obviously at least partially correct. There can be some advantage in having large brains, even outside the hominid line (cf. Edinger, 1962). This must be true because of the regular trends evident in our data toward increasing brain size in various groups of mammals. Such trends could hardly be attributed to genetic drift and must signify a selective advantage for enlarged brains. The significance of the "experiment" that has just been reported is that it identified at least one aspect of that selection pressure as possibly associated with a predator–prey relationship. Its results suggest, however, that different groups of animals have different potentials for increased brain size and that some, like the marsupials, may be limited to levels significantly below those attainable by placental mammals.

Chapter 15

Special Topics

I have titled this chapter intending two senses for "special." I develop "special" topics that did not fit easily into the general framework adopted elsewhere, and I emphasize the "specialization" in the diversity of species. There is a group of mammalian endocasts that, for one reason or another, could not be covered before, and these should at least be listed and described in a book on the evolution of the brain. In most instances not enough material is available about these endocasts for a quantitative analysis of the sort presented earlier, except for the kind of speculative analysis presented for the Mesozoic mammals. Such speculations are justifiable for the Mesozoic mammals, which are uniquely important in evolutionary history, but they would be presumptuous for the present data, which do no more than support the general trends demonstrated in previous chapters.

The second meaning of this chapter title is the implication of speciation: the diversification of species into adaptive niches, which characterized the evolutionary process. Elsewhere in this book I have, perhaps, over-emphasized the apparently oriented directions and rates of evolution of animals and their brains. The generally opportunistic character of the evolutionary process can too easily be overlooked, including the fact that the major result of evolution is the proliferation of species, each adapted to a particular kind of niche. In the emphasis on general trends in other chapters, the facts of a specialization have usually been cited as contributions to greater or lesser variance of an assemblage of species. In this chapter some of these special topics (which may also be read, literally, as topics about species) are presented against the background of the general approach followed in this book.

I will not present a modern bestiary. Furthermore, I will not present additional information about the species already considered. Instead, two classes of facts are reviewed. First, the available information about the evolution of the brain is reviewed, in particular, brain size in selected species of mammals that are less well known with respect to the history of their endocranial anatomy. These include sea mammals, such as whales and seals; flying mammals, in particular, bats; and burrowing (fossorial)

Evolutionary Trends

mammals. The review emphasizes relationships among brain structures, behavioral functions, and the ecological niches of a group of species. In most cases there is considerable information about the neuroanatomy of living species in these groups; I review that information mainly by citations, but its general nature is indicated as related to the fossil history of the brain in these species.

In the second part of this chapter I discuss in some detail one essentially modern assemblage of mammals, the Pleistocene fauna of the La Brea tar pits of Los Angeles. With the exception of the "archaic" mammals of South America, which became extinct when progressive ungulates and carnivores reached that region at the end of the Pliocene and during the Pleistocene, the Quaternary mammals had all apparently achieved the modern level of brain evolution. That issue is analyzed by comparing the Pleistocene La Brea fauna with living relatives of similar species.

Most of the La Brea mammals were significantly larger in body size than their living relatives, and a second purpose in considering them as a unit is to analyze the role of body size more carefully. An old problem is involved, one first recognized by Lapique (1907) and later by many others: brain:body relations are different in closely related species from those in a random set of species.

We can consider the evolution of the brain as the evolution of a separate character evolving at its own rate more or less independently of the other organs of the body. We should also consider the evolution of gigantism and dwarfism as occurring more or less independently of the evolution of at least some organs, in particular, independently of the brain as an organ that controls information processing. These topics are easier to discuss within the context of comparisons of La Brea and living mammals than as parts of the more abstract discussions of earlier chapters (Chapters 2 and 3) on methodology and on the meaning of brain size, although some of the same kinds of issues are raised.

EVOLUTIONARY TRENDS

The mammalian groups to be considered in this section can be differentiated according to their organic response to selection pressures, according to the habitat that they invaded or simply by arbitrary taxonomic listings. In the first instance one can dichotomize the groups as showing evolutionary trends toward small versus large brains relative to their body size and toward small versus large bodies independently of their brain size. Organized according to habitat (which may be more appropriate for the groups that are actually considered), they can be divided into burrowing (fossorial) mammals, water mammals, and flying mammals. If we keep these rational schemes in mind, we can then use the taxonomic listings for a straightforward framework. Our topics are the evolution of the brain in elephants, sirenians (sea cows and dugongs), whales, and bats and I also review some of the material of previous chapters on insectivores, rodents, and edentates. We will, briefly, consider the radiation of the primates, which is the major topic of the next chapter (Chapter 16).

One peculiar feature of this chapter is that the actual shapes of the endocasts of some of these widely separated groups are surprisingly similar. This includes the sirenians, the earliest of the proboscids (*Moeritherium*), and some of the large fossil edentates, discussed briefly in the previous chapter. The similarity may not be entirely accidental in the first two cases, although the fossil evidence is from gross shape, which could easily be a morphological accident.

Large Land Mammals: Proboscids and Others

Numerous elephant fossils have been described (Osborn, 1936), but there is little information on the evolution of their brains. The earliest member of the order Proboscidea is usually thought to have been the Lower Eocene and Upper Oligocene *Moeritherium,* which is known in several endocasts (Fig. 15.1). These are curiously, and perhaps not accidentally, similar to those of living and fossil sea cows. From that point the history is blank until the Pleistocene, when endocasts of mastodons, mammoths, and other elephants are known.

The Pleistocene endocasts are practically the same as in living elephants. But the change in the size of the brain during the 35 or 40 m.y. span between *Moeritherium* and a relatively small *Mastodon* at the Los Angeles County Museum, recovered from the La Brea tar pits, is dramatically illustrated in Fig. 15.1. The body sizes of both specimens can be estimated from very accurate reconstructions. *Moeritherium* was 272-cm long, heavily built, and according to Eq. (2.3), its body weight is estimated as 1000 kg. The La Brea mastodon was 283-cm long, stood 264 cm at the shoulder, and I have estimated its body weight from a scale model at about 2300 kg. Its endocast displaced 4600 ml of water; that of *Moeritherium* displaced only 240 ml. The encephalization quotient, thus, increased from $EQ = 0.20$ for *Moeritherium* to $EQ = 2.2$ for the mastodon, a tenfold increase.

There is some uncertainty about the habitat of the earliest proboscid, *Moeritherium,* and even whether it was a land form, or, possibly, lived a life intermediate between the semiaquatic hippopotamus and the entirely aquatic dugong. The reconstruction of its skeleton, displayed at Yale University's Peabody Museum, indicates a heavyset, long-bodied, and

Evolutionary Trends 343

Fig. 15.1. Endocasts of the late Eocene *Moeritherium lyonsi* from the British Museum (BM 9176), at left, and the Pleistocene *Mastodon* from the Los Angeles County Museum (LACM), at right, illustrating the evolution of brain size in proboscids over the span of 40 m.y. Scale shown by 15-cm ruler.

short-legged animal, hippopotamuslike in structure, though smaller than that artiodactyl. The similarity of its endocast to that of the sirenians may indicate a closer relationship than has usually been accepted, and Romer (1968) considers it possible that *Moeritherium* is a late representative of the common ancestors of the proboscids and sirenians.

The endocast of the mastodon is entirely elephantlike in both size and shape. It seems likely that the radiation of the proboscids after the Miocene was generally accompanied by the presence of the characteristically enlarged elephant brain, although this speculation is backed by no data on endocasts that I have been able to locate.

The specialization of the elephants toward large body size was not unique among land mammals. As mentioned in previous chapters, comparable specializations have been known in several other groups of mammals. The largest of the amblypods among the archaic Tertiary mammals were not quite within the size range of most elephants, but the largest of the titanotheres was larger than the mastodon from which the endocast of Fig. 15.1 was taken. In fact, the trend toward evolution of large bodies among land mammals, though most notable among the proboscids, was actually greatest in a particular lineage of Miocene perissodactyls: the rhinocerotid, *Baluchitherium,* was between two and three times as heavy as the largest known elephant, living or fossil. Among the edentates, *Megatherium* may also have outweighed the largest of the elephants. The elephants differ from all other very heavy species or groups of species, however, in having brains that were enlarged even beyond the extent expected of their large bodies. In living and fossil elephants of the Quaternary, $EQ > 1.0$ is the typical situation. An African elephant reported by Crile and Quiring (1940) had $E = 5.7$ kg, $P = 6700$ kg; thus, $EQ = 1.3$. The value of EQ for the La Brea mastodon mentioned earlier (Fig. 15.1) may be too high, because this may have been an unusually small individual.

In other land mammals that evolved very large bodies—such as the perissodactyls mentioned earlier, the living hippopotamus, and certainly the fossil edentates—the relative size of the brain was noticeably below the expected "average" level of living mammals. One should probably conclude that when selection pressures toward enlargement of the body result in body size above about 1000 kg, they can act more or less independently of those toward enlargement of the brain. The correlated evolution of body and organ size, evidenced by allometric analyses, may have been limited to some extent at the upper extremes of body size.

Another outstanding case, which is consistent with this view, occurred in the whales; they probably evolved extremely large bodies without correlated enlargement of their brains. Thus, small and large whales have morphologically similar brains that do differ in appropriate direction relative to their body size but not to the extent predicted by an exponent of $\frac{2}{3}$; it is more nearly $\frac{1}{3}$. The very largest of the whales may be thought of as mammals that have taken (a selective) advantage of the absence of gravitational stresses that normally limit body size in land mammals, and their body enlargement involved tissues that were not necessarily under the same amount of control by the brain as are most of the tissues of other mammals.

The likelihood of this upper limit on the allometric relationship appears in empirical studies of brain:body data, such as Count's (1947). He found that a parabola rather than a straight line provided the best fit for the mammalian data of the type presented in Fig. 2.3. Although this kind of result must be viewed critically, because one always improves the adequacy of an empirical curve-fitting process by using higher order equations, I am inclined to believe that Count's conclusion resulted from a biological process of the sort just suggested. It should be possible that selection pressures toward enlarged bodies can occur independently of those on the brain, for example, toward accumulation of blubber, and that there are limits to the brain:body allometric relationships found empirically, at the upper extremes of body size.

SELECTION PRESSURES TOWARD SMALL BODIES

The converse selection pressures leading to pygmy forms are also well known in living mammals and other vertebrates, although they can hardly be identified in fossil species. In general, the pygmy forms (e.g., in chimpanzees) have very nearly normal-sized brains, and human pygmies are only slightly, if at all, small brained (Mettler, 1956; Tobias, 1970). The path of evolution should have established a general pattern of relative size among the body's organs and the body as a whole, but the availability of some niches for pygmy forms could have simply selected the smaller-bodied variants of the basic species, with little effect on brain size. As mentioned in previous chapters, the brain tends to be less variable than the body in size (Radinsky, 1967a), and although much of the variation in body size results from nongenetic effects, we know that there are also important genetic differences. In man, these frequently appear associated with race differences; yet, it is notorious and significant that there are at most minor race differences in brain size (Tobias, 1970). These are probably of no significance for intellectual differences—reflecting, more likely, differences in nonneural components of the brain. See, for example, Zamenhof *et al.* (1971).

Among the unusual effects of selection pressures toward small body size, one may probably include the appearance of fissured brains in some very small mammals. The least weasel (*Mustela rixosa*), the smallest of the carnivores, has a typically shaped but miniaturized mustelid brain. Externally, it has a beautifully represented set of gyri and sulci, although in gross size its brain is only slightly larger than that of the smooth-brained laboratory rat.

The effect of selection pressures toward specifically large or small bodies are special cases of evolution and provide many of the exceptions to the general trends that have been identified. It would be a foolish error, however, to use the examples just developed as argument of any force against the use of relative brain size as a general measure with which to compare large assemblages of species. Selection pressures toward unusually large or small bodies can result in species that are "aberrant" in brain:body relations. But these are exceptional, and in our analyses of broad samples of species, the pygmy and giant species merely increase the variability of relative brain size; they contribute to the "error" variance in the analysis.

An important corollary of these considerations is that we should expect much smaller differences in absolute brain size in closely related species than those found in distantly related species. Selection pressures toward changes in body size have been common enough to be identified in good nineteenth century fashion as a rule or "law." The most notable trend, of course, has been toward the enlargement of the body in evolution, since in most niches larger individuals will have some advantage over smaller ones. Evolution of larger species, and the correlation of size with "progress," has been noted by many authors and is usually graced with Cope's name as "Cope's law" (Newell, 1949). This kind of trend will result in "artifacts" in the analysis of relative brain size, and one should be especially careful when a set of comparisons involves fairly closely related species. Some of the expected allometric relations may fail under those circumstances (Lapique, 1907), and one will tend to find brain size more nearly constant as body size increases.

AQUATIC MAMMALS: WHALES

The whales, properly, take up most of our attention as the mammals most completely adapted to an aquatic habitat and with the longest history of that adaptation. Other mammalian groups are mentioned only by citation, including the sirenians (Edinger, 1939; Dechaseaux, 1958), which are entirely aquatic in their adaptations; the pinnipeds among the carnivores, which are largely aquatic in their adaptations and whose endocasts are under study by C. A. Repenning (personal communication); and various other animals with major aquatic adaptations, such as the hippopotamus among the artiodactyls, in which no special aquatic adaptations appear in the endocasts. The most interesting morphological fact about these animals' brains has already been mentioned: the gross similarity of the endocasts of sirenians to that of the ancestral elephant *Moeritherium*.

The history of the brain in whales is a history of early enlargement, dating to the zeuglodonts (suborder Archaeoceti) of the Eocene (Dart, 1923; Kellogg, 1936) and dramatically notable by the Miocene in the ancestors of living porpoises. Endocasts of several species are illustrated in Fig. 15.2, and the reasons for not undertaking a quantitative analysis should be apparent. The endocasts in many whales provide not nearly as adequate a representation of the brain as do the endocasts of other mammalian orders. It is true that endocasts of large-brained mammals are frequently inadequate insofar as the representation of the fissural pattern is concerned. This is true for man, elephants, and other forms, especially when the individuals providing the casts were not very young. But in the cetaceans, the endocasts are more seriously distorted because even the form of the brain may be poorly represented, and the total size of the endocast is significantly greater than the brain (Breathnach, 1955).

Evolutionary Trends

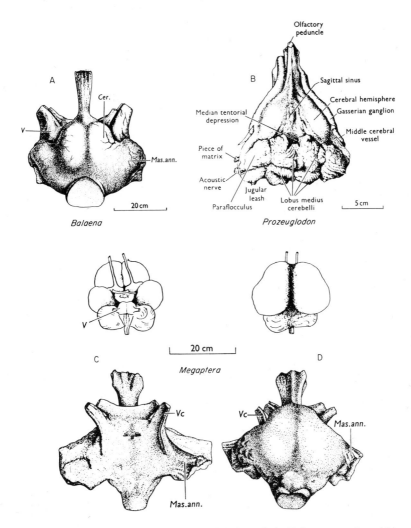

Fig. 15.2. Dorsal views of endocasts of living whale *Balaena mysticus* (A) and Eocene whale *Prozeuglodon atrox* (B). Comparison of brain and endocast in living whale, *Megaptera novangliae,* ventral view (C) and dorsal view (D) illustrate the misleading representation of details of the brain (above) and the endocast (below). Abbreviations: Cer., cerebral hemisphere; Mas. ann., masses annexes (space for non-neural matter); V and V_c, roots of the trigeminal nerve (cranial nerve V). From Breathnach (1955).

The whales, order Cetacea, are grouped into three suborders: the fossil early whales Archaeoceti and two surviving suborders, Odontoceti, or toothed whales, which include smaller species such as the porpoises and dolphins, and Mysticeti, or "whalebone" whales, which include the largest of all mammals and which live by straining plankton through the sievelike baleen.

Endocasts of fossils from the first two groups are known; Eocene endocasts from the archaeocetes such as "Zeuglodon" (*Dorudon*) of the Upper Eocene and *Prozeuglodon* of the Middle to Upper Eocene, first described by Dart (1923), were considered by Edinger (1955a) as differing from endocasts of modern large whales by showing less broadly expanded forebrains and more broadly expanded cerebellums. The latter may be an artifact of the casting of nonneural tissue (Breathnach, 1955). The endocasts of Miocene whales and porpoises, on the other hand, appear quite similar in size and shape to those of the living forms (Figs. 15.3 and 15.4).

Quantitative comparisons are possible, although these are probably more subject to errors when performed for cetaceans than in other orders of mammals. Analyzing endocast volumes relative to body size with the method of convex polygons, makes it clear that the archaeocetes were smaller in relative brain size than either the Miocene or Recent cetaceans. I have estimated their body sizes as ranging from 350 kg (*Dorudon osiris*) to 20,000 kg (*Prozeuglodon atrox*), and their endocasts ranged from 480 ml to 800 ml (Dart, 1923). There are three Miocene cetaceans on which data are available: *Prosqualodon davidi* with $E = 750$ ml and $P = 880$ kg; *?Aulophyseter morricei* with $E = 2500$ ml and $P = 1100$ kg; and *?Argyrocetus* sp. with $E = 650$ ml and $P = 72$ kg. The endocast volume of *Prosqualodon* is from Dart (1923). It and *?Aulophyseter* were small Miocene sperm whales. The data on *?Aulophyseter* and *?Argyrocetus*, a Miocene "dolphin," are from the Los Angeles County Museum, with provisional identifications by L. G. Barnes (personal communication), and are shown here as new data.

When compared to a brain:body polygon of living cetaceans (e.g., using data from Gihr and Pilleri, 1969; Pilleri and Gihr, 1969a,b) the Miocene fossils and the living forms overlap one another. It seems likely that whales, including small species such as porpoises and dolphins, achieved their modern enlarged brains in Miocene times, about 15–20 m.y. ago. It should be noted that *?Argyrocetus* was much smaller than the bottlenose dolphin, *Tursiops*, and the difference in brain or endocast size apparent in Fig. 15.4 is appropriate to the body size difference. *?Argyrocetus* should more properly be compared with the harbor porpoise *Phocaena phocaena* for which Gihr and Pilleri report $E = 500$ g, $P = 61$ kg.

With respect to the details of the endocasts, Edinger (1955a) considered the evidence of the enlarged cerebellum and the impressions of N.VIII of the cranial nerves. She pointed out that the eighth nerve, which

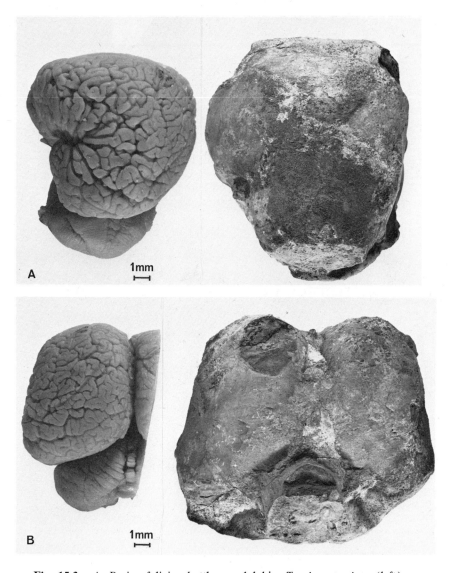

Fig. 15.3. A, Brain of living bottlenose dolphin, *Tursiops tursiops* (left), compared to endocast of small Miocene sperm whale, ?*Aulophyseter morricei* (right), from the Monterey Formation (18 m.y. ago), lateral view. Note that the enlarged brain was characteristic of these small whales. Endocast courtesy of D. Whistler, LACM; Brain courtesy of W. I. Welker, University of Wisconsin. B, Same specimens dorsal view, with left half of *Tursiops* brain placed adjacent to endocast of ancestral species of 18 m.y. ago.

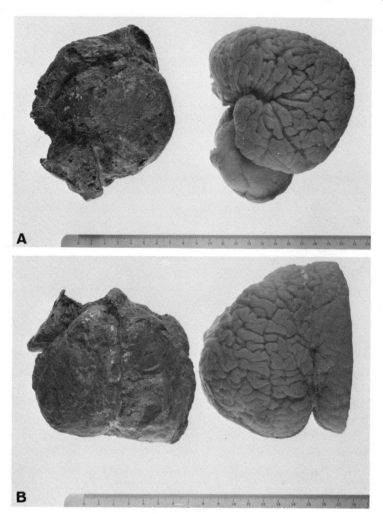

Fig. 15.4. A, Endocast of Miocene "dolphin," ?*Argyrocetus* sp. (left), compared to brain of living *Tursiops* (right). Size comparisons indicated in text. Endocast presented at slight tilt, showing midline; brain presented in exact "normal" aspect giving lateral view only. B, Same specimens in dorsal view. Note similarity in shape but difference in size; latter may be accounted for by body size and special adaptations as discussed in text. Braincast courtesy of W. I. Welker; endocast, courtesy of L. G. Barnes of the Los Angeles County Museum, from U.C. (Berkeley) and is UC 83792. Scale numbers in centimeters.

is mainly auditory in living whales, had its labyrinthine component as the larger one in the archaeocetes. The primacy of hearing in living whales (well known among the porpoises in particular), may, therefore, have

Evolutionary Trends 351

evolved during the interval between the Eocene and Miocene when the earliest porpoise endocasts are known. These endocasts have been described as indistinguishable from those of living porpoises; from Fig. 15.4 I would judge ?*Argyrocetus* as normally delphinoid in this respect. In these endocasts the division of N.VIII in the acoustic and labyrinthine portions is not clear.

In any event, the evolution of the whales and their brains provides interesting, if uncertain and limited, information about our major theme on the evolution of brain size. The materials are not too scarce; a number of Eocene archeocete endocasts are known, and quite a number of Miocene (and Pliocene) cetacean endocasts have been found, though not all have been described. (The two at the Los Angeles County Museum, which were used for Figs. 15.3 and 15.4 still await detailed descriptions by morphologically inclined paleoneurologists.)

The interpretation of the evolution of the whale brain is properly in terms of an adaptation to a niche and morphological changes associated with that adaptation. The evolution of the whales was associated with the evolution of sensory and sensorimotor adaptations for maneuvering in water, and, even more dramatically, the evolution of the auditory and sound producing systems for echolocation and for the perception of objects by their echoes (Kellogg, 1961). The enlargement of the acoustic component of N.VIII, the inferior colliculi of the midbrain (Edinger, 1955a), the entire cerebral cortex, and, in association with the cortex, the cerebellum (Jacobs et al., 1971) were all parts of these adaptations.

The comparative anatomy of the brain of the whale is becoming a more popular topic among neuroanatomists, and one can expect more information on specific adaptations shown by these brains. Recent research has emphasized the fact that the olfactory bulbs are either reduced or entirely absent in some whales (the toothed whales lack them entirely); yet this is a relatively minor response to the water environment, as the fact of the very extensive olfactory brain (pyriform cortex) of these forms would attest (Jacobs et al., 1971).

The enormous expansion of the neocortex in living whales is the most interesting of the adaptations, since relative to the brain as a whole the bottlenose dolphin, for one, exceeds even man in this regard. The evidence from paleoneurology adds a dimension of time to findings such as these because, as far as one can judge from endocasts and estimates of body size, this may have been an ancient adaptation that has characterized the cetaceans during the past 15 or 20 m.y., at least. The evolution of the human brain is a phenomenon of the past few million years; those who compare man and dolphin (e.g., Lilly, 1961) would do well to keep this evidence on the different rates of evolution in mind.

Mammalian Aerial Niches

From adaptations to water, which involved the evolution of unusually large brains, even for mammals, we now move to the adaptation to the bat's world, which was apparently possible with less than an "average" amount of brain for a mammal. As evolutionists, we are concerned with the interaction of niches and biological structures, and it is interesting, but not surprising, that adaptations to two very different mammalian niches would have resulted in radically different adaptations of the relative sizes of the control systems, the brains.

The early history of the bats, like that of the whales, is a mystery. I will discuss only the more familiar types, that is, the Microchiroptera, which have been essentially modern in their skeletal and brain adaptations since the Eocene epoch, and probably evolved from Cretaceous insectivores. The evolution of the fruit-eating bats (Megachiroptera) is not discussed because, to my knowledge, no endocasts from fossils of these larger bats have been described.

The endocasts of the nocturnal, insectivorous, species of microchiropterans have become famous in paleoneurology because of Tilly Edinger's frequent public discussions of the "Tillybat," an animal she has identified as a Paleocene bat on the basis of the endocast, although other views would place that endocast with the Miacid carnivores (Romer, 1968, p. 181).

The issue is interesting because the basis for the identification with bats was in the appearance of the inferior colliculi as outstanding structures of the endocast. The only living mammals among which this occurs are the bats, and it was on this basis that Edinger attempted her identification (Fig. 15.5). As a neural adaptation in living bats, it is certainly correlated with echolocation as a distance sense (Griffin, 1958; Grinnell, 1970). With increased knowledge of the endocasts of early mammals, it was, perhaps, inevitable that other orders would be discovered with comparably exposed and enlarged midbrain structures. This has been shown, definitely, by Gazin (1968) for the Eocene condylarth *Hyopsodus*, the endocast of which was illustrated previously (Fig. 11.4).

Although a relatively large number of chiropteran endocasts are known from Eocene times and the later Tertiary, these are small, and their absolute size and the associated body sizes have not been determined. It is, therefore, impossible in their case, as it was for the whales for other reasons, to analyze the evolution of relative brain size in this highly successful and peculiarly adapted order of mammals. The bats are presently among the most diverse of mammals. According to Walker (1964), the order Chiroptera is divided into 17 families, 178 genera, and a total of

Evolutionary Trends 353

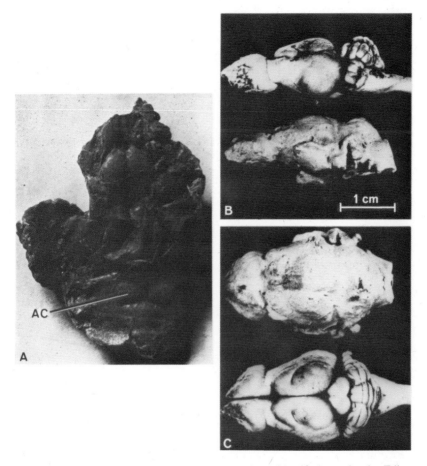

Fig. 15.5. A, Endocast of the Paleocene specimen identified as a bat by Edinger, showing excellent representation of inferior colliculi (AC) of the midbrain. Although similar to a bat endocast, the identification of this specimen remains an open issue. Edinger gives "length of brain" as 2.1 cm. From Edinger (1964a). B, Lateral view of brain (top) and endocast (bottom) of living insectivore, *Tenrec ecaudatus*. C, Same specimen, dorsal view of endocast (top) and brain (bottom). Note midbrain exposure (as in bats), which is not reflected in endocast. From Bauchot and Stephan (1967).

almost 800 species. The important lesson from paleoneurology about bats is that they apparently found their basic adaptive zone early in the Tertiary (or even late Cretaceous, though no fossils from that Mesozoic period have been recovered), and the brain structures necessary for that adaptation are present in the oldest bats with known endocasts. The lesson is that brain structure evolves along with behavioral function associated with a particular niche. Adaptations for the specialized niche of the night-flying bats, which catch insects on the wing by the use of a highly evolved auditory and sound-producing apparatus and midbrain adaptations in the enlarged inferior colliculi, were extremely successful. Such adaptations, once established, remain. Success maintains itself unless a clearly improved system is "discovered" by evolutionary experiments. The post hoc lesson from the bats is that no better adaptation has appeared during the 50 or 60 m.y. or more of their life within their niches.

Burrowing Mammals

Many mammals habitually live underground or live large fractions of their lives underground. Such fossorial niches may have been among the earliest invaded by mammals, although there is a broad range of brain adaptations among the varied known fossorial species. Among the really remarkable burrowers one would include the spiny anteater, echidna, which is a monotreme; a number of marsupials, such as the marsupial "mole" *Notoryctes;* the true moles, that is, insectivores of the family Talpidae; and the lone member of the order Tubulidentata, *Orycteropus,* the aardvark. The remarkable thing about these fossorial animals is that they can effectively navigate below the ground, burrowing rapidly through apparently normal soil, with appropriate adaptations of the nose and eyes to protect their soft tissue from exposure to the soil. Many other mammals, including many rodents and carnivores, such as foxes and badgers, include such behavior in their repertoire.

There is no clear correlation between the fact of a fossorial habit and brain evolution, except in cases in which the habit has an unusual role. For example, the moles have much reduced visual systems and almost vestigial eyes, but this hardly distinguishes them endocranially from their nonfossorial relatives among the relatively small-brained insectivores (Bauchot and Stephan, 1966). The echidnas are truly enigmas in mammalian evolution, with some recent judgments considering them as essentially direct descendants of mammallike (therapsid) reptiles of early Mesozoic times. They are, however, relatively large brained compared to most primitive living mammals. Their encephalization quotients, $EQ \cong$ 0.50 to 0.75, are in the range of "progressive" species of marsupials (e.g., kangaroos) rather than the primitive marsupials (e.g., opossums) or basal insectivores (e.g., tenrecs). For the platypus (*Ornithorhynchus*), $E \cong 10$ g, $P \cong 1.2$ kg; for the echidna, or spiny anteater (*Tachyglossus aculeata*), $E \cong 19$ g, $P \cong 4.2$ kg [brain data from E. J. Dillon (personal communication); body data from Walker (1964)].

When species of a progressive order, such as foxes or badgers among the carnivores, have partially fossorial habits, their brain sizes are appropriate to their order (e.g., as carnivores) rather than to the fossorial habit. As a generalization, one may consider the fossorial habit as a specialization without necessary consequences for relative brain size, although it is a frequent specialization in relatively small-brained orders of mammals. The typical brain size status of the groups of species that include both fossorial and nonfossorial forms determines the degree of encephalization of the fosssorial forms.

SPECIAL SELECTION PRESSURES TOWARD BRAIN ENLARGEMENT

As we noted earlier, proboscids have been an unusual group of mammals with respect to relative brain size because their brains, at least in the Late Tertiary elephants, are unusually large even if body size is taken into account. A similar statement may be made about the smaller whales, which are consistently larger brained than one would expect. In these orders of mammals there may be a kind of "Cope's law" for the brain. For some presently unknown reason these groups, at least in their living representatives, live in niches to which they respond characteristically by brain enlargement beyond the degree required of an enlarged body.

Of all the mammals the most unusual order in this regard is certainly the order Primates. Although the fossil evidence on the primates is only slightly more satisfactory than that on whales and elephants with regard to the evolution of relative brain size, their importance to us makes it appropriate that we undertake a more extended treatment of the evolution of the primate brain. In that treatment (Chapter 16) we will have to make comparisons among relatively closely related species, and it is easier to discuss such comparisons if we complete our review of "special topics," with a consideration of an important Pleistocene fauna, and compare it to its relatively close living relatives and descendants.

LESSONS FROM THE PLEISTOCENE

In our discussions thus far, as well as in our quantitative analysis, we have recognized, repeatedly, a basic conservatism in brain evolution. The brain tends to be about the same size in different individuals within a species in which there are great differences in body size. Adaptations, once achieved, tend to be maintained. It is only under extraordinary selection pressures of the sort that accompany the invasion of new adaptive zones that we see major changes in the external configuration of the brain and its organization. Most adaptations are minor. If one group of insectivores (e.g., the moles) adopts a fossorial habit, there may be a notable deemphasis of vision (e.g., complete covering of the eyes by skin). But the changes will not be revealed dramatically as phenomena of the brain in an order of mammals that is basically nonvisual in the first place.

It is usually difficult to make major distinctions among brains on the basis of the adaptations within an order. Distinctions that are made may rely on questionable homologies of fissural patterns, as von Bonin (1963) has pointed out; Radinsky (1968b, 1969) has, perhaps, been most successful in this difficult effort. When a species invades an environment in which it is advantageous to evolve unusually large bodies, the brain may still maintain a conservative size. In such instances, one may not find the expected brain:body relationship, Eq. (2.1), that occurs among distantly related species. Rather, the older brain pattern of both size and shape might tend to be preserved, with only minimal modifications to accommodate the larger bodies. Thus, whale and porpoises, both large brained, are not as different in brain size as one would predict from Eq. (2.1). The slope, α, in Eq. (3.1), connecting their brain:body points on log–log coordinates, is about $\frac{1}{3}$ rather than the typical value of $\frac{2}{3}$.

Our ability to evaluate, quantitatively, the evolutionary changes in the brain over relatively short time spans (of the order of a few million years) is much reduced by conservatism of that type because the species to be compared will tend to be more similar than in more remote comparisons. One important faunal group of Pleistocene animals, the fossil mammals of the La Brea tar pits of California, illustrates the analytic problems quite well, and I discuss that problem as we consider the culmination of the evolution of the mammalian brain during the last million years.

The La Brea Pleistocene Fauna

When southern California was discovered by the first Europeans with a scientific tradition, among the more impressive discoveries were open tar pits in which animals and people could occasionally be mired. Explorations of these tar pits early in the twentieth century yielded an unusually varied mammalian fauna (Fig. 15.6), which is probably between 10,000 and 50,000 years old and in which skeletons are remarkably well preserved, although the soft tissue has been destroyed. This fauna has been the subject of numerous scientific monographs and has been summed up in an excellent semipopular exposition (Stock, 1956).

Many individual fossils have been obtained, and the material has been studied carefully. There was no need for concern for the hazards of the destruction of individual specimens by the use of special methods such as sectioning and so forth. As a result, many of the specimens are known by one or more endocasts as well as by complete reconstructions of the body. Some of these endocasts have been described in the literature (e.g., Mer-

Lessons from the Pleistocene 357

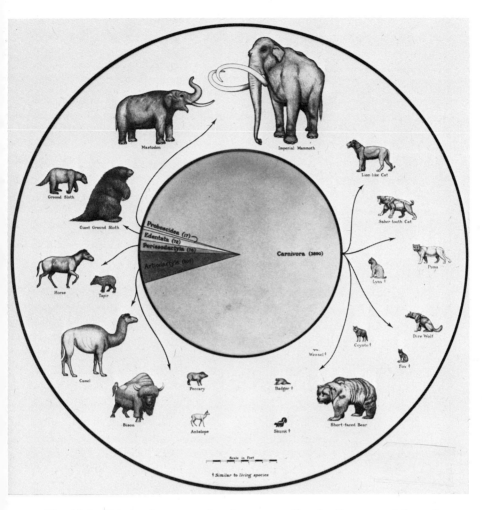

Fig. 15.6. Relative frequency of various mammalian fossils recovered from the La Brea tar pits of Los Angeles; reconstructions are to scale (10 ft = 305 cm).

riam and Stock, 1932), and I have been able to study the complete set of endocasts, made available through the courtesy of the Los Angeles County Museum. The endocast of the mastodon in Fig. 15.1 is from that collection.

The population indicated in Fig. 15.6 is peculiar for a number of reasons, not the least of which is the unusual number of carnivores compared to ungulates and herbivorous mammals generally (cf. Schaller, 1967). This is undoubtedly an artifact of the nature of the entrapment that occurred when animals were mired in the tar. One should expect, first, that many mired animals did, in fact, escape. Second, it was unlikely that herbivorous mammals would seek out the tar any more than other regions with adequate vegetation. For the carnivores, on the other hand, the tar pits would have been an easy source of food because of the likelihood of at least some entrapped prey. The utility of Fig. 15.6 is mainly to indicate the size and variety of species; data on relative numbers of individuals are of interest mainly in suggesting that some large samples are available for study. It must have been the case that herbivores then, as now, far outnumbered the carnivores.

Both in brain and in body the Pleistocene fauna, as demonstrated by the La Brea assemblage, is entirely modern morphologically. There are a few exotic species: the sabretooth, the mammoth and mastodon, and the giant ground sloth, in particular. But in no case do we have the sense of a fauna that would be remarkably unusual for a modern human adventurer. Of the animals in this group, it was perhaps the ground sloth that would have excited the most wonder among hunters of large game because no living mammal is quite comparable to it, certainly not its very small living relatives such as the tree sloths.

The lack of wonder would also characterize one's view of the endocasts of the Pleistocene fossils. They are entirely appropriate to the various species, excepting again the ground sloth, which had a relatively small and unusually shaped brain and to which we can bring no appropriate experience either with the gross or microscopic analysis of the brain. The "elephant" brains from the tar pits are huge and elephantlike, the "lion" brain is identifiably felid, and the artiodactyl and perissodactyl brains are also clearly from members of large ungulate species. The endocast of the sabretooth, while also obviously felid, is remarkable in its unusually deeply convoluted surface, but at no point does one have a sense of being in a truly exotic environment, beyond the exotic environments available to living explorers who would hunt game in the major forests of Africa and South America.

I emphasize the fact that the La Brea fauna looks modern because under quantitative analysis a somewhat different result appears. That result, I believe, is largely an artifact of the relatively large body sizes that characterized many mammals during the Pleistocene radiation. It was a period of gigantism, and one may expect closely related species to appear relatively smaller brained in their giant form than either in their normal or dwarfed forms.

BRAIN:BODY RELATIONS

A number of the La Brea species are so similar to living species in both brain and body size and configuration that there is little point in

Lessons from the Pleistocene

including them in a formal analysis. These include foxes, coyotes, and timber wolves. The ground sloths, which were discussed in the previous chapter, are also omitted. Those species in which notable differences in body size between living and fossil relatives could be shown are presented in the quantitative summary in Table 15.1. The point to note is the regularly lower values of EQ for the La Brea specimens compared to their living relatives.

Table 15.1
Brain:Body Data and Values of EQ for Giant La Brea
Mammals and Their Living Relatives[a]

Specimen	Brain (g)	Body (kg)	EQ
Sabretooths			
Hoplophoneus primaevus (Oligocene)	52	49	0.32
Smilodon californicus (La Brea)	220	330	0.38
Lions			
Panthera leo (Recent)	260	200	0.63
Panthera atrox (La Brea)	340	420	0.50
Wolves			
Canis lupus (Recent)	150	30	1.29
Canis dirus (La Brea)	190	56	1.08
Bears			
Helarctos malayanus (Recent)	390	45	2.57
Tremarctotherium simum (La Brea)	750	720	0.78
Camels			
Camelus dromedarius (Recent)	760	400	1.17
Camelops hesternus (La Brea)	990	1100	0.77
Proboscideans			
Loxodonta africana (Recent)	5700	6700	1.34
Mastodon americanus (La Brea)	4630	2300	2.21

[a] La Brea specimens at Los Angeles County Museum (Stock, 1956). Comparison of sabretooths is of two fossils—*Hoplophoneus* of 30 m.y. ago and *Smilodon* of less than 1 m.y. ago.

To illustrate the problem graphically the data of Table 15.1 are presented in Fig. 15.7, and Eq. (3.6) is also drawn on the graph. It is apparent that lines connecting two related species usually have slopes less than $\frac{2}{3}$, a situation of the type discussed earlier when the whale and porpoise were compared.

As indicated earlier, there is no real mystery about this effect, although it is one that must be considered in brain:body analysis very carefully.

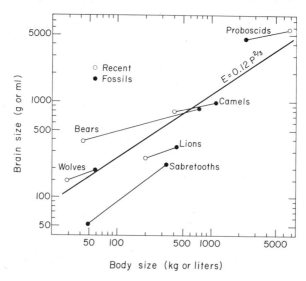

Fig. 15.7. Brain:body relations in La Brea fossils and their living relatives. Since the sabretooths have become extinct, the relationship to the ancestral form *Hoplophoneus primaevus* of the Oligocene (Chapter 13) and *Smilodon* of La Brea is indicated. Data from Table 15.1. Note that in all comparisons among close relatives from the Quaternary the slopes are less than in Eq. (3.6). In the sabretooth line, according to the "additive hypothesis" (see text), there was increased cephalization over the 30 m.y. period that separated the two species, hence the steep slope (see also Chapter 3 and Fig. 15.8).

It has been known for many years, without its significance being fully appreciated. The history of brain:body analysis and especially some of the controversies about allometric equations are often the result of this effect. It is worth a brief digression (see also Chapters 3 and 16).

The basic allometric equation as proposed by Snell (1891) was a simple power function, Eq. (3.1), of the form $E = k\,P^{\alpha}$. Because of the surface:volume relationship and some speculations about the brain as involved in heat exchange, Snell (1891) proposed that the exponent $\alpha = \frac{2}{3}$. Although we now use that value, our reason is empirical and has nothing to do with Snell's reasoning. The immediate effect of Snell's work was to suggest to Dubois (1897) an approach to the problem of encephalization, and Dubois approached it by what he considered to be empirical methods. He chose pairs of species that he considered equally "intelligent" and which differed considerably in body size, plotted their brain:body points on log–log paper, and determined the equation of the line connecting the points. With a limited sample of pairs he found that the average slope was 5/9, a number less than $\frac{2}{3}$, and in retrospect one may say that the shallower slope was due to the fact that he chose species that were more similar than

Lessons from the Pleistocene

those from a random sample of mammals. Lapique also became interested in the problem at about that time, and he found that when he compared subspecies that differed considerably in body size (e.g., breeds of dogs) the exponent was $\alpha = \frac{1}{3}$ or even less (Lapique, 1907). The slope of $\frac{2}{3}$ was not "rediscovered" until von Bonin (1937) chose to restate the problem in completely empirical terms and simply fitted curves to data of a haphazard but very varied collection of mammals. One should now recognize that the slope of approximately $\frac{2}{3}$ will appear and reappear in such data, whether they are sampled from living or fossil species, if the species are different enough, although the exact significance of that slope is still uncertain.

I presented a "theory" of brain size (Chapter 3; also Jerison, 1963), based on the assumption that $\frac{2}{3}$ was a fundamental constant, but we should recognize that this is not a strong assumption. The important point is that it is approximately correct as a number to characterize slopes of arrays of data, and individual points can be related to it by the computation of indices such as the encephalization quotient EQ. In the theory of brain size, however, it was pointed out that one could assume that the brain increased or decreased relative to "average" (better, increased relative to "minimum") size for a particular body size in one of two ways. There may have been a multiplication jump in which the entire brain increased by a particular fraction for a given body size. Alternatively, there may have been an additive jump, in which a certain additional number of neurons was added (presumably by the selective enlargement of some nuclei or systems of nuclei in the brain) and had to be packaged into extra brain tissue. If there were multiplicative steps, then comparisons among assemblages with respect to brain size would produce a series of parallel lines on log–log paper, each with the same slope [$\frac{2}{3}$ according to this approach, which follows von Bonin's; 5/9 according to Dubois and his students, such as Brummelkamp (1940)]. If the steps are a result of an additive process, a very different result obtains, as illustrated in Fig. 15.8. The "equally cephalized" species that have "jumped" above the base level always lie on a curve with a slope less than $\frac{2}{3}$; the exact slope depending on the body size range.

The additive assumption enables us to comprehend Lapique's and Dubois' results, as well as those of von Bonin and those presented here, within a single system. We may consider the general vertebrate situation for randomly selected species of a given class of vertebrates as resulting in a slope of approximately $\frac{2}{3}$ in an allometric function. Whenever species are compared that are evolutionarily close to one another, we may expect that slope to be less than $\frac{2}{3}$, sometimes considerably less. In fact, we should expect some instances in which the slope is 0, that is, in which brain size is a constant independent of body size. This is approximately

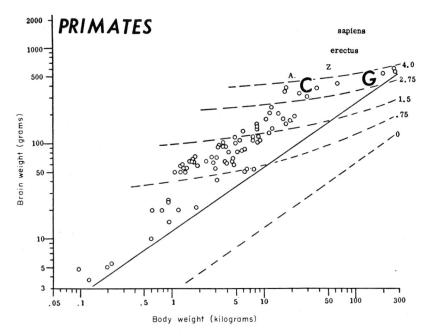

Fig. 15.8. Brain:body relations in living primates and three fossil hominids (A, *Australopithecus africanus;* Z, *A. boisei;* erectus, *Homo erectus*). Dashed curves illustrate the "additive hypothesis," with each curve being the locus of all points in "brain:body space" characterized by the same number of "extra neurons"; parameter multiplied by 10^9 is the number of extra neurons required by the theory presented at the conclusion of Chapter 3. **C** and **G** are assemblages of chimpanzees and male and female gorillas. Solid straight line is Eq. (3.6); dashed straight line is Eq. (3.17). Groups differentiated by parameter are (top to bottom) hominids, pongids, baboons, a diverse group that includes most monkeys and gibbons, and, in the bottom group, squirrel monkeys, marmosets, and prosimians. Further analysis and newer data are included in Chapter 16. From Jerison (1963).

true for mankind. Although there are minor sex and race differences in absolute brain size, within a sex or race there is essentially no correlation between brain and body size for adult human beings. It is a situation to which I devote some attention in the next chapter.

The La Brea fauna can properly be thought of as equivalent to its living relatives and descendants, somewhat larger brained (absolute brain size) as expected from an additive approach to the "theory" of brain size, although not as large brained as they would have to be to be consistent with a multiplicative theory of brain size. Those diverse Pleistocene mammals did, in fact, represent the culmination of the evolution of brains and bodies in all the mammals, excepting only the primates.

Chapter 16

The Primates and Man

By devoting only a single chapter to the topic that is undoubtedly of the greatest immediate interest to us I am emphasizing the limits imposed by my perspective. The history of the human brain is part of the history of the brains of mammals and other vertebrates, and the evolution of mind is being correlated with the evolution of the brain as revealed by the fossil record. It is in the nature of the fossil record that the evidence on man and his ancestors, though increasing because of the successful collecting activities of recent years, remains a relatively small part of the total evidence available from fossil material. Furthermore, the specifically human or primatelike features of the history of our brain are hardly more dramatic, in my judgment, than the facts of early brain enlargement in the first birds and mammals or the facts of progressive enlargement of the brain during the great radiation of Tertiary mammals. This does not minimize the evolutionary significance of the expansion of the human brain, but it emphasizes the importance of other evolutionary trends that have been identified. The real achievement would be to treat all these trends in a unified way and to show, somehow, that they are inherent in the way the neural control systems evolve to cope with the worlds animals live in (Chapter 17).

Some of the most important peculiarly human or advanced mammalian features in the human brain are not accessible to a direct evolutionary analysis. It is impossible, for example, to do more than speculate about the evolution of speech if one tries to base these speculations upon the fossil record. I find analyses based on the shapes of endocasts (e.g., Lieberman and Crelin, 1971) unconvincing. Similarly, the evolution of play, social behavior, maternal behavior, and aggressive behavior, though necessarily associated with the evolution of the brain as well as special skeletal structures, is, basically, inaccessible to studies by the methods that I have used. This is not to deny the importance of speculating about the evolution of such behavior (Howell, 1967; Washburn, 1965), generally by extrapolations from comparative studies of living species. Such speculations are necessary if we are to achieve any understanding of the selection pressures that forced the evolution of the primate brain to go in the several directions that can presently be identified, including the hominid direction.

The speculations scattered throughout this book but developed most fully in Chapters 1, 12, and 17 (see also Jerison, 1963, 1969, 1970a, 1971a) are concerned mainly with cognitive processes, the way one understands and knows, the nature of one's consciousness or awareness, in short, with the evolution of intelligence. These are speculations about the origins of the human mind, and, as such, they concern the work of the brain of one species of primate. In that sense the present chapter is supplemented by many other sections of this book; man and his mind have been used as the standard and measure of the evolutionary status of various other vertebrates. At some point the evidence about the history of the brain in primates should be considered, and that is the purpose of this chapter.

EVOLUTIONARY BACKGROUND

The earliest primates were Paleocene or Cretaceous mammals that were almost indistinguishable from the earliest insectivores, and several famous fossils have been shifted from one order to the other and back as more evidence was uncovered and newer criteria for assignment were developed (e.g., McKenna, 1963). The situation is not resolved even for some living species of mammals. The tree shrews (tupaiids), long considered insectivores, were included among the primates in Simpson's (1945) classification, and most neuroanatomists would prefer them there (e.g., Bauchot and Stephan, 1966; LeGros Clark, 1962), although Campbell (1966a,b) has argued that primatelike features in the tupaiid brain are due to parallel evolution of visual systems in diurnal mammals. The weight of paleontological evidence is presently that tree shrews are insectivores (McKenna, 1966; Romer, 1966; Van Valen, 1965).

McKenna's well-presented discussion of tupaiid affinities may help temper an undue concern with the issue of exact taxonomic status of some of these "intergrades."

> Until a fossil record is well-known, the fine details of tupaiid affinities and taxonomic allocation will continue to be debated on the basis of inadequate evidence, no matter how thoroughly the anatomy of Recent animals is investigated . . . . It seems reasonable to me at present to regard the tupaiids as lepticidlike insectivores . . . with special similarity among primates to Malagasy lemurs, *Adapis,* and *Notharctus.* Among living non-primates the tupaiids are apparently the closest primate relatives, and these conclusions in no way lessen the value of tupaiids to primatology [McKenna, 1966, p. 9].

Evolutionary Background

The separation of the earliest primates from the insectivores, it is generally agreed, was associated with the availability of a diurnal arboreal adaptive zone. In invading that zone, the early primates became more visual than their insectivore cousins and also developed anatomical and physiological adaptations suitable for the active life of mammals that can scamper among branches, feed on the hard fruits of trees, and navigate the unusual environment of forests, leaves, branches, and so forth, in which the life space is in trees rather than on the ground. Among living mammals, the tupaiids conform most closely to one's idea of these early primitive primates. They are highly visual animals, well adapted to the world of diurnal tree dwellers, and appear in life to be like hyperactive squirrels. Their retinas and brains are appropriate for their niches (Campbell, 1966b; Polyak, 1957), so much so that, as noted earlier, Campbell considers them to have evolved the appropriate neural systems in parallel with the primates.

The living tree shrews are extremely important animals for the analysis of the evolution of brain–behavior relations. They are skeletally among the most primitive of living mammals (Romer, 1966). Yet in their habits and brains they mimic the condition of the earliest primates. In relative brain size they are distinctly above the level of the insectivores, and in an analysis by convex polygons they fall in the lower range of the prosimian polygon (Fig. 16.6). But they are not in the primate lineage; at best they may be thought of as models of the earliest primates.

Primate Phylogeny

Primate affinities and lines of descent were presented lucidly and authoritatively a few years ago by Simons (1964), and portions of his chart can serve as an appropriate reference. I have indicated on an adaptation of his chart (Fig. 16.1), the positions and relationships among the species whose endocasts are discussed and included in the quantitative analyses.

The differentiation of the primates from the insectivores occurred no later than the Lower Paleocene, over 60 m.y. ago, and there is evidence of an even earlier, Cretaceous, separation (Van Valen and Sloan, 1965). All the known fossil Paleocene and Eocene primates are classified with the prosimians, allied to living lemurs, lorises, and tarsiers. Radinsky (1970) has recently presented a comprehensive review of the known endocasts of all the fossil prosimians, and I rely heavily on his material in considering these "primitive" primates.

The more modern primates of the suborder Anthropoidea[1] arose from

---

[1] Having chosen, here, to use the term "Anthropoidea" to define the suborder of all primate species that are more advanced than the prosimians, there was a dilemma in how to distinguish hominids (the direct lineage of man: family Hominidae) from

366   16. *The Primates and Man*

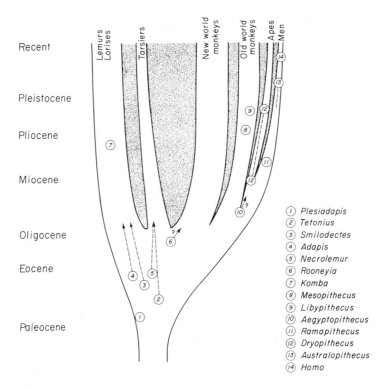

Fig. 16.1. Phylogeny of the primates, with positions of species discussed in this chapter indicated on the phylogenetic tree.

prosimian stock, and this Old World group, to which man belongs, is considerably better known than the New World monkeys. Unless the recently described *Rooneyia viejaensis* (Hofer and Wilson, 1967; Wilson, 1966) is in the line of the South American primates, no fossil endocasts are known for this group. *Rooneyia* is still classified with the prosimians and may have been close to the ancestors of the Old World monkeys rather than

---

other higher primates. The usage in this chapter is imprecise, generally distinguishing hominids from other "anthropoids" without always making the distinction explicit. Part of the difficulty is that taxonomic conventions differ somewhat from the facts of relative brain size. Thus, according to the analysis here, the higher primates may be assigned to at least five groups according to relative brain size in the following descending order: (1) man and his known ancestors, (2) the great apes, (3) the baboons, (4) a diverse group including the gibbons, the remaining catarrhine monkeys, and some platyrrhine monkeys such as the cebus and spider monkey, and (5) the smaller platyrrhine monkeys, such as squirrel monkeys (Jerison, 1963). The groups cut across known lines of descent and suggest that diversification within phyletic lines was sufficiently great to result in the overlap.

to those of the new (Wilson, 1966, p. 246). The specimen remains a solitary find of great importance in our quantitative analysis, and as an early Oligocene primate it may eventually be found to be a later branch in a line of descent from which either the Old World or the South American forms were derived.

Present knowledge of the history of the brains of Old World monkeys is limited to two cercopithecoid specimens from the Miocene and the Upper Pliocene and a rather good assemblage of Pleistocene monkeys and baboons (Freedman, 1960, 1961) that are so similar to living species that they need not be considered separately. Neither of the endocasts of the Neogene cercopithecoids was considered by LeGros Clark (1962, p. 264) as significantly different from living monkeys of that group. These specimens are *Mesopithecus pentelici* (Gaudry, 1862; Piveteau, 1957) and *Libypithecus markgrafi* (Edinger, 1938; Piveteau, 1957).

Although it has been possible to reconstruct the fossil history of the hominoids (apes and men) with some success in recent years (Fig. 16.1), the record is based mainly on noncranial material, and it is only in the hominid line, when we find human fossil remains at the beginning of the Pleistocene, that we begin to have good evidence on the immediate history of our brain. In the earlier fossil record there is presently evidence only from a fragment of the brain of *Dryopithecus* (*"Proconsul"*) *africanus,* a Miocene pongid ("ape"), which LeGros Clark (1962, p. 264) considered to indicate a "still primitive pattern by comparison with the large apes . . . the convolutional pattern is cercopithecoid rather than pongid. . . ."

The relationships among the early hominoids during the late Paleogene and the Neogene have been clarified, recently, by Simons and Pilbeam (1965). They identify three valid genera *Dryopithecus, Gigantopithecus,* and *Ramapithecus* during the Miocene, with the following relationships.

> *Dryopithecus* and *Gigantopithecus* are pongids, *Ramapithecus* a hominid. These two families of Hominoidea were demonstrably distinct by late Miocene time. Species of *Dryopithecus* are regarded as ancestral to those of *Pan* and *Gorilla* and possibly to *Ramapithecus* and species of the latter to those of *Australopithecus.* No unequivocal pre-Pleistocene ancestors of *Pongo* are known at present. It is postulated that canine reduction, facial gracility, and relative decrease in absolute size of the anterior dentition of *Ramapithecus* were associated with greater use of the hands in food-stripping than is the case for chimpanzee and gorilla [Simons and Pilbeam, 1965, p. 141].

There is, finally, one Lower Oligocene primate recently discovered by Simons (1967a), *Aegyptopithecus,* which may be pongid [it was so labeled by Romer (1966)] or in the cercopithecoid line. [An endocast has recently been prepared from this delicate specimen by L. B. Radinsky (personal communication), and it will shortly be available for analysis with respect to relative size.]

In contrast to our meager knowledge of the brains of the precursors of the hominids, the main line of hominid brain evolution is surprisingly well known and should shortly be known from the Pliocene (4–6 m.y. ago) to the present. New discoveries have been made very recently (Howell, 1969; Leakey, 1971) some of which should yield endocasts of late Pliocene australopithecines. There are, perhaps, a half-dozen well-analyzed endocasts of early and mid-Pleistocene australopithecines (Tobias, 1971; Holloway, 1970) and many more from the genus *Homo,* either in its pithecanthropine species, *H. erectus,* or in the variants of the sapient species, such as *H. sapiens neanderthalis.* The precursors of the australopithecines, which Simons has placed about 14 m.y. ago with *Ramapithecus,* have yielded no endocasts.

BRAIN MORPHOLOGY

One need only examine a fossil primate skull of any epoch of the Tertiary period to recognize that these have always been peculiar mammals with respect to their braincases. The oldest well-reconstructed specimen, the Paleocene *Plesiadapis tricuspidens* of about 60 m.y. ago (Szalay, 1971), already showed signs of this peculiarity. Its braincase was appropriate to contain a broadened, flattened and significantly spheroidal brain, as opposed to a cylindrically shaped brain. The tendency discussed earlier (p. 284), toward having the brain more spherical rather than cylindrical in shape, had begun, and these species were, therefore, able to pack a larger volume of brain within a given length of skull. Would an endocast of *Plesiadapis* show features characteristic of the various Eocene and Oligocene prosimians? The answer, if found, would be a major contribution to paleoneurology and to the understanding of human origins; its braincase suggests that it would.

The primates of the Paleogene epoch, comprising the Paleocene, Eocene, and Oligocene, are represented by the five species illustrated by Radinsky (1970) and reproduced in Fig. 16.2. The endocast of *Smilodectes* (S) is sometimes more flexured than indicated in this figure, with the foramen magnum oriented much more ventrally in my cast from another specimen (USNM 17997). One gets a similar impression from the plates prepared by Gazin (1965b) of this genus and *Notharctus.* It is clearly a primate "brain" in many ways. Radinsky emphasized, as had LeGros Clark (1962), that in these specimens one can identify laterally expanded temporal lobes. Radinsky assumed that the development of visual cortex near the midline was the causal factor that forced the lateral expansion of the cortex.

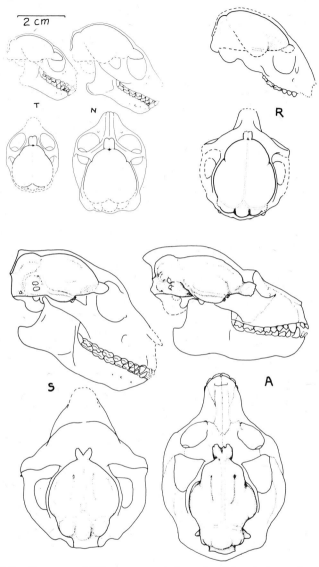

Fig. 16.2. Eocene and Oligocene prosimians showing relations of endocast and skull. See Table 16.1. Modified from Radinsky (1970). Medulla in *Smilodectes* (S) is oriented ventrally rather than dorsally in some specimens, and picture above may indicate too little of a flexure in the brain of this genus. T, *Tetonius;* N, *Necrolemur;* R, *Rooneyia;* S, *Smilodectes;* A, *Adapis.*

In these ancient fossils it is difficult to distinguish the temporal lobe (neocortex) from the pyriform lobe (paleocortex) because the rhinal fissure is not clear in the endocasts. In at least two of the specimens, *Adapis parisiensis* and *Rooneyia viejaensis,* however, there seems to be little doubt that there was a sylvian fissure separating the frontoparietal area from a (neocortical) temporal lobe that had evolved. The evidence is, therefore, in favor of comparable evolution in the uncertain specimens *Smilodectes* and *Necrolemur*. In Gazin's (1965b) original illustration of the endocast of *Smilodectes gracilis* and from personal observations of the endocast and of several partial endocasts from this genus, I have had the strong impression that there was a distinct (neocortical) temporal lobe. But the identification, in any case, cannot be positive.

An interesting issue in the rates of evolution of the parts of the brain is illustrated by this material when it is compared with more recent primate brains. Although it is frequently thought that the frontal pole and prefrontal cortex represent the height of primate brain evolution, occasional strong contrary voices have been heard on this issue. Klüver (1951) and von Bonin and Bailey (1947) have pointed out that "corticalization" among advanced primates, such as Old World monkeys, consists as much or more in the expansion of the temporal lobe as it does in the expansion of prefrontal areas. Their concern was mainly to correct or, at least, ameliorate the effects of the almost gratuitous assumption among neurologists and neuropsychologists of a few decades ago that the frontal lobes were the areas in which higher "psychic" functions were localized.

The evidence of the fossil endocasts could be interpreted as supporting these "gratuitous" assumptions: the earliest prosimian endocasts differed from those of the archaic mammals most notably in two respects. There was, first, the great lateral expansion of the brain in the area of the temporal lobe and, second, notable development of the cerebellum, including the development of lateral lobes and some cerebellar folding. The main feature lacking in these primitive primate brains was a developed or developing frontal and prefrontal area. These facts and the fact that the fossil prosimians also tended to have larger olfactory bulbs than their living descendants are evident in comparisons among the various endocasts (Fig. 16.3), as noted by Elliot Smith (1903), LeGros Clark (1945, 1962), and Radinsky (1970).

There is no real contradiction, however, between the fossil evidence and the possibility that the temporal lobe in living primates is a phylogenetically later development than the frontal or prefrontal area. The fossil evidence concerns the relative expansion of the parts of the brain during the Paleogene epoch. At that time the expansion was in the temporal region, and the frontal region of the brain remained relatively small. Neither the tem-

Evolutionary Background

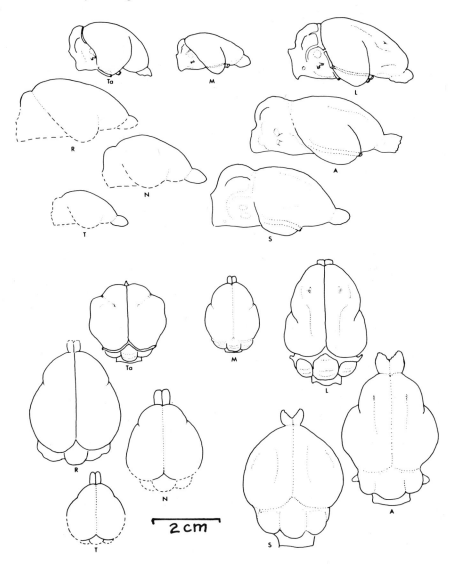

Fig. 16.3. Lateral and dorsal views of endocasts of fossil and Recent prosimians used to perform graphic double integrations to obtain endocast volume estimates of Table 16.1, column (1). Labels as in Fig. 16.2 and Table 16.1. From Radinsky (1970).

poral nor the frontal lobe had achieved the size that each was to reach in some living prosimians or in any of the higher monkeys and apes. The fossil evidence provides relative and absolute brain dimensions as of 50 m.y. ago. Nothing is shown about the later course of development of these parts of the brain. The special development of both the temporal and prefrontal areas in living primates could have occurred during that later evolutionary development.

It is not unlikely that in the evolution of the hominoids the elaboration of the brain began at a lower "cercopithecoid" level with significantly elaborated frontal and temporal lobes at the start but with neither lobe elaborated much beyond the higher lemuroid level. This may have happened during the Oligocene (30 m.y. ago). When the hominids evolved through the australopithecine stage to the *Homo* level, at a much later time (0.5–3 m.y. ago), a significantly greater elaboration could have occurred in the temporal lobes. On the evidence of the living human brain, a region in the tempero-parieto-frontal conjunction is known as the typical sound analysis and speech area, which is perhaps the only unique feature of the human brain (Geschwind, 1965a,b).

EVOLUTIONARY TRENDS

A feature characteristic of the primate brain appeared in the earliest Eocene strata and is visible in the endocast of *Tetonius*. This endocast is relatively broader and more spherical than that of other Paleocene and Eocene mammals. From the external appearance of its cranium the Paleocene primate *Plesiadapis* seems also to have had the characteristically primate brain shape as opposed to the archaic mammalian shape. This was a unique advance at that early stage of the evolution of the mammals and indicates that some of the earliest of the primate specializations were specifically related to the expansion of the brain. The advance is in marked contrast with the generally linear (rather than flexured) brains of either archaic or advanced ungulates and carnivores that were their contemporaries (cf. Figs. 11.3, 11.4, 13.1, 13.3).

It has often been asserted that the specialization of the primates has been a "specialization" towards adaptability, toward being able to survive in a variety of environments. This statement is, perhaps, overly anthropomorphic because of successful human adaptations to a remarkable variety of environments that are normally accessible only to animals with appropriate physiological and anatomical adaptations. It is tempting to make the assertion for the primates as a whole because the brain can be an organ for adaptability, and the brain has been the special domain of primate evolution. As LeGros Clark put it:

> Undoubtedly the most distinctive trait of the Primates, wherein this order contrasts with all other mammalian orders in its evolutionary history, is the tendency towards the development of a brain which is large in proportion to the total body weight, and which is particularly characterized by a relatively

extensive and often richly convoluted cerebral cortex [LeGros Clark, 1962, p. 227].

The remainder of this chapter is devoted to the description and quantitative analysis of that evolution. It is convenient to consider the prosimians separately from the anthropoids (Anthropoidea, that is, New and Old World monkeys, apes, "protomen," and men) in that description because, as we will see, there have been interesting and important differences in the patterns of evolution of relative brain size in these two major suborders of the primates.

RELATIVE BRAIN SIZE IN PROSIMIANS

Radinsky (1970) has published an almost complete study of the available data on fossil prosimian endocasts and has analyzed these data with methods comparable to those used elsewhere in this book. I review his results here, adding a bit of material on the Pleistocene fossil and extinct subfossil lemurs of Madagascar and supplementing his analysis by performing one similar to the analyses used in the preceding chapters.

Let us recall, first, the time spans that are involved in the history of the prosimians and what was happening among the nonprimate orders of mammals with respect to the evolution of the brain. We have data on prosimian endocasts from the Lower Eocene to the Lower Oligocene, that is, from about 55 to 35 m.y. ago. As we have seen in several previous chapters, the archaic mammals of the first portion of that period had brains that were about 20% of the size expected of living mammals of similar body size ($EQ \cong 0.20$). The progressive carnivores and ungulates had achieved considerable advances over this level, at least by Lower Oligocene times (Jerison, 1961), with brains about half the size of their living equivalents ($EQ \cong 0.50$), although some Lower Eocene forms, such as *Hyracotherium*, and Lower Oligocene forms, such as *Archaeotherium*, had brains that were in the same relative size range as those of the archaic forms. There is no evidence that any group of mammals that has been considered thus far in this book, including the earliest ancestors of the elephants and whales, had approached the level of average living mammals, as given by Eq. (2.1). The earliest evidence of such an achievement is in the Upper Oligocene, about 25 m.y. ago, when several ungulate species probably reached that level.

The oldest known primate endocast, that of *Tetonius homunculus* of the Lower Eocene, was approximately 1.5 ml in volume, according to computations by graphic double integration (Table 16.1). I have estimated its body size as being at the lower limit of the range for the living galago, as reported by Walker (1964), which is 80 g. This results in $EQ = 0.68$, which is several times as great as that of any of the other mammals of that period. This picture of the primates as mammals with relatively large brains is characteristic, as LeGros Clark pointed out, and can be verified with quantitative methods.

RADINSKY'S ANALYSIS

By using a method related to one suggested by Anthony (Chapter 3, p. 56), Radinsky (1967a) has been able to analyze relative brain size by comparing brain (endocast) volume to the cross-sectional area of the minimum rectangle within which the foramen magnum of an animal can fit. The method has the advantage that it enables one to analyze relative brain size by using data on the skull alone, and, in the case of the Paleogene prosimians to be considered here, this is all that is known of several specimens. (Radinsky's measure of the foramen magnum was actually the area of an ellipse inscribed in the rectangle. In my analysis of Fig. 16.5 and Appendix III, I used the total area. The measures are equivalent, differing by a factor of $\pi/4$.)

Radinsky's results are presented in Fig. 16.4. They indicate that the radiation of the prosimians from the earliest times, or at least from middle Eocene times (*Smilodectes*) about 50 m.y. ago, involved an expansion of the brain to about the level achieved in Recent species. Radinsky's own statement about this result is somewhat more conservative. He considers "*Smilodectes* to fall slightly outside of, and *Adapis* within, the range of modern prosimians" [Radinsky, 1970, p. 233). We may be able to do more with this analysis if we review the data on which it was based, and Radinsky has provided me with his estimates of brain size for the fossil species that are included in his analysis. These are presented as part of Table 16.1, which also presents my data on some of the same species and other data on prosimians that are useful for quantitative studies.

The main criticism of Radinsky's method is that not enough is known about his basic measure of the independent variable: the cross-sectional area of the foramen magnum. One has the impression that it should be strongly correlated with the general "size factor" that was estimated by gross body size in most of the analyses of this book, and such an impression would be consistent with Anthony's results mentioned in Chapter 3. It would also be consistent with Sacher's result in which he found the size factor heavily represented in the volume of the medulla. But we do not have the mass of evidence of the sort available from body size, which would be needed to verify this conclusion for Radinsky's measure. It would be necessary to undertake an analysis of brain:foramen magnum relations,

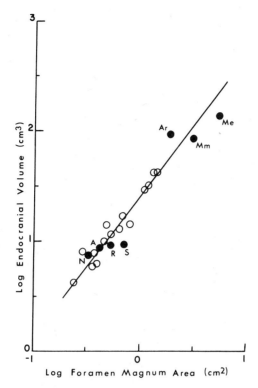

Fig. 16.4. Analysis of endocast:foramen magnum relationship as presented by Radinsky (1970). Abbreviations are Ar, *Archaeolemur fosteri;* Mm, *Megaladapis madagascariensis;* Me, *Megaladapis edwardsi,* all of the Pleistocene; and N, A, R, S, as in Table 16.1. Open circles represent all families of living prosimians; filled circles represent fossil prosimians.

which would have to be as extensive as the published analyses of brain:body relations. Some of the questions that should be raised are answered in the graphic analysis of foramen magnum data to be discussed presently (Fig. 16.5) and by the quantitative analysis of these data included in Appendix III.

Among the methodological problems, it is not always clear just how to measure the foramen magnum in an imperfect fossil, and there also exist specimens (such as *Tetonius*) in which this measure cannot be obtained. I have not been able to obtain measures comparable to Radinsky's for *Rooneyia,* and the exact measure he used for *Smilodectes* would vary depending on the specimen chosen for the measurement. The only one of his four Paleogene prosimians in which I have no question about the cross-sectional area of the foramen magnum is *Adapis* because of the nature of the endocast that is available (LeGros Clark, 1945). I make these statements with casts of three of these fossils, *Smilodectes, Adapis,* and *Rooneyia* on my desk as I write; I have not seen *Necrolemur.* Any investigator who has worked with the occasionally badly crushed material available to the paleontologist would know that such problems can be serious. I emphasize this point because there is an air of objectivity about a pair of simple linear measures of the foramen magnum, whereas one has a sense that estimates of gross body size from skeletal materials such as the skull and long bones (Chapter 2) are more subjective. This is not necessarily the case.

FORAMEN MAGNUM AND THE SIZE FACTOR

In order to achieve better insight into the measure of the foramen magnum (and incidentally review the problems of measurement more adequately), let us consider some data that are taken from photographs of brains and endocasts of living insectivores and prosimians, in which information on body size was also available (Bauchot and Stephan, 1967). These are presented as a set of three graphs in Fig. 16.5, in which the foramen magnum measure taken from the endocast is compared with the equivalent measure of the medulla of the brain itself and in which these measures are correlated with body size. The results should give one pause. The correlations are not bad, and they show that all the measures are associated with a "size factor." We are interested in measures like these or the other measures of body size discussed in Chapter 2 because the size factor is a major determinant of brain size.

A more detailed analysis of the data of Fig. 16.5 is presented in Appendix III, including correlation coefficients, partial correlation coefficients, and inferences about the relationships among the several measures. In that analysis I found a surprising problem in the use of Radinsky's method: the size of the foramen magnum is apparently determined by absolute brain size as well as by body size. In other words, it is not a "pure" measure of a size factor (which I have considered body size to be). A large-bodied but small-brained mammal would tend to have an enlarged foramen magnum, as it should for the measure to be interpretable as Radinsky interprets it. But unfortunately, a small-bodied species would also have an enlarged foramen magnum, according to the statistical evidence, if it were relatively large brained, even if the cross-sectional area of its medulla were relatively small. In short, the statistical analysis indicates that if the cranium is enlarged in association either with body enlargement or with independent enlargement of the brain, there tends to be an enlargement of the foramen magnum. (This apparently is not the case for the cross-sectional area of the medulla as measured on the brain

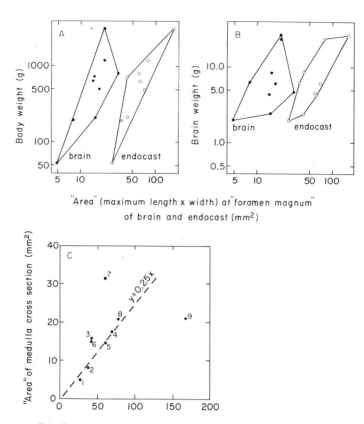

Fig. 16.5. A, Correlations between body weight and area of foramen magnum (open circles) and between body weight and area of medulla (filled circles). Terms "brain" and "endocast" used to designate minimum convex polygons determined from measurements made on brain (medulla) and on skull (foramen magnum). B, Correlations between brain weight and areas of foramen magnum and medulla (key as in A). C, Area of medulla as function of area of foramen magnum. Note that in label of abscissa in A and B, "foramen magnum" refers to medulla for brain measures. Data from Bauchot and Stephan (1967). See Appendix III for statistical analysis.

itself, which is more nearly related to the "pure" body-size factor, as hypothesized by Anthony and others, and should be a valid measure of that factor.)

The effect of this probable confounding of the "body-size factor" with brain size in the measure of the foramen magnum is to decrease differences among species in an analysis like Radinsky's. Species will seem more similar in relative brain size than they actually are, when analyzed as in Fig. 16.4. These are difficult methodological issues probably inherent in Radinsky's method, and they should be examined more carefully with more extensive data than were used in Fig. 16.5 and Appendix III. In the following analysis the "size factor" is estimated not from the foramen magnum but from the more traditional measure (or estimate) of body size.

Brain:Body Analysis of Paleogene Prosimians: Procedures

There are three independent parts in the following analysis. The first and simplest was to estimate brain size in the Paleogene prosimians. This was done by using the method of graphic double integration (Chapter 2) on the endocasts as reconstructed by Radinsky (1970). The data for this analysis are the dorsal and lateral views of the endocasts in Fig. 16.3 and include three living species as well as the five fossil species. The second and much more difficult part of this analysis was to estimate the body sizes of the specimens. The final part simply required more data on living prosimians to provide a kind of base line against which to compare the species of Fig. 16.3. All the analysis could be performed using only published reports of fossil and Recent species, although I have supplemented the first two parts by personal observations.

To estimate the body sizes of the eight species of Fig. 16.3, I have relied more than in previous chapters on analogies to living species. In the case of the fossils my procedure was always to seek accurate illustrations of the skeletal remains and to compare them with some of the better known fossil or Recent prosimians. I assumed that when forms were similar, the weight or volume of the specimen could be determined as proportional to that of the model. Gregory (1920) and Gazin (1958) provided much of the necessary information about the skeletal dimensions of *Adapis parisiensis* and *Smilodectes gracilis*. Tables of body length, body weight, and brain weight, in particular Weber's tables (1896), were also useful.

To illustrate the way in which body size was estimated, let me review my entire procedure as I made an unusually difficult estimate of the data for *Adapis parisiensis*. The problem began with establishing the identity of the specimen that provided the original endocast. The endocast, incidentally, is at the British Museum (Natural History), and the mold is still part of the collection of the anthropology department (specimen M.20192, endocranial cast of M.1345). LeGros Clark (1945) first described the endocast as *Adapis parisiensis,* a rather variable species of small Upper Eocene prosimian. I have been unable to locate the original skull from which the endocast was made, but I believe it may have been the one described by Filhol (1883) and illustrated by Gregory (1920, Fig. 50) as 8-cm long. The issue was complicated by the fact that Piveteau (1957) reproduced LeGros Clark's figure but described the specimen as *Adapis magnus,* a significantly larger species (skull length about 13 cm). Since the original skull was from the phosphorites of Quercy, an important French geological site, it seemed possible that an error had been made in earlier identifications. It was only after ascertaining LeGros Clark's recollection of the original (personal communication) that I definitely placed it as *A. parisiensis.*

The next step was to determine the size of the skull and, if possible, of other limbs of the original specimen that provided the endocast. The problem of identification is still not completely resolved because Filhol's (1883) figures of the humerus and femur of *A. parisiensis* as reproduced by Gregory (1920, Figs. 6, 7, 15) are not completely satisfactory with respect to scaling. In several instances the statement "natural size" is made, but in one key instance Gregory gave the scale factor as "$\frac{-}{1}$" omitting the numerator. Although this may have been a typographical error, other inconsistencies in the figures suggest that he was not certain about Filhol's original illustrations. Faced with this kind of situation, I have generally, but not always, tried to resolve the discrepancies by referring to the original data. This could not be done here. My approach, instead, was to consider other material on *A. parisiensis,* in particular the specimen at the Yale Peabody Museum (YPM 13888). This consists of a small skull, 6.7-cm long, but its cranium was large enough to contain the endocast of the specimen that LeGros Clark studied. (I could verify this by juxtaposing the copy of the endocast and the cranium.) At that time I was also able to examine the skulls of other lemurs, of which *Lemur albifrons* (YPM 995) appeared to be most comparable to *Adapis* with respect to the skull length devoted to the cranium, although the total skull length of the living species was 9 cm, due mainly to a longer snout.

With these various considerations in mind, the reconstruction of *Notharctus tenebrosus* (=*osborni* of Gregory, 1920, Plate XXV) seemed to represent an animal of about the same size and shape as *A. parisiensis.* From Gregory's illustration I was able to calculate a most probable head-and-body length of 44 cm for *A. parisiensis.* The Recent *Lemur mongoz* had regularly been used by Gregory as the comparison species to illustrate similarities and differences among the Eocene and Recent lemurs, and my problem was considerably simplified by the fact that Weber (1896) had presented length:weight data on two specimens of *L. mongoz.* In one specimen, $L = 46$ cm; $P = 2.1$ kg; $E = 28$ g. In Weber's second specimen, $L = 42$ cm; $P = 1.3$ kg; $E = 21$ g (data to two significant figures). Since the specimen of *Adapis* was considered similar in body shape to this lemur, it was necessary only to use the data of Weber's two specimens in order to estimate the body size of my fossil. By using the principle of similitude and basing the computation on the larger of Weber's specimens, I could estimate a body size of 1.8 kg for *Adapis*. From Weber's smaller specimen I could estimate 1.5 kg. This led to a final estimate of 1.6 kg for *Adapis*, to two significant figures.

A much less tortuous route led to exactly the same figure for *Smilodectes*. I merely noted Gazin's comparison between *Smilodectes* and *Notharctus* and his views about the approximate comparability in size of *Smilodectes gracilis* and *Notharctus tenebrosus* (Gazin, 1958, p. 47). The other body size estimates of the fossil prosimians presented in Table 16.1 were developed in similar ways.

The most useful source of information on the sizes of the brains and bodies of living prosimians were the excellent recent compendia published by Bauchot and Stephan (1966, for insectivores and prosimians; 1969, for the anthropoids or "simians"). Summaries of these data are included in Stephan *et al.* (1970). Whenever possible, I also referred to standard sources such as Walker (1964) for evidence on expected ranges of adult body sizes. There remain many uncertainties in this procedure, such as the incorrect identification of species, erroneous reporting of data (including typographical errors), and the repetition of errors by those using the literature.

Brain:Body Analysis in Paleogene Prosimians: Results

The basic brain:body data on the species of Fig. 16.3 are summarized in Table 16.1. This includes Radinsky's estimates of brain size, as well as my estimates in which the method of graphic double integration was used. We will first undertake a nonnumerical analysis using minimum convex polygons (Chapter 2). The results are summarized in Fig. 16.6, in which the fossil data are graphed as open squares and the living specimens of Table 16.1 as open circles. The other 20 data points in Fig. 16.6 are from Stephan *et al.*, (1970, Table 2). The analysis in Fig. 16.6 is based on the minimum convex polygon drawn about the data on living prosimians of Stephan *et al.*, and the polygon is the "map" of the region in "brain:body space" where we would expect prosimian data. The line is Eq. (3.6), the allometric function for brain:body relations in "average" living mammals, and it is interesting to note the symmetry of the polygon for the living prosimians about the independently derived allometric function.

There are a number of specific results obtained by Radinsky that are not supported in this analysis, but his general result that the fossil prosimians were remarkably close to their living descendants in relative brain size was confirmed. Three species were below the polygon: *Adapis parisiensis, Smilodectes gracilis,* and *Tetonius homunculus.* One species fell

Relative Brain Size in Prosimians

Table 16.1
Brain and Body Size Data and Encephalization Quotients (EQ) for Tertiary and Recent Prosimians[a]

Genus and species[b]	Brain size[c] (g or ml)		Body size[d] (g)		EQ[e]	
	(1)	(2)	(3)	(4)	(5)	(6)
Eocene and Oligocene fossil species						
Tetonius homunculus (T) AMNH 4194	1.50	—	80	—	0.68	—
Smilodectes gracilis (S) YPM 12152 & USNM 17997	9.04	9.5	1600	—	0.55	—
Adapis parisiensis (A) BM 20192 & AMNH 11045	9.60	9.0	1600	—	0.58	—
Necrolemur antiquus (N) YPM 18302	4.14	7.5	150	—	1.22	—
Rooneyia viejaensis (R) UT 40688-7	7.17	9.5	200	—	1.75	—
Recent species						
Tarsius spectrum (Ta)	3.76	4.65	120	199	1.29	1.14
Tarsius syrichta	—	3.36	—	92.2	—	1.35
Microcebus murinus (M)	1.67	—	65	—	0.86	—
M. murinus (male)	—	1.87	—	52.4	—	1.11
M. murinus (female)	—	1.72	—	45.5	—	1.04
Lepilemur ruficaudatus (L)	7.15	—	550	—	0.88	—
L. ruficaudatus (male)	—	6.94	—	912	—	0.60
L. ruficaudatus (female)	—	7.79	—	890	—	0.70

[a] Data on Recent prosimians in this table indicate reliability of methods: compare brain sizes in columns (1) and (2) and body sizes in columns (3) and (4) for Recent species.

[b] Letters are labels used in Figs. 16.2 and 16.3. Museum identification numbers are from AMNH, American Museum of Natural History (New York); YPM, Yale Peabody Museum; USNM, United States National Museum (Washington, D.C.); BM, British Museum (Natural History); UT, University of Texas.

[c] Column (1) endocast volume by double integration of data of Fig. 16.3; column (2) endocast volume of fossils as estimated by Radinsky for Fig. 16.4, and median brain weights of Recent species collected by Bauchot and Stephan (1966). Radinsky's estimate of *Adapis* was by water displacement; other fossils by subjective impression.

[d] Column (3) body weights: estimates of fossils discussed in text; Recent species from Walker (1964); column (4) median body weights of Recent species collected by Bauchot and Stephan (1966).

[e] Column (5) EQ from columns (1) and (3); column (6) EQ from columns (2) and (4).

within the polygon: *Necrolemur antiquus*. And one species, *Rooneyia viejaensis,* was slightly above the upper margin of the living prosimian polygon. Two of the three "control" species (the living species of Fig. 16.3 and Table 16.1), *Tarsius spectrum* and *Lepilemur ruficaudatus,* fell within

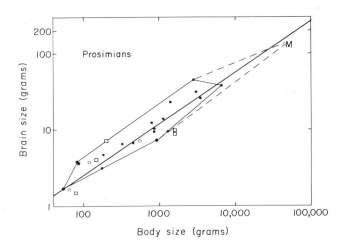

Fig. 16.6. Brain size as a function of body size in the prosimians. Polygon extended as dashed line to include *Megaladapis edwardsi* (M), the giant Pleistocene fossil lemur. Small squares are Eocene and Oligocene fossils of Table 16.1. Note that prosimians are presently "average" mammals, since line, Eq. (3.6), is nicely centered through the polygon. Three of the fossil prosimians are below the polygon. Open circles are data on living prosimians from Table 16.1, columns (1) and (3). Filled circles are data of Stephan *et al.* (1970) on living prosimians.

the polygon, but the third, the mouse lemur, *Microcebus murinus*, fell below the polygon.

The result with the mouse lemur should not have occurred if the data were valid and the "map" complete, and it is worth discussing as an illustration of a limiting feature of this analysis. The "error," if it should be considered that, resulted in part from our use of Walker's (1964) compendium on the mammals, in which a weight range of 45–85 g is given for the mouse lemur. On that basis the midrange value of 65 g was assigned to the specimen in Table 16.1. In their extensive report of quantitative data, Bauchot and Stephan (1966) reported body weight ranges of 46.8–54.0 g in their male specimens and a range of 29.8–101.3 g in the females. Males and females did not differ significantly in brain size, with brain weights ranging from 1.54 to 2.00 g. In averaging the same data, Stephan *et al.* (1970) were surely selective and must have eliminated some specimens as emaciated or as gravid females. In any event, it is clear that, as the smallest lemur, *Microcebus* would inevitably be near one corner of a polygon such as Fig. 16.6. Such a specimen could easily be out of the bounds of the polygon merely because it is close to a natural edge and end point, and the polygon makes a very acute angle at that

Relative Brain Size in Prosimians

point. A more precise analysis using the distribution of encephalization quotients, EQ's, will clarify the underlying relationships.

To perform the more precise numerical analysis the values of EQ were computed for the 20 specimens listed by Stephan et al. (1970) as prosimians (including two tree shrews that were kept in this analysis and that fall within rather than at the edges of the polygon of Fig. 16.6). The frequency distribution of EQ was determined as in previous applications by graphing on cumulative normal probability paper in the usual way, and it is presented in Fig. 16.7. The data on EQ from Table 16.1 were added to Fig. 16.7 by placing the points on the (visually fitted) line.

The computations for the prosimians in Table 16.1 resulted in values of EQ that placed the three Recent species at the eightieth and twenty-fourth percentiles, that is, within one standard deviation of the mean. The analysis of the fossil data can be undertaken directly in the same way by examining their location on the graph.

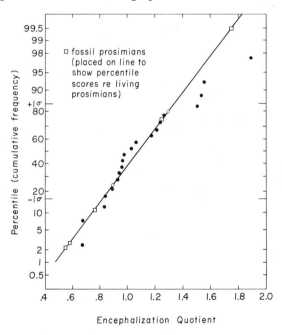

Fig. 16.7. Cumulative frequency distribution of relative brain size in the prosimians. Paleogene fossils were placed on the line (squares) to indicate their "percentile scores." The two lowest specimens are the Paleogene fossils, *Adapis* and *Smilodectes*. Open circles are living species of Table 16.1, column (5); *Microcebus* and *Lepilemur* cannot be distinguished on graph and are represented by the lower open circle. Filled circles are from data of Stephan et al. (1970) on living prosimians.

The two least cephalized specimens had lower values of EQ than any of the 20 prosimians in the data of Stephan et al. These were *Smilodectes*, which fell at about the second percentile, and *Adapis*, which was below the third percentile. If we think of the line in Fig. 16.7 as defining a probability distribution, we would conclude that these two prosimians of the Middle and late Eocene were significantly smaller brained than living prosimians "at the 2 or 3% level." Although this may be questionable statistical thinking, it does nevertheless express an appropriate judgment. Two of our fossils were small brained, so much so as to encourage us to think of their living relatives as having evolved from them into larger-brained species. These "small-brained" fossils had brains that were slightly more than half the size of the brains of typical living prosimians. The living prosimians are "average" living mammals with respect to relative brain size (see Appendix II).

Our two other fossil prosimians, the Lower Eocene *Tetonius* and the Upper Eocene *Necrolemur*, are more like the living prosimians, falling at about the tenth and seventy-fifth percentiles, respectively. A superficial analysis would, therefore, consider these animals as offering evidence of relatively little progressive evolution in the prosimian brain during the last 40 m.y. This is really an inappropriate evaluation of the positions of *Tetonius* and *Necrolemur*. Both were related to living tarsiers, and if we want a true comparison group for them, they should be compared with living tarsiers rather than with the entire assemblage of prosimians. The only tarsier in the basic Recent sample had $EQ = 1.53$, and, if the visually fitted line were extended as shown, that specimen would fall at about the ninety-fifth percentile.

The evolution of *Tetonius* and *Necrolemur* was part of the evolution of the tarsiers, suggesting that an increment in relative brain size took place in that group. They rose (as represented by the individuals under study) from the tenth to the seventy-fifth percentiles of the entire prosimian range during the Eocene epoch, and by Recent times the tarsiers had reached the ninety-fifth percentile of that range. In summary, there was an increase in relative brain size in the tarsioids, and they may have reached the brain size range of living tarsiers by the end of the Eocene.

The case of *Rooneyia* seems to me especially unusual. Although I would prefer not to rely on the evidence of relative brain size in the discussion of an individual specimen, in this instance the evidence seems particularly compelling, buttressed as it is by one's qualitative impressions of a generally "advanced" skull and brain: a conventional interpretation of this species hardly seems warranted. The only living prosimian in the comparison group of Figs. 16.6 and 16.7 in which the brain was relatively larger was the generally aberrant aye-aye (*Daubentonia madagascariensis*).

Rooneyia had actually achieved a relative brain size that was within the living anthropoid (simian) range, comparable to that of marmosets. For a Lower Oligocene species (35 m.y. ago) this was something special. Its brain was actually 3 to 4 times the size typical for the mammals of its time.

Present judgments, based on the overall pattern of characters, tend to place *Rooneyia* with the prosimians, acknowledging some tendencies toward the simian grade. Without overemphasizing the single character (actually a mosaic of characteristics) of relative brain size, it seems that a stronger judgment than that is possible. The failure of the prosimians, generally, to progress in brain size beyond the level of "average" living mammals [see Appendix II, Fig. 16.6, and Eq. (3.6)], as well as the evidence that this suborder had achieved its present level of relative brain size by the late Eocene times and certainly no later than the Miocene epoch, suggests that *Rooneyia* represented either a grossly atypical prosimian condition with respect to the brain or that it should be considered as a harbinger of things to come for the primates as they evolved away from the prosimian condition. As such, *Rooneyia* may be too late or too specialized to represent the ancestral advanced primate suborder, Anthropoidea, but it might more adequately be considered as a conservative descendant from the common ancestor of the simians and the group within which *Rooneyia* is placed. On the basis of the present analysis, *Rooneyia* must have represented a group with the genetic potential for further increases in relative brain size, a potential that no longer characterized the prosimians of its time.

Neogene and Pleistocene Prosimians

I do not present an analysis of the relative brain size of the Miocene lemuroid *Komba* ("*Progalago*") *robustus* (LeGros Clark and Thomas, 1952; Radinsky, 1970) or of most of the giant fossil and subfossil extinct lemurs of the Pleistocene. *Komba* was at the same level of brain and body size as *Galago senegalensis,* which is one of the species in our comparison group that determined the shape of the minimum convex polygon in Fig. 16.6. Radinsky's quantitative analysis of the giant Pleistocene prosimians indicates that they would be comparable to living prosimians, falling approximately on the allometric line of "average" living mammals, Eq. (3.6).

This procedure for the Pleistocene group may be justified on the grounds that it is unusually difficult to estimate the body sizes of the giant Madagascar lemurs because there are no living models with which to compare them. I have, in fact, performed the analysis with the largest of these species, *Megaladapis edwardsi,* because in that instance a restoration of the body form has been attempted, which provided sufficient clues for the estimation of its size (Zapfe, 1963). Furthermore, since this was the largest of the Madagascar lemurs, I could enter its data as maxima in Fig. 16.6, thus determining whether anything unusual happened to the "map" of the relative brain size of the lemurs when that range of body sizes was extended.

According to its reconstruction, *Megaladapis* was heavily built; it apparently looked like a bear with oversized tarsierlike fingers and could have been adapted to a bearlike niche. It was about 100 cm long, and as an animal of heavy habitus it would have weighed about 50 kg, according to Eq. (2.5). L. B. Radinsky (personal communication) indicated that the endocast of the specimen described in his report (1970) displaced 138 ml of water. This permitted me to add the data on *Megaladapis* as point M in Fig. 16.6 and an extension of the minimum convex polygon by the dashed lines on the figure. It is clear that the trend of the prosimians was maintained in this species.

The analysis in the cumulative frequency distribution of EQ (Fig. 16.7) can be used for a finer evaluation of the status of *M. edwardsi* with respect to relative brain size. Its $EQ = 0.85$, a low value for a Quaternary prosimian, placed it at about the fifteenth percentile of living species. Without presenting a complete analysis, it seems likely that this is an instance of gigantism of the type discussed previously (Chapter 15) in the review of the data on the Pleistocene mammals of the La Brea tar pits of California. As selection pressures toward gigantism were experienced, the evolutionary systems that responded did not have to include the brain and sensory systems to the same extent as, for example, skeletal and muscle systems. The increase in brain size necessary to accommodate a larger body in which the basic plan of neural and sensorimotor organization follows a species-specific pattern would be less than that necessary to accommodate a larger body of a randomly selected mammalian pattern.

The most interesting evolutionary feature of the radiation of the Madagascar fossil and subfossil lemuroids is in the variety of niches that they filled. They did not respond to the selection pressures of those niches in the typical primate way of evolving relatively large brains. In fact they were essentially "average" mammals according to quantitative criteria of relative brain size. Their adaptations were accomplished in the typical mammalian fashion of evolving special skeletal adaptations such as large bodies or special finger pads (for tree climbing). The secondary role of the brain in these adaptations supports my earlier contention that the post-Paleogene prosimians no longer responded to selection pressures by the enlargement of the brain.

This has important implications for our general analysis. It suggests that, although the primates as an order had, and used, the potential for brain enlargement in their adaptive radiation, this potential was characteristic of only some of the groups at any given time. It could be, but was not necessarily, the response to the selection pressures of a particular evolutionary stage. We always have to consider the various selection pressures and the response by the brain to them (when such a response occurred). The earliest selection pressures to which the prosimians responded in characteristic primate fashion by the enlargement of their brains were very likely different from the selection pressures that later led to the enlarged brains of the monkeys and apes. These, in turn, must have differed from the selection pressures to which the hominids responded by their remarkably rapid increase in relative brain size, accomplished almost entirely within a million years or so in the Pleistocene epoch.

RELATIVE BRAIN SIZE IN ANTHROPOIDS

The quantitative analysis of the evolution of the anthropoid brain as revealed by the fossil record could be presented in a few pages, although more space is used in order to present several alternate analyses. The evidence that the nonhominoid anthropoid endocasts are essentially the same in living and Neogene fossil forms and that there has apparently been no progressive increase in relative brain size, at least in the cercopithecoids (Old World monkeys), is important for the analysis. It suggests that one may use data on living forms, without much risk of error, to represent the probable brain:body relationships of the primate stock from which the hominids and pongids were derived. It is unfortunate that there are no significant fossil data on pongid brain evolution. [LeGros Clark's comment on "*Proconsul*" (see p. 367) is essentially all that can be said.] It, therefore, is assumed that the pongids may also be represented by the living species.

More on the Meaning of Brain Size

It is appropriate to devote a few paragraphs now to review and reemphasize previous discussions (Chapter 3) on the meaning of brain size. The use of gross brain size as a measure of human brain evolution has had an unusual history. Essentially everybody concerned with the problem refers to the measure and does use it; yet it has become a popular critical exercise among those concerned with human evolution to discuss the irrelevance of brain size (e.g., Dart, 1956; von Bonin, 1963; Holloway, 1966, 1968). Although this measure is not easy to defend, because it is a genuinely "desperate" measure, it is the only direct measure that can be applied to the evolution of the brain as an information-processing system. For that reason gross brain size must be used as the primary datum, and all its possible implications must be exploited. The facts mentioned in Chapter 3 help ease one's scientific conscience in connection with the use of brain size: it is a natural biological statistic that enables one to make reasonably well-educated guesses about the likely morphological status of the insides of the brain.

To use the phrase of one recent critic (Holloway, 1966) brain size or cranial capacity is not a "suitable parameter" for studies of the evolution of the brain. Rather it is a "statistic" that enables one to estimate a variety of suitable parameters, such as the size of diencephalic nuclei (p. 74), the number of cortical neurons, and perhaps even the degree of connectivity of neurons in the brain.

Brain size, relative or absolute, often has been used uncritically in the study of human evolution. It is usually an easy measure to take, and changes in brain size in hominid evolution have been so dramatic that they excite the interest of the most casual student. Among the questioned usages for this measure we should mention, first, its use as a kind of "Rubicon" (Keith, 1948), a criterion to differentiate the sapient from the non-sapient, or to quote Keith, "apehood from manhood." The "error" here is more philosophical than morphological. Keith's Rubicon of 750 ml is a useful "decision criterion" if one recognizes that the data being decided are statistical rather than deterministic: if we use Keith's criterion to decide on a difference among fossil finds, we also accept certain likelihoods of errors with every decision. Neither Keith nor his critics thought in these probabilistic terms, and the idea that a scientific judgment could be demanded and justified in situations where the probability of error was part of the judgment was strange to them.

A more serious error or misuse has been not of brain size or endocast size but in the treatment of the endocast as if it were a brain and the attempt to extract detailed information about the development of various parts of the brain, including development of sulci and gyri, as well as major lobes. One can have no sympathy with this misuse of data. It has been amply demonstrated that endocasts and brains do not correspond perfectly in any animal and correspond rather poorly to one another in all respects except size in most large-brained mammals, including man. The limits of the proper use of hominid endocasts for this purpose have been discussed in some detail by von Bonin (1963), although he is unnecessarily skeptical about the use of gross brain size in the analysis. As I indicated earlier, in some orders of mammals, in particular the carnivores, the endocast mirrors the brain's surface rather well, and valid studies of the fissural pattern can be undertaken (e.g., Radinsky, 1969).

Another "error" that has produced some of the skepticism with regard to the use of brain size has been a psychological one. The user of brain size as a measure tends to act as if the grams of brain do the work of the brain, neglecting the fact that a complex instrument is involved. It is almost sacrilegious to put a brain on a scale, weigh it as so much meat, and then speculate about structural and functional relations. With regard to this "error" I am more sympathetic with the perpetrators than with the critics. It is true that grams of brain and, particularly, milliliters of endocast (often plaster or stone) do not do the work of the brain. But it is also probably true that the important integrative work is not done by individual neurons or molecules inside neurons. In fact, we have so little genuine knowledge of how the work of the brain is done to produce really complex behavior in vertebrates that one might properly be diffident about criticizing any technique of analyzing the brain that results in significant correlations with dimensions of behavior. It is probably true that the functional properties of the brain will finally be understood to result from the interaction of neuronal systems, rather than the action of single units. The neuronal systems that are likely to be significant for the most interesting mammalian behaviors may be very extensive, involving millions of neurons or perhaps, essentially, all the brain (Magoun, 1963).

This preamble repeats some things said elsewhere in this book because we are in sensitive territory when we begin the analysis of the evolution of the human brain. Let us, nevertheless, go on to the data and the analysis.

There will be three omissions. I have no quantitative section on the evolution of the brain in monkeys. That was discussed earlier, when the two known Neogene fossils were mentioned (p. 367). I have no section on the evolution of the brain in apes. And I have no section on the evolution of the brain in the pre-Pleistocene hominids. Such information should eventually be forthcoming, in particular on the earliest hominids, as new fossils are found. Most important for the next section would be an endocast of *Ramapithecus,* a genus now generally accepted as the earliest known hominid (Simons, 1967b). The gap in the hominid skeletal record is about 10 m.y. long, from the earliest *Ramapithecus* (14 m.y. ago) to the earliest *Australopithecus,* about 4–6 m.y. ago.

Australopithecene Endocasts and Bodies

Recent discoveries by the Leakeys (M. D. Leakey *et al.,* 1971; R. E. F. Leakey, 1971) and work on the paleoanthropology of *Australopithecus boisei* (Tobias, 1967) provide us with the necessary basic information on the body sizes of the small *A. africanus* and the larger *A. robustus,* and Holloway's (1970) work in Tobias' laboratory has provided a new and more accurate set of measures of the volumes of the endocasts than had ever been available previously. These data are summarized in Table 16.2.

The estimates of body size followed Tobias' (1967) judgment considering the possibility of a 25–35 kg body range for the gracile species and 50 kg for the robust species.

The first step in the analysis is to establish a comparison group against which to set the data of Table 16.2. Such a group is available in the re-

Table 16.2
Brain:Body Data for Australopithecines[a]

Specimen	E (ml)	P (g)	EQ	N_c
Taung	440	25,000	4.29	3.93
		30,000	3.79	3.87
		35,000	3.42	3.82
Sts 60	428	25,000	4.17	3.85
		30,000	3.69	3.79
		35,000	3.33	3.74
Sts 5	485	25,000	4.72	4.24
		30,000	4.19	4.18
		35,000	3.78	4.13
Sts 19	436	25,000	4.25	3.90
		30,000	3.76	3.84
		35,000	3.40	3.79
Sts 71	428	25,000	4.17	3.85
		30,000	3.69	3.79
		35,000	3.33	3.74
MLD 37/38	435	25,000	4.24	3.90
		30,000	3.75	3.84
		35,000	3.39	3.78
"Zinj" and SK 1585	530	50,000	3.25	4.29

[a] Endocast volumes and codes for identifying specimens are from Holloway (1970), and EQ is encephalization quotient according to Eq. (3.7). E is endocast volume; P is body weight. Factor of 10^9 omitted in listing of "extra neurons" under N_c. This column gives the number of "extra neurons" according to Eqs. 3.13–3.17. See Jerison (1963, p. 282) for a computational example (cf. Figs. 15.8 and 16.10).

cently published data of Stephan et al. (1970, p. 292), in which averaged brain:body data from 21 species of living Anthropoidea are presented. These data are plotted in Fig. 16.8, along with the graph of the allometric equation for average living mammals, Eq. (3.6). A minimum convex polygon about the living anthropoid data of Fig. 16.8 was drawn, exclusive of *Homo sapiens,* as the "map" in brain:body space, within which one would expect to find nonhominid anthropoids.

The australopithecine data may be evaluated by adding appropriate points to Fig. 16.8 from the data of Table 16.2 on the fossil hominids. This was done as in previous chapters (see Chapter 2) by placing a rectangle in the position of the gracile australopithecines of Table 16.2 such that the indicated values of brain and body size would fall within the rectangle. The lower limit of the rectangle, for example, is at a brain size of 428 ml and a body size of 25 kg. In the case of the Zinj specimen (Z), in which $E = 530$ ml and $P = 50$ kg, I took account of the fact that this, too, is basically an estimate of values for a population, and I represented the information as an open circle rather than a point, to suggest a probable error of estimate. Data in Fig. 16.8 on *Homo erectus* and *H. sapiens* are discussed later in a different context.

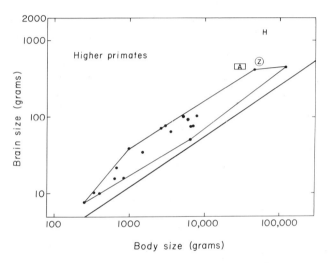

Fig. 16.8. Brain:body relations in "higher primates" (monkeys, apes, and men); line is Eq. (3.6). A, *Australopithecus africanus* range; Z, *Australopithecus boisei* ("Zinjanthropus"); H, *Homo sapiens*. Note that entire higher primate polygon is above the line of "average" living mammals, and that hominids are above the polygon. Straight line is Eq. (3.6), $E = 0.12P^{2/3}$. Data on monkeys and apes (filled circles) from Stephan *et al.* (1970); data on hominids from Table 16.2.

If we accept the usual interpretation of data according to the method of minimum convex polygons, we recognize that the australopithecines were above the living nonhuman primate polygon. The smallest brained of the australopithecines of the gracile species would have had to weigh 60 kg to reach the upper margin of the polygon. No one has suggested so heavy a body for *A. africanus*. The obvious conclusion from this first analysis is that the earliest known australopithecines were already above the typical nonhominid primate level of relative brain size. We may note, in addition, that they were still far below the pithecanthropine (*Homo erectus*) level.

Further Analysis of the Anthropoid Polygon and the Australopithecines

The analysis is not entirely satisfactory, because the base group is inadequate in several respects. There have been gorillas with brains weighing more than the 530 g indicated for *A. boisei,* for example, and the polygon should be larger. One "error" resulted from the fact that Stephan *et al.* (1970) chose to ignore the well-known sexual dimorphism of gorillas and took an average value of specimens of both sexes that they considered to be typical adult animals. A more adequate base group can easily be constructed, and this is done in the next two exercises. We will first undertake to draw a more representative brain:body map for living primates, based on Bauchot and Stephan (1969). For this exercise the largest male specimen of each species was used (see Chapter 2, pp. 42–43), and a single point was graphed for each species. To illustrate an important trend among related species several genera (e.g. *Cercopithecus*) are indicated in a common code to enable the reader to note the "regression" or allometric function that would be most suitable for the species of each genus. The new "basic" data are summarized in Table 16.3, in which values of brain weight and body weight for each species are supplemented by the computed values of EQ and N_c (the number of "extra neurons" computed according to the methods described on pp. 78–80, Chapter 3). These are the basis for further discussions of the status of the australopithecines as well as for methodological discussions later in this chapter and elsewhere in the book. The new data were graphed for the revised analysis by convex polygons (Fig. 16.9), and it is apparent that the enlarged base group does not change

Table 16.3
Brain and Body Data for Higher Primate Species[a]

Genus and species	E (g)	P (g)	EQ	N_c
Saguinus tamarin	9.5	410	1.43	0.24
S. midas	10.4	350	1.74	0.28
S. oedipus	9.8	413	1.47	0.25
Callicebus moloch	17.6	670	1.92	0.40
C. cupreus	14	514	1.82	0.34
Alouatta villosa	65.5	7,824	1.38	0.88
A. seniculus	46.8	3,560	1.67	0.75
Cebus capucinus	73.8	3,765	2.54	1.10
C. albifrons	80	1,640	4.79	1.28
C. apella	75	2,400	3.49	1.18
Saimiri sciureus	24.8	630	2.81	0.54
S. oerstedii	26.4	893	2.37	0.55
Ateles paniscus	106.4	7,400	2.33	1.39
A. belzebuth	118.4	8,890	2.30	1.49
A. geoffroyi	117	7,787	2.48	1.50

Table 16.3—(cont.)

Genus and species	E (g)	P (g)	EQ	N_c
Cercopithecus aethiops	73.2	4,819	2.14	1.06
C. pygerythrus	72.6	5,670	1.90	1.03
C. lhoesti	93	8,500	1.86	1.21
C. hamlyni	72.2	6,000	1.82	1.02
C. mitis	81.5	8,250	1.66	1.08
C. mona	69.3	5,300	1.90	1.00
C. ascanius	71.5	4,500	2.18	1.05
C. talopoin	41.1	1,380	2.76	0.76
Macaca sinica	84	8,392	1.70	1.10
M. nemestrina	122	8,610	2.42	1.53
M. fascicularis	80.5	7,080	1.81	1.09
M. mulatta	106.4	8,719	2.09	1.36
Cercocebus albigena	116	10,500	2.02	1.43
C. torquatus	140	8,680	2.76	1.72
C. galeitus	118.5	10,700	2.03	1.45
Papio cynocephalus	213	22,220	2.24	2.19
P. anubis	222	35,000	1.73	2.12
P. papio	193	21,800	2.06	2.02
P. ursinus	181	24,490	1.79	1.87
P. hamadryas	179	16,000	2.35	1.97
Mandrillus sphinx	179	32,000	1.48	1.76
Presbytis entellus	119.4	21,319	1.29	1.29
P. cristatus	81.7	16,500	1.05	0.93
P. obscurus	64.7	7,030	1.47	0.89
Hylobates lar	105	5,700	2.74	1.42
H. moloch	97.4	6,228	2.40	1.32
H. agilis	87.5	7,372	1.93	1.17
Symphalangus syndactylus	133	12,744	2.03	1.57
Pongo pygmaeus (male)	395	90,720	1.63	3.07
Pan troglodytes (male)	440	56,690	2.48	3.62
Gorilla gorilla (male)	570[b]	172,370	1.53	3.86
Pongo pygmaeus (female)	287.5	44,452	1.91	2.59
Pan troglodytes (female)	325	43,990	2.17	2.89
Gorilla gorilla (female)	426	90,720	1.76	3.30
Homo sapiens (male)[c]	1,361	55,500	7.79	8.83
Homo sapiens (female)[c]	1,228	51,500	7.39	8.21

[a] Most of these data are reviewed in Bauchot and Stephan (1969) and are of the heaviest specimen reported; data given only when at least two species are reported, with the exception of man and the anthropoid apes. Parameters as in Table 16.2.

[b] From Schultz (1950), based on Glydenstope (1928). Schultz gave a volume of 607 ml. In other references he indicated that he used a 6.2% addition to weight as a correction for a weight–volume "error," and for the present it is assumed that this was done in this case. The datum appears to be for the heaviest wild mountain gorilla. Heavier weights are for zoo animals, which get notoriously obese.

[c] Bauchot and Stephan cite Gjukic (1955) for these as the means of 721 male and 935 female individuals.

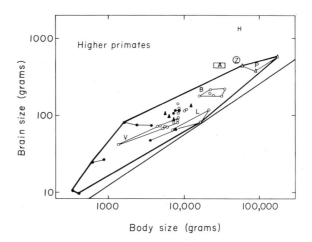

Fig. 16.9. Brain:body relations in larger samples of higher primates as given in Tables 16.2 and 16.3. Groups of single genera indicated: V, Vervets (*Cercopithecus*); L, langurs (*Presbytis*); B, baboons (*Papio* and *Mandrillus*); P, great apes (*Pongo, Pan, Gorilla*). Note that orientations of smaller polygons and connected points, which represent congeners, are not parallel to Eq. (3.6), but represent more flat slopes, as required by the "additive hypothesis." Points coded as circles and triangles to differentiate groups as in Table 16.3.

our previous conclusions about *Australopithecus;* it stands as a genus intermediate in relative brain size between the living primates and *Homo erectus*.

Among the living primates as shown in Fig. 16.9, there are trends within the several genera that differ from the overall trend shown by the slope of the polygon and Eq. (3.6). Such trends are reflected in the data of Table 16.3 as a negative correlation between *EQ* and body weight within a genus. A particularly clear example occurs in the case of *Cercopithecus,* which was represented by eight species. In these species an allometric exponent (slope) of about $\frac{2}{5}$ rather than $\frac{2}{3}$ would have been called for to fit the data. This means that heavier species tended to be closer to the allometric line for average mammals, which are all smaller brained than monkeys, and lighter species were relatively further above that line. Perhaps the meaning of *EQ* is clarified by these considerations, but they should also lead to some skepticism about the adequacy of an allometric analysis when it is applied to closely related species. *EQ* as a measure tends to be biased against species with heavy bodies.

The straightforward analysis with convex polygons also has limitations, as noted in the consideration of the prosimians and the unusual case of the mouse lemur that fell outside a polygon because it was too close to the lower tip. Considering the implications of allometric analyses or the analysis by convex polygons, the empirical nature of the operation is disquieting, and most students of the problem have felt some need to analyze the underlying relations that produce the orderly graphs such as the ones under discussion. In my effort to analyze this issue I developed the "extra neurons" measure, N_c, which was computed for the primates in Table 16.3 and for the australopithecines in Table 16.2. Details of the analysis are discussed in Chapter 3. At this time I will consider its implications for the present data and especially for the analysis of the status of the australopithecines.

"Extra Neurons": The Additive Hypothesis

The discussion presented in Chapter 3 should be supplemented at this point as we apply the basic concepts to our data. The anatomical issue is: what happens when a brain evolves to larger size? If the answer were that every part of the brain in the "basic" animal becomes proportionately larger, this would mean a multiplicative increase in total brain size. Such a position has occasionally been argued, but it is now clear that the differential enlargement of the brain in living species usually involves the development of specialized neural systems. For example, the variations among the procyonids in their use of their forepaws as hands is reflected in the variations in the amount of brain—cortical surface—that is identified as involved in the use of the forepaws (Welker and Seidenstein, 1959; Welker and Campos, 1963). In general, one would expect any novel behavioral capacity to be reflected in the development of some novel neural structures. Human language and its localization in the brain is, perhaps, the oldest example of this principle of localization and proper mass (Chapter 1) at work.

If the enlargement of the brain results from the selective enlargement of only some areas, then this is equivalent to adding tissue to the brain as a whole. In that case the expected effect on total brain size and on the total number of neural elements in the brain is essentially of the sort described in my theory of brain size (Jerison, 1963; also Chapter 3). In my discussions of this issue I would have preferred to avoid indicating where in the brain additional neurons would be. I described them as cortical, but that is almost for didactic convenience. The "extra neurons" need not actually be cortical, and the main reason for describing them as such is that the equations used in estimating their number were derived from counts of cortical neurons and the cortical mass.

The significance of this approach is that it implies a particular effect that would be found in an allometric brain:body analysis. Similar species should have similar brains, and the size of the brain in similar species should tend to be independent of body size. Within closely related species one should, therefore, tend to find brain:body data in which large species and small species would be connectable by lines with slopes less than $\frac{2}{3}$, perhaps considerably less than $\frac{2}{3}$. Within a species there should be essentially no correlation between brain size and body size, except in the case of sexual dimorphism, where the sexes may differ the way closely related species differ. Another implication of the additive hypothesis, when it states that closely related progressive species may have the same number of "extra neurons," is that the differences in brain size among such progressive (as opposed to "primitive") species will be due mainly to parts of the brain that control generalized functions in all mammalian species. These parts are only a small fraction of the brain in progressive species, although they may be a significant fraction in primitive species.

In mathematical terms the equations imply that a true brain:body function exists that is different from the allometric equation of the type of Eq. (2.1). The overall brain:body relationship will consist of a set of relatively flattened curves, such as the one that could be drawn to relate the species of *Cercopithecus* in Fig. 16.9 (see Fig. 15.8). Taken together, these curves, which exist for different species with different specializations, will be contained within a convex polygon with an orientation or slope of approximately $\frac{2}{3}$. The convex polygon and its contained points are a fact. One implication of this fact, if the additive hypothesis is true, is that larger species will usually have more "extra neurons" and hence will be specialized in more spectacular ways than smaller species. This is a theoretical justification of sorts for Rensch's (1956) contention that larger species are more "intelligent" than smaller species.

The implication of this point of view for the evolution of the primate brain is twofold. First, it implies that within any group of similar primates, brain size will tend to be constant. Second, it implies that it should be possible to discover specialized structures in the brains of primate species that are unusually enlarged in some species as opposed to others. Specifically, in the case of the australopithecines and in the evolution of the hominids it suggests that it would be appropriate to hypothesize brain enlargement associated with the development of new skills (such as language in the very large-brained genus *Homo* or elaborate social control or sensory integrative systems in other groups). In short, we must seek neurobehavioral responses to unusual selection pressures that would have resulted in significant enlargement of the brain. For this purpose, we should keep in mind that many of the brain's systems that are currently under active investigation from a neurobehavioral point of view are notoriously small, and their identification as the substrata of a particular behavior pattern would not imply a very much enlarged brain. This is true, for example, of the subcortical and paleocortical systems that are under active study for the analysis of aggression and emotional behavior.

AUSTRALOPITHECINES AND THE ADDITIVE HYPOTHESIS

To return to the morphological analysis, let us review the data on the australopithecines from the point of view of the "extra neurons" hypothesis exactly as stated in Chapter 3. The issue would be: can we or should we separate the australopithecines from living primate assemblages that are assumed to have achieved comparable levels of behavioral or brain evolution? Specifically, therefore, we will compare brain-body relations of the australopithecines as presented in Table 16.2 with those of baboons (which represent a community of living primates that are ecologically in a niche that has been considered similar to that of the australopithecines) and with those of great apes, which are similar to the australopithecines in gross brain size.

The baseline data are those of all the mature baboons and apes in the Bauchot and Stephan (1969) review. These individuals varied considerably in body size, in particular because of sexual dimorphism. I graphed these data (Fig. 16.10) and placed the australopithecine data on the same graph. I then derived the curves of equal values for the "extra neurons" parameter to separate these groups from one another, and these are the curves drawn on the graph (cf. Fig. 15.8).

The results demonstrate several of the effects just discussed. First, all the pongids (as related species) fell into a single group and the several kinds of baboons were in another group. Within each group the absolute brain size was independent or almost independent of body size. It should be kept in mind that many individuals of each species are graphed. It was possible to indicate the species and sex of the pongids, and the contentions of the previous paragraph are clearly supported by their data. These related pongid species, and the individuals of these species, were of various brain sizes, but the variations were independent of body size and almost independent of species. In the one pongid species in which an unusually large body size is characteristic (male gorillas only), the curves giving equal numbers of "extra neurons" predicted the gross brain size of the large-bodied group.

The analysis did not succeed in completely separating the baboons from the great apes or the great apes from the australopithecines. It is likely that the overlap of the first two groups was an artifact of the unusually small body size of one of the specimens taken by Bauchot and Stephan (1969) from Kennard and Wilner (1941) and recorded as *Papio papio* with $E = 237$ g and $P = 12.1$ kg. Walker (1964) indicates a lower

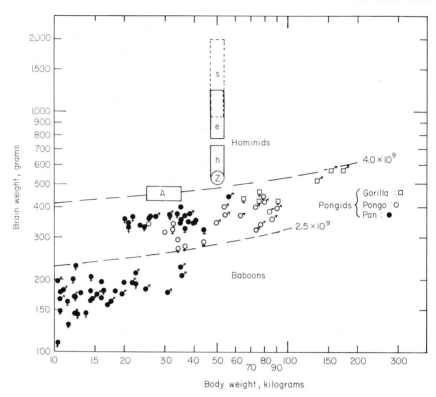

Fig. 16.10. Brain:body relations of populations of baboons, apes, and hominids to indicate pattern of increase in relative brain size among these groups. The "extra neurons" parameter is drawn at $N_c = 2.5 \times 10^9$ and at $N_c = 4 \times 10^9$. Note that groups are reasonably well differentiated by those values of this parameter. The relative constancy of brain size within each group is reflected by the shape of the graphs of N_c as a function of body weight. Baboons and chimpanzees, both shown by filled circles, are entirely differentiated by $N_c = 2.5 \times 10^9$; all baboons were below that value and all chimpanzees were above that value. Hominids indicated are *Australopithecus africanus* (A), *Australopithecus boisei* (Z), "*Homo habilis*" (h), *H. erectus* (e) and *H. sapiens* (s). Data on hominids from Holloway (1970), Tobias (1971), and Leakey *et al.* (1971). Other data from Bauchot and Stephan (1969). See Tables 16.2 and 16.3 and previous figures in which data points represented species rather than individuals as in the present graph. This figure is comparable to Fig. 15.8, with more individuals included and more hominid fossils represented.

limit of 14 kg and a range of 14–41 kg for the genus, and the specimen should probably be ignored. The incomplete separation of the australopithecines from the great apes occurs as a result of the overlap in N_c of the smaller brained of the gracile australopithecines (assuming the indicated body sizes) as shown in Table 16.2 with the larger brained among the male mountain gorillas; both appear to be in the range of 3.8×10^9 "extra neurons."

It may be just as well that this slight overlap appeared in the computations of the australopithecine data because in no case was the separation extremely great, even when the parameters were computed on the basis of different brain and body weights for the fossil hominids (Tobias, 1967). This acknowledges the fact that if the australopithecines were beyond the range of living great apes they were just barely beyond that range. The exercises by the method of convex polygons, which were nonnumerical atheoretical analyses, indicated that the australopithecines fell above the range of the living pongids, and one would judge that at the very least they were at the upper extreme of the living pongid range.

Let me comment on the nature of the additive ("extra neuron") theory of brain size. In my view, even if the theory of "extra neurons" as the basis of differences in brain size among the species of primates were exactly true, there would be no reason either to expect or insist on the absence of overlap among groups of animals to be classified with the help of the parameter. There is no reason to consider the gorilla (or chimpanzee or baboon, for that matter) as using their store of "extra neurons" in the same way or in the way that australopithecines used theirs. The "extra neurons," it should be recalled, represent all the neuronal material developed for functions beyond those required to maintain generalized (but progressive) mammals at the behavioral level of their most primitive ancestors. All advances beyond that level are presumed to have been accumulated by added numbers of neurons to different structures in the brain. The specializations of apes and baboons will only partially overlap those of the hominids. The gross amount of specialized tissue could, therefore, be exactly equal in two different phyletic lines, although the areas that were enlarged could have been in different parts of the brain.

The australopithecines, according to this view, may simply not have gone as far as they were to go in their hominid specializations, which the facts of subsequent hominid evolution support. Their marginal status, albeit at an upper margin, suggests that one may properly view the australopithecines as slightly larger brained than the typical living pongid. This was certainly true for the case of living pongids in the same size range as the australopithecines, and it is well to keep in mind that we have no exact idea of how to treat changes in brain size associated with changes in body size. I proposed the theory of "extra neurons" as a model of how a correct theory would probably work, but it is reasonable as no more than a first approximation to such a theory.

Brain Evolution in the Genus *Homo*

It has already been noted that the *Homo* stage involved a considerable enlargement of the brain. That enlargement took place over a period that may have covered less than a million years, unless Tobias (1966) is correct in his analysis of *Homo habilis*. This species is currently the subject of some controversy, although it is apparently being accepted as a valid species and as a probable precursor to *H. erectus* by some workers (Howell, 1969). With minimal materials Tobias has estimated a brain size of *H. habilis* as between 643–724 ml. The specimen was from the same level as zinjanthropus, from about 1.75 m.y. ago. If the cranial capacity was indeed that calculated by Tobias, it indicates a clearly enlarged hominid brain. A later find, also assigned to *H. habilis* and also from that level, had more complete cranial material, and its endocast was described as having a volume of 560 ml (Leakey *et al.*, 1971). It would appear that the major expansion of the brain in the hominids occurred after *H. habilis* had evolved.

The points in time when the enlargements occurred will never be known exactly. The first small step probably occurred at least 5 m.y. ago, and the advance to the pithecanthropine (*H. erectus*) and then to the sapient levels probably occurred much later, probably about 1 million and 250,000 years ago, respectively. (This is discussed in more detail in the next chapter.) These later stages in hominid evolution have been reviewed by Coon (1962) in his book "The Origin of Races," which is a mine of information on the evolution of the hominids from the pithecanthropine level to modern man. The data summarized in the remainder of this section on *H. erectus* and *H. sapiens* are from Coon (1962, Appendix Table 37).[2]

Brain:body analysis is not required because all the species were about the same in body size; the males probably weighed from 50 to 80 kg,

---

[2] Through correspondence I have learned of the work of V. I. Kochetkova (1960) on endocasts of Pleistocene *Homo sapiens* in Soviet collections. Her work, which was also summarized by Shevchenko (1971), appears to be consistent with the position developed and cited here on the basis of Coon's (1962) treatise. Tobias' (1971) recent summary of the status of fossil hominid endocasts is also consistent with Coon's. A startling new report by R. E. F. Leakey (1973) appeared as this book went to press. Leakey describes a fossil hominid skull ("1470") from Lake Rudolf in Kenya, in which the endocranial volume was determined to be greater than 800 ml. The deposits in which it was found are dated as 2.8 m.y. ago. If the brain size and age are valid, this would put the enlargement of the hominid brain to the *H. erectus* level almost 2 m.y. earlier than previously supposed. This new find does not change any of the fundamental points raised in the discussion in the remainder of this chapter. It may now be necessary, of course, to identify selection pressures appropriate for this most important enlargement in the hominid brain in the earliest Pleistocene environments.

much as modern men, and the females were somewhat smaller. All the analysis can, therefore, be restricted to absolute brain size.

Accepting Coon's judgments on the species to be assigned to *H. erectus* and *H. sapiens* and dividing the latter group into the neanderthals and the remaining sapient types, it is not difficult to present the data by preparing cumulative frequency distributions of absolute brain size in four samples. These are from Coon (1962, Tables 37A, 37D, 37E, and 37F). This has been done in Fig. 16.11. Although slight differences among the groups

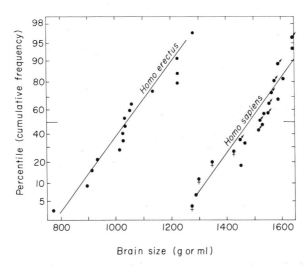

Fig. 16.11. Brain size data in the two species of the genus *Homo*. *H. sapiens* specimens identified by sex are neanderthals, others are *H. sapiens sapiens*. Data from Coon (1962).

are apparent (the sexes are not separated), the important distinction is clearly between the *H. erectus* assemblage and the others.

The classic pithecanthropines, including the original specimen discovered by Dubois (1898), are three of the lowest four values (775–935 ml) in the *H. erectus* cumulative curve. The fourth is one of the Sinanthropus specimens, which had an endocranial volume of 915 ml.

The *Homo sapiens* stage is present in the fossil record beginning about 250,000 years ago with the Steinhelm and Swanscome skulls to which endocranial volumes of 1150 and 1275 ml have been assigned (Coon, 1962, pp. 492–497). Later sapient groups, such as the neanderthals, had typical endocast volumes of the order of 1500 ml or more, that is, somewhat above living human averages (about 1350 or 1400 ml). The differences are unimportant, as far as one can tell, and in fact differences in brain size in living "subspecies" or "races" of men have not actually been studied with sufficient care to assure one that significant differences other than those due to sex differences exist even among different races (Tobias, 1970). It may be to the point to note some very recent results of the analysis of the constituents of the brain (in chicks and rats rather than men and women), in which it was found that sex differences in brain weight in these vertebrates (which are significant) reflect differences in supporting tissue and not in the number of neurons and glial cells (Zamenhof et al., 1971). The males are slightly more fat headed than the females, but the sexes are equal with respect to neuron and glial counts. [Sex was not taken into consideration in computations of N_c, and it is one of several additional variables that should be part of an exact theory of brain size. There are other variables, aging (including loss of neurons), degree of lamination of the cortex, and the relation of "projection" to "association" cortex, that probably play major roles but were excluded in the interest of simplicity.]

The earliest brains of sapient man were probably organized in much the same way as they are today. We should recognize that in the functioning of the brain of progressive mammals considerable plasticity is inherent in the structure. This means that, in fact, the functional capacities of an adult brain may vary considerably among individuals within a species, depending on the way the brain has been used. In man the first months and years of life are critical. The period amounts to weeks or months for rhesus monkeys (Harlow and Harlow, 1962) and days or weeks for rats (Denenberg and Rosenberg, 1967). The capacity for plasticity, and in a larger sense, for culture, evolved with the evolution of the brain. The ways in which brains are actually used by assemblages of mammals, especially human beings, are not completely predictable from the structure that evolved.

ENLARGED BRAINS IN PRIMATES: SELECTION PRESSURES

It should be possible to reconstruct the adaptive zones that involved each of the major advances in relative brain size in the anthropoids. A good deal of effort has gone into the question of the australopithecine zone, although considerably less (in recent years) has been devoted to the transition from *Australopithecus* to *Homo*. I have seen no interesting speculations on the transition from the prosimians to the anthropoids. In this section I will sum up current opinions and add a few speculations leaving much of the final analysis to the next chapter. Howell's (1967) review is particularly helpful as background for this summary.

From Prosimians to Anthropoids and Hominids

The transition from the prosimians to the anthropoids probably took place at the end of the Eocene or in early Oligocene times. If the Lower Oligocene (35 m.y. ago) endocast of *Aegyptopithecus* were available, we could guess whether the evolution of the enlarged brain of the anthropoids —about 2 to 3 times the size of the prosimians (cf. Fig. 16.8 and Fig. 16.6)—occurred then or later. The evidence of *Rooneyia,* which may have been a transition form, shows that primates maintained their characteristic responses to selection pressures by brain enlargement and concurrent behavioral, rather than major skeletal, adaptations. This has also been the long held view of most primatologists (e.g., LeGros Clark, 1962).

Most of the living prosimians are in Madagascar and have evolved there in relative isolation during much of the Tertiary. There they occupy typical primate diurnal niches of other parts of the world as well as many nocturnal niches. Prosimians in other parts of the world (Africa, Southeast Asia, Philippines) are nocturnal animals. This may be a clue to the adaptive zone of the early anthropoids, that is, that they were merely the natural continuation of the lemuroid line for adaptations to diurnal tree living. The increase in brain size in the anthropoids would have had to reflect the changing selection pressures on the nonpredacious species of that zone, possibly associated with the development of more advanced carnivores as predators, even in the trees.

The hominid line, which Simons (1967b) and Pilbeam and Simons (1965) would initiate with *Ramapithecus* (a known fossil), has probably evolved as a result of different selection pressures. It is interesting to quote Simons on the nature of those selection pressures.

> It is possible but would be most surprising if the hominid dental similarity of *Ramapithecus* did not indicate correlation in feeding behavior with man. It is more logical to assume that the animal used the hands in food apprehension more like tool-making man and his Pleistocene relatives than that this ancient creature fed like apes or monkeys when it does not exhibit their dental specializations [Simons, 1967b, pp. 252–253].

A more general argument has been developed during the past two decades. It began with the role of bipedalism. In a thoughtful theoretical essay Bartholomew and Birdsell (1953) outlined the probable ecology of an environment that would have produced the selection pressures toward hominid development. Among the features they emphasized was the possibility that the normally brachiating primates might have invaded an earthbound niche, such as the savanna country presently occupied by many species of baboons. In that case, there would have been great selective advantages for bipedalism. This would be especially the case for species with a long history of brachiating (but not overspecialized for brachiation, for example, by major reduction of the thumb as in spider monkeys). In tall grass the bipedal gait would provide a higher platform from which to survey the range continuously and at the same time free the hands for other purposes. The latter freedom was interestingly interpreted by Bartholomew and Birdsell (1953), who suggested that the branches of a tree for a brachiating animal are genuine tools, involved in intricate mechanical activity. The hands would, thus, have been adapted for the use of tools in a bipedal ground-living "protohominid," and genuine tools could become parts of the normal way of life for these creatures.

The second basis of the argument, which also puts the hands and teeth together, is the assumption that the hominids [*A. africanus* rather than *A. robustus*, according to Robinson (1967)] became more carnivorous than other primates. (Many living primates are entirely herbivorous, though meat eating by baboons and chimpanzees has been observed fairly frequently.) In omnivores the teeth would become apparently unspecialized and hominidlike, enabling the animal to chew hand-prepared material as well as to cut meat with sharp premolars, chisellike canines, and incisors. The human bite is capable of inflicting serious wounds, and the potential for the use of a manlike mouth by a predacious mammal, even without the sharp, stabbing canines of, for example, the baboon, may have been seriously underrated (Every, 1965).

In its most extensive elaboration (Washburn, 1965) the argument combines these features of a land-living anthropoid, adapted to bipedalism (perhaps running bipedally and walking on all fours, but at least standing when observing the terrain), with carnivorous and hence predacious habits. Such habits in skeletally generalized animals imply social organization for the hunt and adaptations for socialization, and they also imply the requirement of a much greater range and lower population density than for herbivorous animals. Washburn pointed out that the greatest range of living primates other than man may involve about 15 square miles and that the individuals of many species live out their lives in much narrower bounds. As hunters, however, they would need to find prey and would have to be far-ranging animals. The early australopithecines would have had ranges up to hundreds of square miles, according to this analysis.

The role of the hunter without stabbing teeth also would have put selection pressures on hunting en masse and by stratagem and on the development of tools, or the ability to learn to use natural materials such as bones or rocks as tools, to stun or kill potential prey. The "osteodontokeratic" culture envisioned by Dart (1957) is entirely consistent with such pressures. This argument, also developed by Bartholomew and Birdsell (1953), completes a picture of an adaptive zone with strong selection pressures favoring variants with more elaborate nervous systems, capable of behavioral plasticity, that is, capable of surveying a broad vista and choosing a path among many possible ones on the basis of memory and specific cues that signal the possibility of game.

The later transition to *Homo* would have had to occur as a natural development from the same selection pressures, with some development of primitive language as a social signaling system, or as a means of specifying a "real" world (Chapter 17). This is the case whether one takes *Homo habilis* as the beginning of a new genus or if one considers the evolution of the new genus as first represented by pithecanthropines evolving in Africa or Southeast Asia.

We need not look far for special selection pressures toward the major enlargements of the brain within the genus *Homo*. The socialized life of a predacious primate is so obviously benefited by linguistic skills, and language is so manifestly the peculiarly human development, that changes in the brain to permit that advantageous supplement to perception and communication would have had obvious selective advantages throughout the period of hominid evolution (cf. Andrew, 1962; Sebeok, 1965).

There is more than language, of course, and we should consider more general analyses of the responses of the organism to selection pressures by the evolution of brain systems. Some of the possibilities are considered in the next chapter, in which I try to unify our results on the advances in the brain as it evolved in the vertebrates.

Chapter 17

The Significance of the Progressive Enlargement of the Brain

The brains of the earliest birds and mammals became larger relative to those of their immediate ancestors among the reptiles because that was the appropriate response to a new set of selection pressures. These pioneers among the "higher" vertebrates had entered adaptive zones during the early and mid-Mesozoic that could only be coped with successfully if certain demands on the processing of information about events at a distance, or about events in the foreground as distinguished from the background ("figure:ground" relations), could be met. We were able to see that, at least in principle, the only way for them to process the information was by adding to the capacity of the central nervous system to do its job (Chapter 12). The selection pressures inherent in the new adaptive zones placed larger-brained variants at a selective advantage.

The explanation was adequate for the initial departure of mammals and birds from the reptilian level, but we have not yet given an account of selection pressures that could have demanded the later expansion of the brain. We have not yet discussed fully the causes of the progressive expansion of the brain in mammals that was characteristic of the evolution of the brain during much of the Tertiary period, after the long period of stability of Mesozoic mammals. The evidence on the birds is less of a problem for analysis than that on the mammals because the bird brain evolved somewhat more rapidly to its present size (Chapter 9). If we consider *Archaeopteryx* as a transitional form representing a period of rapid evolution from the reptilian to the avian grade, and if *Numenius* or the living gallinaceous birds represent the basal avian grade, we need account only for the relatively small enlargement of the brain in birds that has occurred since the late Eocene, about 40 m.y. ago.

In the mammals, however, the situation is more complex and more difficult. The evidence is that the evolution of the mammalian brain may have been rapid during the late Triassic transition from the reptilian to the mammalian grade, about 180 m.y. ago, but a steady state was then maintained for a period of about 120 m.y. The Mesozoic and most of the early Cenozoic mammals had uniformly small brains during that entire long geological interval. I have interpreted this as evidence of their having been in a fairly stable adaptive zone in which there were no unusual selection pressures toward further brain enlargement.

At the end of the Mesozoic, perhaps in late Cretaceous times, one group of mammals, the primates, showed a new spurt in relative brain size. Primate species evolving before the end of the Eocene had become comparable to "average" living mammals in relative brain size. Although the primates were precocious in this regard, the same kinds of trends eventually occurred in almost all the mammalian species on which we have evidence. The trend was dramatic in the carnivores and ungulates (Chapter 13), but there was evidence of it even among the insectivores (Chapter 10), which are usually considered to be the most primitive living placental mammals. The only mammals in which the trend may have been entirely absent were the didelphids among the marsupials.

THEORETICAL BACKGROUND

I will try to explain these trends in this chapter, which may be thought of as a review and theoretical essay on the general problem of the progressive evolution of brain size and its relation to the evolution of behavior. This chapter continues the discussion of Chapter 12, and some of the ideas that are developed were introduced in Chapter 1. After considering, briefly, several empirical constraints on the analysis, I will review a few concepts that I consider basic to understanding why brains of some vertebrates have evolved to larger sizes. I will then discuss the specific responses by the brains of mammals (and, to a lesser extent, birds) to the selection pressures of the Cenozoic era that led to the progressive enlargement of their brains. That analysis concludes with the remarkable evolution of the hominid brain during the Pleistocene epoch, and I will try to show that the evolution of the hominid brain can be comprehended as a biological phenomenon of the same general type as that involved in the evolution of the brain of other mammals.

Empirical Constraints on Theorizing

The evolution that has taken place, as we have been able to reconstruct it, sets natural limits on our theories, although some of these limitations are more frequently points of departure for elegant speculations than restraints on fantasy. The facts that the early primates were tree dwellers and that the early anthropoids appear to have been brachiators, for example, were put to good use in the speculations of Bartholomew and Birdsell (1953), discussed in the last chapter on the probable ecology of the "protohominids."

A more significant limitation on our ability to build theories may be less obvious and is associated with the fact that our concern is with gross brain size. Most of the identified functions of the brain that are at all well understood are functions of very minute amounts of tissue, and knowledge of the brain in the large is surprisingly limited. We know that extensive brain damage can have almost trivial effects on measured behavior (Hebb, 1949) and that the activation of very small regions of the brain can be surprisingly effective in eliciting elaborately defined behavior patterns (Valenstein et al., 1970). When we assert that evolutionary changes in the size of the brain were associated with certain behavioral or physiological adaptations, we should have some evidence that large masses of nervous tissue were actually required as parts of the mechanisms of those adaptations.

An interesting example of this theoretical problem has to do with the evolution of homoiothermy discussed in Chapter 12 (see also Nyberg, 1971; Jerison, 1971a). Adaptations for the intrinsic control of body temperature are certainly among the more dramatic of the differentiating features of the vertebrates. Most living lower vertebrates, as defined in this book, are poikilotherms, basing their control of body temperature on extrinsic mechanisms. The homoiotherms, that is, the birds and mammals, are considered more advanced in their control of body temperature because of the intrinsic mechanisms that they have evolved. Yet Heath (1968) has shown that there are many grades between purely intrinsic and purely extrinsic controls, and it is quite clear that reptiles as well as birds and mammals are generally "warm blooded" when they are active. They differ in the way they maintain their warm bloodedness and in how long they typically maintain it. From the point of view of the role of the brain in these processes, we might expect, a priori, that animals with extrinsic temperature control, such as lizards, which must make gross behavioral responses moving into or out of sunny and shady areas when they sense appropriate changes in body temperature, would use more neural tissue than homoiotherms in handling the control of body temperature. From studies of brain mechanisms for homoiothermy it seems to be the case that only a small amount of neural tissue in the hypothalamus is actually involved in the control (see, for example, Myers and Veale, 1970). We have nevertheless, not concluded that poikilotherms have to be larger brained than homoiotherms because there is evidence to the contrary, but we should be at least as reluctant to accept the conclusion that homoiothermy was a cause of brain enlargement. The control of body temperature is probably an inappropriate dimension of adaptive change to consider as a cause of significant changes in brain size.

Theorizing about the adaptations that led to or arose from increases in brain size is usually concerned with human evolution and with appearance of various human traits. Behavior patterns associated with hunting, learning, aggression, motivation, and, of course, language are the ones most usually discussed. We must, nevertheless, remain wary with respect to all these categories (excepting only the evolution of language) for the same reason that we cannot accept homoiothermy as an explanation for the enlargement of the brain. There are many species of mammals, such as the wolves, that are effective hunters, even hunting socially in packs, yet their brains are typical in size for "average" living mammals. Learning as a "function" (see Chapter 1, pp. 10–11, 24) should probably not be dealt with as if it were a unitary adaptive capacity, a simple behavioral function. If one were to invoke the evolution of learning capacity to explain the evolution of enlarged brains, it would be in the face of our knowledge of systems consisting of only a few neurons in a nerve network, which seem to be capable of simple "learning" (Bullock, 1967).

The same kinds of restrictions arise when we think of aggression or motivation as abstract entities. These are particularly good examples for this exposition because of the now well-known facts of brain stimulation and its effects on behavior. Electrodes implanted in very restricted parts of the brain, subcortically rather than cortically, can serve to excite limited amounts of neural tissue yet elicit behavioral aggression or the evidence of motivation. Animals of many species will learn and extinguish their learning under suitable schedules of reinforcement in which the reinforcement consists of this kind of brain stimulation, and trivially small amounts of brain are actually in the area under stimulation or even in the excitatory or inhibitory pathway. The point is not so much to deny the importance of learning or motivation or aggression in the evolution of brain size but to point out that the role of these categories should be considered in ways in which a good deal of neural tissue would have to be involved to control their participation in behavioral adaptation.

I will discuss the problem of speech and language at greater length later in this chapter, but it is appropriate to note here that this category is an exception to the rule. As far as is presently known, very extensive amounts of brain are involved in the accomplishment of simple useful speech, and the involvement may extend far beyond the so-called speech areas (Penfield and Roberts, 1959). I bring this category up at this point, however, in spite of our ignorance of the nature of brain mechanisms for speech and language.

As a behavioral category, language is of special concern to us as human beings, and our intuitions about its role and importance are compelling. But there is a long distance between our intuitions, or even the logical analysis of language and language mechanisms, and a really useful picture of brain:behavior relations that underlie the human use of language. The evolutionary origins of language are considered later in terms of the principles used throughout this book. From an evolutionary point of view one may do better to think of language and speech in association with the development of perception and imagery rather than with the development of communication and social controls, although both sets of factors undoubtedly contributed to the utility of language as a behavior mechanism in the evolving hominids.

Basic Concepts

During the preparation of this book I found that I brought a few prior notions to the analysis and that other notions were derived from the data as I tried to understand my results. The propositions that seem most basic have been considered before, especially in Chapters 1 and 12. Let me list three.

1. "Reality" or a "real world" is a construction of the nervous system. It is, in fact, a model of a possible world which enables the nervous system to process the mass of incoming and outgoing information in a consistent way. It is a trick, as it were, to enable an organism with a large nervous system to handle almost inconceivably large amounts of information that are usually thought of as nerve impulses or states of membranes of single cells.

2. When a particular aspect of the model (reality) is created in a particular way (presumably by the action of complex and large-scale neural networks), that method of building models will be maintained, insofar as possible, even if there evolve new ways of receiving information about the outside world. The specific point was raised when the problem of using auditory information in place of visual information to create a "real world" was considered. Stebbins apparently had the same issue in mind when he defined "the conservation of organization."

> Whenever a complex, organized structure or a complex integrated biosynthetic pathway has become an essential adaptive unit of a successful group of organisms, the essential features of this unit are conserved in all of the evolutionary descendants of the group concerned [Stebbins, 1969, pp. 124–125].

3. The enlargement of the brain has always been a result of selection pressures and did not occur merely as a matter of, for example, genetic drift. When there is an instance of evolutionary enlargement of the brain,

it should be possible to identify the selection pressures that led to it. Within this principle, two subordinate principles can be considered:

(a) The evolutionary enlargement of the brain was usually a secondary consequence of other trends, in particular those toward increased body size—examples of Cope's law (Newell, 1949). These were relatively trivial effects from the point of view of this book and are handled by allometric equations or by related methods involving the idea of an "expected" brain size for a given body size. It may be, however, that the allometric increase in the absolute size of the brain in the evolution of species of larger body size was one method of increasing the total neuronal mass, until a "critical mass" was achieved. The effects discussed as associated with "extra neurons" could be consequences of such allometric size relationships. According to the theory of brain size developed here, in which an additive hypothesis was used, larger species of mammals should have many more "extra neurons" than related smaller mammals. The point was clearly demonstrated in the data of Table 16.3. It is similar to Rensch's (1956) conclusion on the "intelligence" of larger animals, although he reached his position through a different analysis.

(b) When the enlargement of the brain has been a primary evolutionary effect, as in the case of the progressive evolution of the brain in ungulates and carnivores or as in the case of the primates, one may be concerned with the evolution of higher levels of organization or of intelligence. It is this category of brain enlargement that interests me most, although it appears to have been a relatively unusual event in the evolution of the vertebrates.

Other principles have been discussed in the first chapter, and I have also presented summaries of my point of view there and in Chapter 12. Some of the analysis that follows will be, necessarily, repetitive. But it is my present purpose to present an overview of the entire evolutionary story of changes in brain size as the vertebrates evolved. I emphasize, of course, the progressive changes that occurred mainly among the mammals and mainly during the past 50 m.y. In this context the great challenge is to analyze the rapid evolution of brain size in the later hominids, using the same principles that apply to other vertebrates, and, specifically, to other mammals.

SELECTION PRESSURES TOWARD ENLARGED BRAINS

The evolution of enlarged brains was a novel solution to an adaptive problem when it first occurred. The vertebrates could have evolved and radiated into all the niches in which they are known today, at least insofar as the appropriate skeletal adaptations can be read from the fossil record, without evolving brains enlarged beyond the requirements of the body sizes for those niches. There is nothing in the fossil record of lower vetebrates to indicate that their lives were less "rich" than those of fossil birds and mammals that lived in similar environments. And we know that, despite their relatively small brains, fish and reptiles have not become extinct as a result of the radiation of larger-brained mammals and birds that appear to live in similar niches. We may emphasize therefore, that vertebrates do not live by brains alone, and, although large brains may signify certain styles of life, the selective advantage associated with an enlarged brain has rarely been a major one. It is really anthropocentricity that leads us to emphasize the brain and its evolution.

I analyzed the first expansion of the brain in mammals and birds with this point in mind. The idea was to consider how and under what circumstances it would be possible for the earliest mammals to live as normal "reptiles" (the immediately ancestral group), differing from their relatives only in the "irrelevant" fact that their brains were somewhat different in size. I sought to maintain conservation principles, which would not imply a general advantage for an enlarged brain, but only a special advantage in a restricted niche or in a novel adaptive zone. The point of suggesting that the role of the brain was specific to nocturnal mammals is that an enlarged brain could permit them to lead entirely normal "reptilian" lives, and they would differ from normal reptiles only in that they could use nonvisual sources of information about events at a distance. The analysis of the enlargement of the brain in birds was somewhat more difficult in that it was less obvious how to identify an available adaptive zone that could be occupied by "reptiles" only if they had evolved feathers and large brains. One answer was to suggest a primatelike niche (Chapter 9, p. 199; Chapter 12, pp. 276–278).

PROGRESSIVE ENLARGEMENT OF THE BRAIN IN TERTIARY MAMMALS

What kind of adaptive zone entails selection pressures that would make of the early archaic and "progressive" Tertiary mammals merely larger-brained counterparts of their immediate ancestors among the small-brained Mesozoic mammals? The obvious zone is that for diurnal visual animals that could live as the extinct ruling reptiles had lived. When such a zone was to be invaded by mammals, however, it is clear that they could not simply revert to their ancestral reptilian adaptations. The adaptations had to be mammalian; the sensory systems that would evolve would be mammalian, and so also would be the central representation of the system in the brain.

The history of the early Tertiary mammals indicates that they evolved in two stages. First, they merely extended their range to include the possibility of large body size, and the earliest Cenozoic mammalian radiation may not have involved the habit patterns of living mammals. Rather, the ancestral forms could be considered as essentially enlarged insectivores or didelphids with relatively minor adaptations for lives comparable to those of the then recently extinct archosaurs. They were probably relatively nonvisual animals that could live out more of their lives by day, but they presumably evolved no special sensory adaptations for diurnal lives.

Under the circumstances any species of mammal that evolved a more adequate visual distance-sensing system would have more selective advantages. Since the typical mammalian distance-sensing system at that time was an encephalized auditory and olfactory system, with capacities for storing information over time, one would expect, according to the conservation principle, that a newly visual species of mammal would have its new visual system modeled after the, by then, normal mammalian auditory distance-sensing system. Such a neurosensory system would be encephalized to a much greater extent than reptilian visual systems, and this may have been the basic element in the further expansion of the brain in Tertiary mammals.

The evidence is that such an expansion occurred first in the primates, and we should recall now the discussion of the evolution of the avian brain. In that discussion (pp. 275–278 *et seq.*) I have noted that life in trees based on visual information would have that information in the form of mottled figures against mottled background, with natural camouflage inhibiting one's capacity to form an accurate picture of events at any distance. Under these circumstances there would be many advantages to the use of color and a constructed "real" space with objects and things in it. It would be very difficult to move about freely in such an environment on the basis of prepotent cues from certain patterns of stimulation that act as effective stimuli for fixed action patterns because the stimulus-patterns would be unstable and changing.

The earliest primates, hardly more than modified insectivores entering an adaptive zone of diurnal tree-dwelling animals, probably evolved such a more adequate visual system as their major distance sense. That system, according to conservation principles, would be modeled after the long-established mammalian auditory system with its major central nervous system representation. The new visual system could be significantly superior to the auditory system, however, because much of the useful spatial information about events at a distance could already be encoded by retinal elements. These could be reencoded as cortical information and integrated with other information from other sense modalities represented in the cortex.

At this point let us recall that the auditory world of the immediate ancestors of the primates, according to our approach, should have been, at best, the equivalent of a reptile's visual world. Responses should have been tightly bound to stimuli, and the typical behavior mode should have been that of the "fixed action pattern" released by a pattern of stimulation of the sort made familiar by the ethologists' analysis of the behavior of birds. Although there is no obvious reason to assume that such tight stimulus binding did not occur in the early Tertiary primates with their hypothesized diurnal, encephalized visual systems integrated with their previously organized and retained (in the evolutionary sense) distance-sensitive auditory and olfactory systems, I suggest that there were important divergences either already present or in the offing. One major aspect of the new visual system is that it would be represented as a temporal as well as a spatial system. A second aspect, resulting from the integration with other cephalized sensory systems, is that events sensed from one perspective could be correlated with a set of events from the same source seen from another perspective. This means that they could respond to stimulus configurations as "objects" with particular positions in space and durations in time. The response to a pattern of stimulation by a fixed action pattern response would be less likely if for no other reason than that the adequate pattern of stimulation was no longer as fixed.

It is not easy to speculate about when in the course of evolutionary history this kind of shift took place, but I would guess that in addition to its appearance in early primates it occurred to some extent in all the large-brained vertebrates (birds and mammals) at a very early point in their history and was correlated with the achievement of a particular brain size. It does not have to be an all-or-none affair, however, and one can imagine the persistence of the fixed action pattern approach to behavior in most birds for most purposes, while there were more flexible or plastic responses available to some species for which stimulus configurations could be redefined as objects rather than as releasers of responses. I would suggest that "object definition" was typical in the behavior pattern of the early primates and that it became increasingly common and the basis of more plasticity in the behavior patterns of other mammals evolving as large-brained forms during the Tertiary. (The direction of behavioral and brain evolution of birds, on the other hand, appears to have been toward the perfection of the fixed action pattern and, only secondarily, toward plasticity of behavior in many circumstances.)

In summary, the first stage of the enlargement of the brain in progressive Tertiary mammals may be associated with the evolution of visual systems modeled after the then existing mammalian auditory and olfactory systems rather than the earlier reptilian visual system that had become reduced in mammals. As such, there would be two important features of the newly evolved visual system attributable to the fact that it would be a corticalized rather than a retinal system. First, there would be a significant temporal element to visual experience, since the contribution of time is a major difference between auditory and retinal–visual information processing. Visual information would be labeled by duration as well as extent. (There is also a requirement for temporal integration—memory—if smell were to be used as an accurate distance sense; see p. 269.) Second, since similar information would be coming from the visual, auditory, olfactory, and tactile senses, there would be a tendency to integrate the data from all these sensory systems by cortical integrating networks. The effect of that integration would be to identify a pattern of stimulation with an "object" at a particular position in space, independently of the exact pattern of stimulation received from the object at a given moment. A spatial background against which the object would be placed as "figure" may also be assumed as part of this construction since one of the major characteristics of the background is the fact that it remains fixed in time. Together these capacities are moves in the direction of creating a perceptual world by the brain as a "model," following Craik's (1943) usage of modeling as "explaining," to make sense of the incoming stimulation to the brain.

QUANTITATIVE ANALYSIS OF VISUAL AND AUDITORY SYSTEMS IN MAMMALS

The analogies between vision and hearing that are discussed in this section recall some of the quantitative analysis of the visual and auditory systems discussed previously (Chapter 12). We can compare the relative extents of the neural representations of the two systems in living mammals, following cell counts published by Lashley (1950) and by Chow (1951). These data are summarized in Table 17.1. We note that the mammalian visual system, beyond the fact that it seems to have more peripheral elements, is even more heavily represented than the auditory system in the cortex of primates.

We should also recall that these systems, though described as sensory, are in fact integrative systems in important ways. It has been recognized for some time that defects in visual or auditory performance following lesions in both the primary and secondary portions of these projection systems can be understood as associated with attentional failures, failures to inhibit responses, and other higher order activities (Killackey and Diamond, 1971). (Attention and inhibition are, basically, primitive behavior functions, but when these are correlated with specific visual or auditory inputs and specific response patterns that had been learned in connection with

Table 17.1

Number of Neurons or Other Elements at Various Levels of Visual and Auditory Systems in One Hemisphere[a]

Visual system				Auditory system	
Level	Rat	Rhesus monkey	Man	Level	Rhesus monkey
Acuity "units,"[b] visual field	32,000	—	338,000	Cochlear nuclei (III)	88,000
Retinal ganglion cells (III)	292,000	—	—	Superior olivary complex (III)	34,300
Optic nerve (fibers) (III–IV)	—	1,210,000	1,010,000	Lateral lemniscus (III–IV)	38,100
Lateral geniculate neurons (IV)	34,000	1,084,000	1,200,000	Inferior colliculus (IV_R)	392,000
Striate cortex neurons	655,000	145,285,000	—	Medial geniculate parvocellular (IV)	364,000
				Auditory cortex	10,200,000

[a] Roman numerals refer to levels indicated in Fig. 12.2. Data from Chow (1951) and from Chow et al. (1950), rounded to nearest thousand neurons.

[b] "Acuity units" is a measure relating behavioral acuity to ganglion cells in the rat and a direct behavioral measure of "discriminable intervals" in man.

those inputs, it becomes a total behavior pattern that may truly be described as "higher order.") It is apparent from Table 17.1 that the corticalization of vision in living mammals is at least comparable to that of hearing.

BEHAVIORAL ADVANTAGES OF ENLARGED BRAINS IN UNGULATES AND CARNIVORES

At this point let us consider the results described in Chapters 13 and 14, in which we traced a pattern of evolutionary expansion of the brain in carnivorous and herbivorous species of mammals that may have been in a predator–prey relationship. How would enlargement of the brain, and the kind of model of the effect of enlargement that has been emphasized so far, be relevant to that evolution? The answer is not immediately apparent because one should not accept, without question, suggestions about aggression and flight as functions requiring additional central neural tissue in significant quantities.

I suggest instead that the actual functions that developed involved the perceptual equipment necessary for flexible response to predators and prey. To the extent that the small Oligocene horse *Mesohippus* appeared to a predacious sabretooth *Hoplophoneus* as an object that moved about and changed its appearance as it moved, yet remained potential prey, the Oligocene sabretooth could maintain its tracking pattern and probably inhibit its attack response until it was in a position to spring and kill. Similarly, that ancestral horse's capacity to escape its fate as potential prey was undoubtedly improved by whatever capacity it had to integrate olfactory, visual, and auditory information from its predator as it was being stalked, identify ("name") the source of danger, and adopt an appropriate strategy for escape.

Romer presented this kind of analysis when he discussed the replacement of the archaic carnivores, the creodonts, by the more progressive true carnivores. His statement can be applied generally to the evolution of advanced habits as related to the evolution of brain size.

> The brain of the creodonts was generally of relatively small size and their intelligence presumably low. This may have been a main cause of the early extinction of almost all members of the group; for with replacement of the slow-footed and stupid herbivores of early Tertiary time by the swifter modernized ungulates, intelligent group pursuit (as in the wolf pack) or clever stalking (as in the case of cats) became necessary for the capture of prey.[1]

The specific analysis that I would add to Romer's statement would have to do with the definitions of "intelligent" and "stupid" and the mean-

---

[1] Reprinted from "Vertebrate Paleontology," 3rd ed., by A. S. Romer, The University of Chicago Press, Chicago, Illinois, copyright 1966, pp. 229–230.

ing of "clever pursuit." I would argue that all these terms can be basically perceptual and cognitive and relate to the constructed "real" worlds of both predator and prey. The problem of pursuit is one that emphasizes accurate information about the position of the potential prey and some capacity to retain information about previous patterns of motion of the prey. The evolution of the capacity to handle such information would have been at least as important as the perfection of sensorimotor coordination to enable an animal to be swift as well as bright. And considerably more neural tissue is probably required for the evolution of improved perceptual and cognitive capacities than for capacities associated with more rapid and accurate motion.

ENLARGEMENT OF THE ANTHROPOID BRAIN

A difficult problem for a speculative analysis is the source of selection pressures that resulted in a doubling of the size of the brain in the primates between the basic condition of the prosimians and the appearance of the anthropoids (the New and Old World monkeys and the apes; in previous chapters the suborder Anthropoidea was treated as including the hominids). The prosimian brain had rapidly reached the size of the "average" of living mammals early in the radiation of the primates (about 40 m.y. ago), but the typical anthropoid brain is between 2 and 3 times the size of the brain of prosimians and other "average" mammals (Table 16.3 and Appendix II). Furthermore, the fossil evidence, though limited, suggests that the increase in the size of the brain in the anthropoids occurred relatively early in their radiation, perhaps in the Oligocene, about 30 or 35 m.y. ago. It is important for our analysis of the evolution of brain size in the hominids that we achieve some understanding of the position of the anthropoids because it was from an anthropoid base that the progressive further increase in brain size in the Pleistocene hominids occurred.

Among the behavioral correlates of increased brain size in the anthropoids, we could consider social behavior, play, parental care, and so forth. But all these categories are often seen in other mammalian species without notably enlarged brains. The particular way in which they occur in anthropoids may be different, but, again, there appears no necessary a priori relationship that would enable one to identify a specific selection pressure on the anthropoids from which the prosimians were spared.

There is a possible resolution of the problem that involves an analogy with the enlargement of the brain in the ungulates during the Tertiary period. Living prosimians are relatively restricted in their distribution. Most living diurnal species are on the isolated island of Madagascar, and the prosimians elsewhere are in nocturnal niches. The prosimians and the anthropoids are generally herbivorous animals, and their present and past conditions are, therefore, analogous to those of the Tertiary ungulates. The suggestion is that the natural predators of the tree-dwelling primates of the Eocene and Oligocene may have become efficient enough to place the prosimians at a selective disadvantage relative to larger-brained anthropoids, and the evolution of larger brains, analogous to that of the Tertiary ungulates, was also appropriate for the primates. Prosimians that found their way to Madagascar were not under the same selection pressure since almost no large carnivores reached that island until the Pleistocene. The Madagascar lemurs could then accomplish a significant adaptive radiation along dimensions independent of brain size, and they even filled niches of bear-sized herbivores. Elsewhere, however, the prosimians either retreated to more obscure nocturnal niches or were replaced by larger-brained species, the anthropoids, which could cope with the predation of the more advanced carnivores and other specially adapted predacious species, such as the snakes, among the reptiles.

The adaptive advantages of the enlarged brains of the anthropoids may have been associated with superior forms of vision, including more complete stereopsis permitting better depth perception and more adequate use of color information. The diurnal prosimians have an appropriate retina for color perception (Wolin and Massopust, 1970), and at a retinal or subcortical level these adaptations must characterize primates as a class. Such adaptations, with elaboration of the cortical representation of vision, hearing, and motor behavior in the anthropoids, would have been associated with superior locomotion and superior perception, including the development of brachiation as a form of movement. This would have resulted in a much more mobile arboreal mammal than had previously evolved, one that could be more sensitive to the presence of predators and could move gracefully through trees with significantly finer accuracy than the essentially quadrupedal prosimians. There are also more subtle dimensions of behavior, involving social vigilance (Hall, 1960) and defensive communication, that may have evolved at that time to enable the herbivorous anthropoids to avoid or escape their predators.

Within the Anthropoidea too little is known about the evolution of the brain during the transition from monkeys to apes to make valid evolutionary judgments. An analysis according to the "extra neurons" hypothesis clearly placed the living apes above the monkeys (including the baboons) with respect to the number of neurons devoted to "intelligent" behavior (see Fig. 16.10). Behavioral experiments are equivocal at this time. So many uncontrolled factors can enter into the testing of animal "intelligence" that all conclusions must be hedged to the point that they become almost meaningless. Most students believe that the pongids are capable of more elaborate and more plastic behavioral adaptations than are the monkeys either from New World or Old World stock. But the current evidence with delayed reaction or learning-set tasks, though suggestive that the apes are advanced, is insufficient to permit conclusions about a transition between monkeys and apes to be applied to evolutionary analyses (Warren, 1965). It is as easy to imagine the precursors of the hominids as derived from a simian as from a pongid stock if we are restricted to evidence of the present behavioral capacities of these groups.

HOMINIDS AND THE HUMAN BRAIN

The evolutionary progression in primates that culminated with the achievement of the brain size of *Homo sapiens* covered over 10 m.y. according to the most recent evidence. In considering the selection pressures that determined this evolution, it is probably appropriate to imagine the condition of the ancestral primate in the hominid line as being close to the anthropoid condition presently exemplified by the baboons and forest apes. Speculations on this protohominid condition were reviewed in the previous chapter and are considered again later in this section.

The most important new result is the discovery that the relatively small-brained australopithecines were probably the typical hominids for most of the last 5 m.y. So long an occupation of a hominid adaptive zone without much enlargement of the brain has important implications for the kind of picture that we construct of the selection pressures that resulted in the enlarged human brain (*Homo erectus* and *Homo sapiens*), when this finally occurred.

When *Australopithecus africanus* was discovered by Dart (1925), as the Taung child with its 400–500 ml brain, there was little question among evolutionists that the specimen was in the lineage of the apes. A proper "missing link" for the evolutionary consensus of 50 years ago had to be an apelike animal with a manlike brain, and the Taung brain was too small. The questions at issue were whether man had, indeed, evolved from apelike ancestors, and those who accepted evolution assumed, without question, that man's mind and intelligence were the key to his evolution. That was the basis of the 1913 Piltdown forgery ("Eoanthropus"), in which a human skull was associated with an ape's lower jaw (Coon, 1962).

But even before the fraudulence of "Eoanthropus" had been exposed, an alternate view of human evolution had been proposed by Washburn and his students (Washburn, 1951). Washburn recognized that different evolutionary characters have frequently evolved at different rates and that the evolution of the human brain may have been a later response by hominid species to a situation in which other phenotypic characteristics had already reached a hominid grade. It was in this atmosphere that the scientific community eventually accepted the Taung child as a hominid; the crucial statistical evidence was based on dental and skeletal features (Ashton et al., 1957; see also LeGros Clark, 1967).

The picture is that of the early hominids as a group of moderately large primates (20 kg or more), living in a niche similar to that of the living baboons. These animals were supposed to have ranged much more widely than do the tribes of baboons studied by Hall, Washburn, Devore, and others (Crook, 1970). The major achievement of the past decade of field studies in primatology has been to define the behavioral responses to such an ecological niche. These have lent strength to the conjectures of Bartholomew and Birdsell (1953) about the nature of such a niche, presented in the previous chapter. They indicated, furthermore, that the enlargement of the human brain could have followed rather than preceded the invasion of that hominid niche.

The fact that the australopithecines had survived as a genus (and perhaps as a species) for about 4 or 5 m.y., a normal and not unusually short species "life span" (Simpson, 1953), is the critical point. It demonstrates that the selection pressures of the australopithecine niche were not, of themselves, sufficient to produce the enlarged brain that one associates with true men, although they probably forced some enlargement of the brain. The probable adaptations of the australopithecines must have served as precursors for later hominid adaptations by permitting hominids to explore the niches that were selectively advantageous for a species with a significantly enlarged brain.[1a]

ADAPTIVE ZONE OF THE AUSTRALOPITHECINES

In speculations on the adaptive zone of the early hominids it was emphasized that the initial "protohominid" condition may have been that of a semicarnivorous grasslands ape. Brachiation, though normal as an adaptive mode (the animal would have had the neural and motor capacities

---

[1a] As more fossils are found it should be possible to extend, and if necessary, modify this and later speculations about selection pressures toward larger hominid brains. For example, R. E. F. Leakey's (1973) latest find, described in Chapter 16, footnote 2, could alter the details of this discussion, especially with regard to the timing of the enlargement of the brain to the *Homo erectus* level and the probable ecology of the geographical area where this occurred. The role of glaciation discussed later under "Changes in the Hominid Adaptive Zone," might have been to alter the climate globally, making it possible for some early Pleistocene hominid tribes to live in a seasonally variable climate with very harsh winters, without necessarily being in a glacial region. Selection pressures of permanent residence in such an area would be the same as those discussed later in this chapter when considering possible early pithecanthropine environments.

associated with brachiation), had not produced skeletal specialization to the degree that would inhibit the use of the arms, hands, and thumb to hold, handle, and, eventually, mold tools. The major skeletal adaptations of the early hominids permitted the bipedal gait (perhaps not fully developed in the early forms with the adaptation for running rather than walking). Early bipedal hominids could survey, continuously, a broad scene that would be obstructed by tall grasses if their locomotion and posture were like that of quadrupeds of their body size.

The change from a basically herbivorous to a carnivorous diet would have been coupled with the development of active predacious habits. In view of the relatively generalized skeleton this would have led to the use of the free arms and hands as tool-grasping organs (already adapted to "handling" branches, which have been viewed as tools of brachiators). Tools, haphazardly collected, could serve in the place of talons, claws, or canine teeth to bring down small herbivores as prey. Furthermore, the predacious habit would have led to social organization in which a broad geographic range was covered by small numbers of social predators.

Important in understanding the early hominid adaptive zone are the sets of niches of social predators such as wolves, the niches currently occupied by savanna-dwelling baboons, and the very different niches of the forest apes: the chimpanzees and gorillas. It is generally assumed that the baboon's niche, rather than that of the great apes, is close to that of the early hominids (Devore and Washburn, 1963). Baboons lead complex social lives in which the troop and its organization determine the lives of individual members of their basically herbivorous society. The early hominids would have been more loosely organized, more carnivorous, and more individualized, with less specialization toward defense by ritually established troop leadership for specified males or for the almost military structure that is consistent with the adaptive modes of the baboon.

If we imagine a niche intersecting that of a loosely structured group of gorillas, as described by Schaller (1963), of the baboons with their more tightly knit organization and lesser capacities to withstand predators when they are away from the group (Washburn and Howell, 1960; Washburn, 1965), and of wolves as broad-ranging social predators (Crook, 1970), we may have some picture of an early hominid population. The carnivorous lives of the early hominids would imply a much wider geographic range and more thinly concentrated population than that of any of the living monkeys as well as a social organization more appropriate for attack than for defense. There would also be a clearer advantage to auditory signaling systems and visual and other neural systems for scanning and retaining information about broad vistas perceived by the vigilant primate (Hall, 1960).

The oldest hominids of reasonably well-known brain and body size, *Australopithecus boisei* ("Zinjanthropus") described by Tobias (1967), and *"Homo habilis,"* recently described by Leakey *et al.* (1971), with 530 and 560 ml endocasts, respectively, were only slightly beyond the relative brain size range of living pongids. These were brains sufficient for the adaptive zone just described. There must have been other selection pressures that later led to the remarkably enlarged brain of *Homo erectus* and *Homo sapiens*. We should be able to visualize the potential advantage of an enlarged brain, however, if species like the australopithecines began to explore new niches, and we should be able to identify such niches as having been entered during the last million years or even longer ago.

CHANGES IN THE HOMINID ADAPTIVE ZONE

The selection pressures that led to our enlarged brains must have appeared at the stage of human evolution represented by the pithecanthropines (*H. erectus*). The research of the past two decades on the australopithecine niche defines a point of departure in human evolution. But it is in the contrasts between the australopithecines and the pithecanthropines, rather than between the australopithecines and living forest apes or savanna baboons, that we may identify the selection pressures that led to man. As I have mentioned before, the australopithecines were long-lived as a species. There is no reason to assume that there was significant organic evolution of their brains during their span of 5 m.y. or so. The spectacular transition occurred with the appearance of the pithecanthropines of the genus *Homo* during early or mid-Pleistocene times. To illustrate the kind of environmental challenge required, let us examine the implications of a well-known Pleistocene phenomenon for the hominization of the brain.

One outstanding ecological fact of Pleistocene hominid evolution was certainly the climatic changes associated with glaciation. It is not difficult to picture the situation of a semitropical, omnivorous australopithecine species sufficiently broad ranging to have explored the geographic extremes of its adaptive zone. During parts of the Pleistocene, tribes at the periphery of the total geographic range would have faced the challenge (over millennia) of changing climates. They could remain in such areas only by restricting their daily ranges and developing new adaptations for successful hunting and for protection from the cold. Adaptations useful for communication within the tribe, originally devised for use over long distances, would have to be modified for communications within a much restricted normal home range, with only a limited number of shelters and unusually close and continuous proximity to other members of the tribe.

Problems of crowding may thus be much older than we realize. It may have been the adaptation to crowding and cold that imposed the special selection pressures upon that broad-ranging dominant predacious species of man–apes that led to the human condition. It was probably under such selection pressures that the evolution of the genus *Homo* took its brain beyond the long-persistent australopithecine level. The evolution had to be rapid, and it should have occurred near the edges of the Pleistocene glacial zones. The potential was there in the broad-ranging apelike hominid who survived as a species from the Pliocene to the mid-Pleistocene in semitropical climates. The adaptations for social yet carnivorous habits in such a broad-ranging species, challenged by an environment in which there was seasonal narrowing of the range and the need for an artificial climate (through fire and clothing) and artificial prey (through domestication or semidomestication of some animals), were accomplished by brain evolution rather than skeletal evolution and resulted in the broad and flexible behavior repertoire of the human species.

Speculations like these were familiar during the past hundred years, since the evolution of man was first recognized. In view of recent concern with the ecology of the australopithecines and the deemphasis of the pithecanthropines, it is important to restate the major point: the basic evolutionary changes probably occurred at the transition between *Australopithecus* and *Homo* (but beyond the level of *"Homo habilis"*) rather than with the advent of the australopithecines. Those concerned with the subsequent evolution of human behavior should reemphasize the role of that transition if they are to appreciate what were probably the last significant organic changes in man's brain as it evolved to its present status. These evolutionary changes occurred between 0.5–0.25 m.y. ago. Since that time, there have probably been only minor developments of the brain and intelligence that can be attributed to organic evolution.

THE WORK OF THE HOMINID BRAIN

We must now consider in more detail what the brain does that made it selectively advantageous for the later hominids to have enlarged brains. A satisfactory statement of how the brain's work produces the essence of human (or animal) behavior is still not possible and remains one of the great challenges to the neurosciences. It is appropriate, however, to consider some of the adaptive dilemmas that were faced by the hominids and were solved through behavioral means for which an enlarged brain was probably essential.

I have stated some of the adaptive problems in the previous section.

How could the species "conserve" ("the conservation principle") the primate adaptations to a life in forest and trees, in which the geographic range was limited, while entering an adaptive zone in which the daily range could encompass tens of square miles? How could the behavior patterns adequate for defensive activities of herbivorous brachiators be converted conservatively into the patterns of bipedal, social, ground-dwelling predators? Finally, how would a species with adequate adaptive responses to the two challenges just mentioned cope with new demands of a predator species in the harsh climates of a glacial area?

The answers, at the level that these can be presented at this time, are of the type suggested in the discussion of biological intelligence in Chapter 1, in the discussion of gross brain functions in Chapter 3, and in the general view of brain:behavior relations suggested at many points in this book. I prefer to emphasize the role of the evolution of cognitive capacities such as consciousness, imagery, or the higher order perception made familiar by the Gestalt psychologists. One reason for this is that most other behavioral functions that can be described as being selectively advantageous to a species are known in small-brained as well as large-brained vertebrates and in many invertebrates. This includes such common categories as learning, tool using, social communication, and so forth, which are known in insects, mollusks, and many small-brained vertebrate species. The saving feature for this analysis is the possibility that similar behaviors may result from different underlying mechanisms in different species. Human speech may be mimicked by birds, yet, although this raises some interesting questions for the analysis of language as well as for the analysis of bird behavior, there is no question that the brain mechanisms involved in normal human speech are different from those in the speech of, for example, a myna bird. It is with the nature of the brain mechanisms in the hominids that we should concern ourselves, and I would continue the cognitive approach as outlined in the introductory chapter in discussing the effect of the action of these mechanisms.

The basic concepts presented earlier in this chapter, especially the point that reality as one perceives it is a construction of the brain to "explain" the mass of neural information that it processes, are the core of this approach. We may then ask: in what sense did reality in australopithecines differ from that in their smaller-brained relatives among the primates? Similarly, we may ask in what sense the reality of the pithecanthropines had to be modified from that of their australopithecine ancestors in order to make survival easier in their niches. I suggest some answers at the conclusion of this section, after first reviewing the nature of imagery and its implications for the evolution of language.

IMAGERY

In this discussion I will present as fact what is really a chain of speculations about the structure of consciousness. This is done for didactic reasons because it becomes burdensome to repeat the caveats and disclaimers that would normally be used to preface each hypothesis. The statements are, in short, a presentation of what is possibly the case, consistent with information about cognitive and perceptual activity, and with what is presently known about brain:behavior relations.

In Chapter 1 I suggested that the functional organization of the brain in early hominids tended toward the modeling of auditory information and tactile and kinaesthetic information after the model of corticalized vision that had been achieved in the mammals generally and had been perfected to an unusual extent in the anthropoids. In other words, the hominids tended toward the construction of a perceptual world in which the information from the various sensory domains was bound together to provide a consistent picture of a spatially extended world, filled with objects that could move and emit sounds, be touched, smelled, and seen, and in which "constancy" of objects over time was guaranteed. Furthermore, it would have been a world in which time could be tremendously extended, to permit the integration of images over seconds, minutes, or even longer time periods. If time may be extended to retain events of the past, why not extend it into the future as a projection of events to come?

A perceptual world organized in this way is a world that is familiar to us. It is not only a world in which we can see and hear and touch, but a world in which we can imagine. And our images are not only of past events but of future events as well.

I would not suggest that all this could be accomplished by the brain of an australopithecine, although there is no immediate reason to suggest that each of the activities was not within the reach of that brain. Rather, I am suggesting that the kind of problems faced by the possessor of such a brain might be handled in essentially human ways. If a tool were being made, it could have been handled not in response to events past but in anticipation of events to come. The tool might, thus, be modified to be more appropriate to the imagery that it can evoke, that is, broken where it protrudes "inappropriately" or chipped, eventually, when the technology reached the point of working stone tools.

In this kind of perceptual world the role of sound could be that of completing the information about the perceptual structures rather than that of establishing and maintaining communications. It is easy to communicate, as we should know from "communication" in invertebrates and lower vertebrates, and most of the information that must be shared with others of a community of vertebrates—even a protohominid community—could be handled by a very elementary code, using relatively little brain tissue. To translate a visual world with extension in both space and time into an even richer auditory world, so that events in space are labeled and thus recognized on their next appearance, hours or days later, places a heavier demand on a nervous system than the requirement for vocal signaling sufficient to "release" a "fixed-action pattern" of the kind known in some primitive vertebrates.

The stage was set with the evolution of the hominids for the construction of perceptual worlds in which it would eventually be possible to evoke a visual image by the use of touch and sound or a tactile or auditory image by the use of other modalities such as vision or even odors. The fact that this also resulted in improved communication should not be confused with the requirements for communication in troops of monkeys or apes. The quality of communication in living troops of nonhuman primates is such that it could be simulated with very little information-processing material, and it appears to be no more effective in the control of behavior than the vocal communication of other species, for example, the vocal communication of birds.

The quality of language that makes it special is less its role in social communication than its role in evoking cognitive imagery, and I suggest that it was this kind of capacity that was evolving in the early hominids. (We need language to tell a story much more than to give directions for an action.) It was for this kind of capacity that central neural structures that involved visual, auditory, tactile, and motor units (including motor systems of the tongue and larynx) had to become more elaborate, and it was appropriate that the structures, the "speech" areas, evolved at an anatomical position near the confluence of the primary or secondary areas associated with the central neural representation of these modalities in other anthropoids.

The use of tools and the use of language should both be considered in the context of the organization and reorganization of perceptual space. Both helped provide a temporally unified past and future, in which an intended state was signaled, and actions were taken to make it likely that the "reality" of the future conformed to the imagined ("predicted") reality. An unshaped stone may be imagined in its shaped state, as mentioned before, and practiced blows can force it into that shaped status.

LANGUAGE AND BRAIN DEFICIENCY

The evolutionary viewpoint leads us to the position that the capacity for language evolved with the evolution of the human brain. There are, of course, difficulties in defining language precisely, and one may demonstrate symbolic, languagelike behavior in various nonhuman species (Premack, 1971). It is, nevertheless, obvious that the use of language is an especially and perhaps, peculiarly, human adaptation and that it is only with difficulty that analogies or possible identities with language may be demonstrated in the behavior of other species.

An issue has occasionally been raised about when in human evolution and the evolution of brain size it became possible for hominids to use language. This is an unanswerable question. For one thing, language in living men is learned, and the evolutionary heritage refers only to the capacity for learning language. There is no reason to correlate that capacity with a particular amount of hominid brain, except in instances of extreme pathology. If it is analogous to other behavioral capacities in other species, language should be identified with a particular type of brain. Thus, a microcephalic living human would not be expected to be completely incapable of the use of language (although in some cases such individuals may not have demonstrated the capability) any more than a decerebrate bird would be expected to be incapable of flight.

A useful way to think of language in evolutionary terms is as a species-specific adaptation, such as the flight mechanism of the hummingbird or stalking behavior in cats. We know that mammals may suffer brain damage in the laboratory or in nature, yet we would scarcely expect them to revert to reptilian or piscine adaptive mechanisms. In the same sense, we should not be surprised when human beings with pathologically small brains retain the capacity to learn speech. The use of language by deficient humans may be unusual, but it must be gauged relative to the normal human use of language, not to the studies on "language" in other species in which ingenious experimental manipulations are used to force us to sharpen our definitions of the nature of language.

Were it not that so spectacular a human capacity is involved, we could omit this discussion. Yet language is what seems to make us most obviously different from other animals, and many of us would go to some length to protect our identities and emphasize the difference. This is a valid aim for the evolutionist because language as we know it is certainly a characteristic that could be used to distinguish man from other species in the taxonomic as well as theological sense. We should conclude that deficiencies in human development will not produce nonhuman individuals, merely deficient human beings.

Plasticity and the Capacity for Culture

Plasticity, or modifiability, is one of the fundamental capacities of nervous systems. In the construction of a complex "real" world, in which each sense modality contributes to the cognitive image, it would be difficult to imagine a prewired central nervous system that was prepared, ready-made, for all the capacities. Among the consequences of the elaboration of functionally modifiable nervous systems, one must include the capacity for acculturation, the development of adult–child relations that lead to social and emotional dependence, the use of artifacts, communication by languages, and so forth. Each of these (or other categories, such as play, duration of infancy, love) can be used as a point of departure for speculations about the sources of hominization. But in every instance, the question of the evolution of the brain has to be tied to the work of the brain.

The simplest intuitive description of the brain's work (for me) is that it creates a "real" world. Within that real world all the events of a lifetime take place. (The realities differ, of course, depending on the nervous systems that do the work of building a real world and on the experiences—the information—to which the nervous systems are exposed during the lifetimes of the individuals.) The perception of others in social roles can often be handled at the level of organization of "fixed action patterns," as discussed in Lorenz's (1935) famous analysis of the companion in the bird's world. The perception of the self, on the other hand, may be a peculiarly human development of the capacity for creating "objects" in a "real" world. The reality of the self as an object (a person) is one of the most compelling of intuitions. Our capacity for imagery and imagination, though still poorly understood, may be related to our models of our "selves" and is clearly a kind of information processing that goes beyond simple model building based on the accommodation to stimulation through a variety of sense modalities.

Individuals capable of constructing elaborate multisensory "real" worlds might construct a reality that seems more fundamental than the immediate information from the senses. The capacity for imagery, in which one manipulates a possible real world in one's imagination, must early have led the hominids, by the time these capacities were well developed, to reach an appreciation of a past prior to one's lifetime and a sense of a future after one's death.

These have become singularly human hallmarks, understandable as cognitive structures with temporal as well as spatial extent and with some symbolic superstructures such as self-awareness or consciousness added to them. We can carry these speculations no further if we limit ourselves to the organic evolution of the brain. But it is well for us to recognize that when it became selectively advantageous to construct a reality beyond any immediate sense modality, the path to hominization could be followed. The capacity for abstraction is inherent in the process of construction of such a reality as a synthesis of information from many sources. The development of language in the process was, for this perspective at least, a reasonable extension of a conscious perceptual world, given the existence of visual and nonvisual imagery and auditory and vocal capacities. We model "reality" linguistically just as we model it visually or tactilely. Languages may be media for communication because different human brains construct essentially the same "reality." And we can share that reality because it is already shared in the linguistic structures we have in common and in the linguistic factors that are part of our fundamental image of the real world. Speech and reading, in this sense, provide a genuine shared consciousness for members of the human species, given a common cultural background.

IMPLICATIONS FOR HOMINID BRAIN EVOLUTION

The questions presented early in this section on the brain's work can now be reviewed. These questions were answered in part as we discussed imagery and its relevance to toolmaking and to language. Our concern now is with the actual events in the history of the hominids that led to the evolution of the australopithecine brain and to the later evolution of the pithecanthropine brain from australopithecine stock. The role of toolmaking, incidentally, is fairly trivial in the analysis of the evolution of the brain (Hall, 1963) and has been discussed, in passing, in the previous pages. The point is that toolmaking could be accomplished with very little brain tissue and of itself would imply little about an enlarged brain. It is only if we assume that imagery was involved in toolmaking that we have reason to expect a selective advantage for a large-brained species. Comparable arguments would also hold for other categories, such as "learning." We may, therefore, devote most of this discussion to the evolution of language, since it is impossible to imagine grammatical and learned language as an accomplishment of a naturally small-brained species.

There has been much work during the past two decades on the definition and formal analysis of language and its biological basis, but the analysis refers to an aspect of language that is only marginally relevant for the evolution of the capacity for language and for the earliest probable use of linguistic structures. It is likely that the evolution of language began with the invasion of the australopithecine niche as we have discussed it here, in which a much broader geographical range than was typical for primates became the norm for the australopithecines and in which the tribe acted as a social unit adapted for predatory activities. Recognizing that primates, generally, are noisy animals with an elaborate set of species-specific calls, either innate or learned early in life (Marler, 1965), and recognizing that primates are well adapted to discriminate sounds, we must begin our appreciation of the evolution of language with a consideration of the potential utility of sound for the broad-ranging predacious australopithecines.

There are two potential uses of sounds in an advanced animal's life. First, there is the relatively straightforward use of sound to maintain contact with others in the group as part of normal social behavior. Little brain tissue is needed for such use, certainly no more than in "average" living mammals such as the social predators (wolves) or social ungulates. The second use of sounds is to encode information with a time-binding neural code. It is likely that the broad-ranging australopithecines, faced with a much extended range and a demand to encode considerably more information about events at a distance, could succeed in their new primate niche because they had evolved additional capacities to encode auditory and vocal information. Other social predators, such as wolves, probably solve some of this information-processing problem by using their well-developed olfactory system, but this system had become much reduced in the diurnal primates and was not available for redevelopment. The essential problem is to form an adequate map of an extremely extended terrain and to retain that map while using it to navigate the terrain and locate objects within it.

The capacity that had to be achieved by the australopithecines was for a kind of "auditory" analysis that would result in the temporal encoding of information about objects in space. As an extension of the typical primate capacity to assign sounds, which they themselves can produce, to events at a distance (a capacity shared with many mammals and most birds) and the presently poorly understood capacities for temporal integration that are inherent in auditory functions, it is likely that the australopithecines developed the capacity to name, in some sense, the important objects in their visual fields. They also must have had the capacity to use these names for easier retention of the nature and location of the objects as they navigated their ranges.

The effect of such activity would have been to extend the perceptual capacities of each individual by adding dimensions to objects and to space itself, in particular, temporal dimensions and the cognitive dimensions contributed by the produced and heard sounds. To the extent that the australopithecines were, in fact, social hunters, this capacity would have been quite useful because it would have enabled the individuals in a tribe to share their experiences, their images, as it were, by producing the sounds necessary to evoke a similar image in another individual as a construction of the sensory systems of the brain. The world of the australopithecines could hardly have been linguistic beyond the level just indicated. The role of language may have been limited to naming, and its utility would have been in enabling individuals to share their experiences with others in a tribe beyond the sharing possible from the specific calls and warning cries normally emitted only in the presence of the object eliciting the call.

Having attained this level, an aberrant, peripheral group of hominids could develop further as part of the adjustment to crowding and to cold if their homes were in glacial areas. Within a tribe attempting to survive in a varying glacial area, there would be clear advantages to being able to communicate from generation to generation the accumulated information from the past. The use of artifacts and artifices in clothing, in the hunt, in fire, and in tools can hardly be imagined if each individual had to invent the full array for himself. Communication would often take the form of creating appropriate imagery in the listener by preparing more and more detailed statements about the information. There is a selective advantage to the capacity for culture, even at that primitive level, in such tribes. The capacity for language, which may have been present in a primitive form in the semitropical species of australopithecines, would have become a major evolutionary lever in the evolution of later hominids, the pithecanthropines, who could then attempt a life at the edges of the normal range of the earlier hominids.

Tribes at the periphery, attempting to adapt to a restricted range, could use language as a surrogate for some of the activities that are in their normal behavior repertoire and for which a broad geographical range is necessary. Language could enable individuals to share real or imagined (but possible) experiences with one another. They could then "displace" some activities, which would normally be appropriate responses to these experiences, by verbal exercises in which all the activities are carried out symbolically. This possible value of language requires its use to tell stories, present dramas, and gratify needs for grand experiences of comedy or tragedy, which are familiar uses today.

The catalog of other values and uses of language is inexhaustible for a species in which the individuals are normally in close social contact and yet maintain a predatory life. The point to emphasize is that the origins of language were as likely in the pressures to create a better model of a real world, that is, in perceptual and cognitive development, as in the pressures toward being able to communicate with one's fellows.

CONCLUSIONS

The vertebrate brain has evolved to control the normal range of behavior within each vertebrate species. It is a specialized structure, of course, but one theme of this book has been to emphasize the possibility of using the

Conclusions

absolute and relative size of the brain to estimate its information-processing capacity. In this way it seemed most reasonable to identify the brain of lower vertebrates as associated with a common level of adaptive behavior mechanisms. These are, typically, the mechanisms of fixed action patterns in response to patterns of stimulation, in which relatively little construction of models of real worlds is assumed to occur. In short, in the lower vertebrates one may be dealing with brain:behavior systems at the Cartesian level, essentially reflex machines with few requirements for plasticity or flexibility.

In the higher vertebrates, the birds and mammals, plasticity and flexibility are evident in all living species. Yet the birds developed along a different direction, perfecting to an unusual extent the fixed action pattern as the basic behavioral response to environmental requirements. It was really within the mammals that more flexible patterns of behavior have been the rule. The evolution of intelligence occurred mainly within the mammals and only in a casual way in birds, if one defines intelligence as the capacity to learn new response patterns in which sensory information from various modalities is integrated as information about objects in space.

These trends reached their most elaborate development in the evolution of the primates, a group of mammals adapted toward adaptability, in which skeletal specialization was minimal and adaptations were more completely determined by the enlargement of the brain and the development of learned behavior mechanisms than in any other vertebrates. The trend culminated in man, and we know it as the capacity for imagery, for language, and for culture.

Bibliography

Abbie, A. A. (1940). Cortical lamination in the Monotremata. *J. Comp. Neurol.* **72**, 429–467.
Ades, H. W. (1959). Central auditory mechanisms. *In* "Handbook of Physiology" (Amer. Physiol. Soc., J. Field, ed.), Sect. 1; Vol. I, pp. 585–613. Williams & Wilkins, Baltimore, Maryland.
Adey, W. R. (1959). The sense of smell. *In* "Handbook of Physiology" (Amer. Physiol. Soc., J. Field, ed.), Sect. 1; Vol. I, pp. 535–548. Williams & Wilkins, Baltimore, Maryland.
Adolph, E. F. (1951). Some differences in responses to low temperatures between warm-blooded and cold-blooded vertebrates. *Amer. J. Physiol.* **166**, 92–103.
Adrian, E. D. (1947). "The Physical Background of Perception." Oxford Univ. Press (Clarendon), London and New York.
Aitken, J. T., and Bridger, J. E. (1961). Neuron size and neuron population density in the lumbosacral region of the cat's spinal cord. *J. Anat.* **95**, 38–53.
Allison, T., and Van Twyver, H. (1970). The evolution of sleep. *Natur. Hist.,* **79**, 56–65.
Andrew, R. J. (1962). Evolution of intelligence and vocal mimicking. *Science* **137**, 585–589.
Andrews, C. W. (1897). Note on the cast of the brain cavity of *Iguanodon*. *Ann. Mag. Natur. Hist.* [6] **19**, 585–591.
Anthony, J. (1951). Existe-t-il un moyen anatomique satisfaisant d'exprimer le degré d'organisation cérébrale des Mammifères? I. Les principales méthodes employées jusqu'à présent. *Mammalia* **15**, 53–68.
Anthony, J. (1951). Existe-t-il un moyen anatomique satisfaisant d'exprimer le degré d'organisation cérébrale des Mammifères? II. Discussion. *Mammalia* **15**, 124–137.
Anthony, R. (1938). Essai de recherche d'une expression anatomique approximative du degré d'organisation cérébrale autre que le poids de l'encéphale comparé au poids du corps. *Bull. Soc. Anthropol. Paris* [8] **9**, 17–67.
Ariëns Kappers, C. U. (1929). "The Evolution of the Nervous System in Invertebrates, Vertebrates and Man." Bohn, Haarlem.
Ariëns Kappers, C. U., Huber, G. C., and Crosby, E. (1936). "The Comparative Anatomy of the Nervous System of Vertebrates, Including Man," 2 vols., Macmillan, New York.
Aronson, L. R. (1963). The central nervous system of sharks and bony fishes with special reference to sensory and integrative mechanisms. *In* "Sharks and Survival" (P. W. Gilbert, ed.), pp. 165–241. Heath, Boston, Massachusetts.
Aronson, L. R., and Noble, G. K. (1945). The sexual behavior of Anura. II. Neural mechanisms controlling mating in the male leopard frog, *Rana Pipiens*. *Bull. Amer. Mus. Natur. Hist.* **86**, 83–140.

Ashton, E. H., Healey, M. J. R., and Lipton, S. (1957). The descriptive use of discriminant functions in physical anthropology. *Proc. Roy. Soc. London, Ser. B* **146**, 552–572.

Augusta, J., and Burian, Z. (1961). "Prehistoric Reptiles and Birds." Paul Hamlyn, London.

Axelrod, D. I., and Bailey, H. P. (1968). Cretaceous dinosaur extinction. *Evolution* **23**, 595–611.

Baerends, G. P. (1957). The ethological analysis of fish behavior. *In* "The Physiology of Fishes" (M. E. Brown, ed.), Vol. 2, pp. 229–269. Academic Press, New York.

Barnett, L. (1955). "The World We Live In." Time-Life Inc., New York (distributed by Simon & Schuster).

Barnett, L. (1960). "The Wonders of Life on Earth." Time-Life Inc., New York.

Bartholomew, G. A., Jr., and Birdsell, J. B. (1953). Ecology and the protohomonids. *Amer. Anthropol.* **55**, 481–498.

Bauchot, R. (1963). L'architectonique comparée qualitative et quantitative du diencéphale des insectivores. *Mammalia* **27**, Suppl. 1, 1–400.

Bauchot, R., and Stephan, H. (1961). Etude quantitative de quelques structures commisurales du cerveau des insectivores. *Mammalia* **25**, 314–341.

Bauchot, R., and Stephan, H. (1966). Données nouvelles sur l'encéphalisation des insectivores et des prosimiens. *Mammalia* **30**, 160–196.

Bauchot, R., and Stephan, H. (1967). Encéphales et moulages endocrâniens de quelques insectivores et primates actuels. *In* "Problèmes Actuèls de Paléontologie (Évolution des Vertébrés)." *Colloq. Int. Cent. Nat. Rech. Sci.* **163**, 575–587.

Bauchot, R., and Stephan, H. (1969). Encéphalisation et niveau evolutif chez les Simiens. *Mammalia* **33**, 225–275.

Beltan, L. (1957). Etude sommaire d'un moulage naturelle de la cavité cranienne d'un Boreosomus de l'Eotrias de Madagascar. *C. R. Acad. Sci.* **243**, 549–551.

Berggren, W. A. (1969). Cenozoic chronostratigraphy, planktonic foraminiferal zonation and the radiometric time scale. *Nature (London)* **224**, 1072–1075.

Berry, M., Hollingworth, T., Flinn, R., and Anderson, E. M. (1973). Morphological correlates of functional activity in the nervous system. *In* "Macromolecules and Behavior" (G. B. Ansell and P. B. Bradley, eds.), pp. 217–240. University Park, Baltimore, Maryland.

Bing, R., and Burckhardt, R. (1905). Das Centralnervensystem von *Cerotodus forsteri* Semon, R. Zool. Forschungsreisen in Australien. Vol. 1. *Ceratodus* Jena. *Med. Naturwiss. Ges.* **4**, 1893–1913.

Binkley, S., Kluth, E., and Menaker, M. (1971). Pineal function in sparrows: Circadian rhythms and body temperature. *Science* **174**, 311–314.

Bishop, S. C. (1943). "Handbook of Salamanders." Cornell Univ. Press, Ithaca, New York.

Bitterman, M. E. (1965). The evolution of intelligence. *Sci. Amer.* **212(1)**, 92–100.

Black, D. (1915). A study of the endocranial casts of *Ocapia, Giraffa* and *Samotherium* (Miocene form), with special reference to the convolutional pattern in the family of Giraffidae. *J. Comp. Neurol.* **25**, 329–357.

Black, D. (1920). Studies on endocranial anatomy. II. On the endocranial anatomy of Oreodon. *J. Comp. Neurol.* **32**, 271–314.

Blinkov, S. M., and Glezer, I. I. (1968). "The Human Brain in Figures and Tables." Basic Books, New York.

Bok, S. T. (1959). "Histonomy of the Cerebral Cortex." Van Nostrand-Reinhold, Princeton, New Jersey.

Bonner, J. T. (1965). "Size and Cycle." Princeton Univ. Press, Princeton, New Jersey.
Brainbridge, R. (1958). The speed of swimming in fish as related to size and to the frequency and amplitude of the tail beat. *J. Exp. Biol.* **35**, 109–137.
Bramwell, C. D. (1970a). The first hot-blooded flappers. *Spectrum* **69**, 12–14.
Bramwell, C. D. (1970b). Those flappers again. *Spectrum* **72**, 7.
Bramwell, C. D. (1971). Flying ability of Archaeopteryx. *Nature (London)* **231**, 128.
Bramwell, C. D., and Whitfield, G. R. (1970). Flying speed of the largest aerial vertebrate. *Nature (London)* **225**, 660–661.
Brazier, M. A. B. (1959). The historical development of neurophysiology. *In* "Handbook of Physiology" (Amer. Physiol. Soc., J. Field, ed.), Sect. 1, Vol. I, pp. 1–58. Williams & Wilkins, Baltimore, Maryland.
Brazier, M. A. B. (1968). "The Electrical Activity of the Nervous System," 3rd ed. Pitman, London.
Breathnach, A. S. (1955). Observations on endocranial casts of recent and fossil cetaceans. *J. Anat.* **89**, 532–546.
Bridgeman, P. W. (1931). "Dimensional Analysis," rev. ed. Yale Univ. Press, New Haven, Connecticut.
Brink, A. S. (1956). Speculations on some advanced mammalian characteristics in the higher mammal-like reptiles. *Paleontol. Afr.* **4**, 77–96.
Brink, A. S. (1958). Note on a new skeleton of *Thrinaxodon liorhinus*. *Paleontol. Afr.* **6**, 15–22.
Brown, B., and Schlaikjer, E. M. (1940). The structure and relationships of *Protoceratops*. *Ann. N.Y. Acad. Sci.* **40**, 133–266.
Brown, M. E. (1957). Experimental studies on growth. *In* "The Physiology of Fishes" (M. E. Brown, ed.), Vol. I, pp. 361–400. Academic Press, New York.
Brummelkamp, R. (1940). Brain weight and body size: A study of the cephalization problem. *Proc., Kon. Ned. Akad. Wetensch.* **39**, 1–57.
Bullock, T. H. (1967). Simple systems for the study of learning mechanisms. *In* "Neurosciences Research Symposium Summaries" (F. O. Schmitt *et al.*, eds.), Vol. 2, pp. 203–327. MIT Press, Cambridge, Massachusetts.
Bullock, T. H., and Horridge, G. A. (1965). "Structure and Function in the Nervous Systems of Invertebrates," 2 vols. Freeman, San Francisco, California.
Burrington, R. S., and May, D. C. (1970). "Handbook of Statistical Formulas," 2nd rev. ed. McGraw-Hill, New York.
Butcher, H. J. (1968). "Human Intelligence." Methuen, London.
Campbell, C. B. G. (1966a). Taxonomic status of tree shrews. *Science* **153**, 436.
Campbell, C. B. G. (1966b). The relationships of the tree shrews: The evidence of the nervous system. *Evolution* **20**, 276–281.
Campbell, C. B. G., and Hodos, W. (1970). The concept of homology and the evolution of the nervous system. *Brain, Behavior, Evolution* **3**, 353–367.
Carter, G. S. (1967). "Structure and Habit in Vertebrate Evolution." Sidgwick & Jackson, London.
Chow, K. L. (1951). Numerical estimates of the auditory central nervous system of the rhesus monkey. *J. Comp. Neurol.* **95**, 159–176.
Chow, K. L., and Leiman, A. L. (1971). The structural and functional organization of the cortex. *In* "Neurosciences Research Symposium Summaries" (F. O. Schmitt *et al.*, eds.), Vol. 5, pp. 149–312. MIT Press, Cambridge, Massachusetts.
Chow, K. L., Blum, J., and Blum, R. A. (1950). Cell ratios in the thalamocortical visual system of macaca mulatta. *J. Comp. Neurol.* **92**, 227–239.
Clemens, W. A., Jr. (1966). Fossil mammals of the Type Lance Formation in Wyoming. Part II. Marsupialia. *Univ. Calif., Berkeley, Publ. Geol. Sci.* **62**, 1–122.

Clemens, W. A., Jr. (1968). Origin and early evolution of marsupials. *Evolution* **22**, 1–18.
Coghill, G. E. (1929). "Anatomy and the Problem of Behavior." Cambridge Univ. Press, London and New York.
Colbert, E. H. (1962). The weights of dinosaurs. *Amer. Mus. Nov.* **2076**, 1–16.
Colbert, E. H. (1965). "The Age of Reptiles." Norton, New York.
Coleman, P. D., and Riesen, A. H. (1968). Environmental effects on cortical dendritic fields. *J. Anat.* **102**, 363–374.
Connolly, C. J. (1950). "External Morphology of the Primate Brain." Thomas, Springfield, Illinois.
Coon, C. S. (1962). "The Origin of Races." Knopf, New York.
Cooper, S. (1966). Muscle spindles and motor units. *In* "Control and Innervation of Skeletal Muscle" (B. L. Andrew, ed.), pp. 9–17. Livingstone, Edinburgh.
Cope, E. D. (1877). On the brain of *Coryphodon*. *Proc. Amer. Phil. Soc.* **16**, 616–620.
Cope, E. D. (1883). On the brains of the Eocene Mammalia *Phenacodus* and *Periptychus*. *Proc. Amer. Phil. Soc.* **20**, 563–565.
Cope, E. D. (1884). The vertebrata of the tertiary formations of the west. *Rep. U.S., Geol. Surv. Territ.* **3**, 1–1009.
Count, E. W. (1947). Brain and body weight in man: Their antecedents in growth and evolution. *Ann. N.Y. Acad. Sci.* **46**, 993–1122.
Craik, K. J. W. (1943). "The Nature of Explanation." Cambridge Univ. Press, London and New York.
Crile, G., and Quiring, D. P. (1940). A record of the body weight and certain organ and gland weights of 3690 animals. *Ohio J. Sci.* **40**, 219–259.
Crook, J. H., ed. (1970). "Social Behaviour in Birds and Mammals." Academic Press, New York.
Cuvier, G. (1835). "Recherches sur les Ossemens Fossiles," 4th ed., Vol. 5 and Atlas. D'Ocagne, Paris.
Dart, R. A. (1923). The brain of the Zeuglodontidae. *Proc. Zool. Soc. London*, 615–654.
Dart, R. A. (1925). *Australopithecus africanus:* The man-ape of South Africa. *Nature (London)* **115**, 195–199.
Dart, R. A. (1956). The relationships of brain size and brain pattern to human status. *S. Afr. J. Med. Sci.* **21**, 23–45.
Dart, R. A. (1957). The Osteodontokeratic Culture of *Australopithecus prometheus*. *Transvaal Mus. Mem.* **10**.
de Beer, G. (1937). "The Development of the Vertebrate Skull. Oxford Univ. Press, London and New York.
de Beer, G. (1954). "*Archaeopteryx lithographica*." British Museum, London.
de Beer, G. (1956). The evolution of ratites. *Bull. Brit. Mus. (Natur. Hist.), Zool.* **4**, 57–70.
Dechaseaux, C. (1958). *In* "Traité de Paléontologie" (J. Piveteau, ed.), Vol. 2, Tome VI. Masson, Paris.
Dechaseaux, C. (1964). L'encéphale de *Neurogymnurus cayluxi* insectivore des phosphorites du Quercy. *Ann. Paleontol.* **50**, 83–100.
Dechaseaux, C. (1968). Le cerveau d'*Archaeopteryx* est-il de "type avien" ou de "type reptilien"? *C.R. Acad. Sci.* **267**, 2108–2110.
Dechaseaux, C. (1969). Moulages endocrâniens d'artiodactyles primitifs. *Ann. Paleontol.* **60**, 195–248.

Dechaseaux, C. (1970). Cérébralisation croissante chez le courlis (*Numenius*) au cours de la période qui va de l'Eocène supérieur à l'époque actuelle. *C.R. Acad. Sci.* **270**, 1–3.

Dendy, A. (1911). On the structure, development and morphological interpretation of the pineal organs and adjacent parts of the brain in the tuatara (*Sphenodon punctatus*). *Phil. Trans. Roy. Soc. London, Ser. B* **201**, 227–331.

Denenberg, W. H., and Rosenberg, K. M. (1967). Nongenetic transmission of information. *Nature (London)* **216**, 549–550.

Denison, R. H. (1938). The broad skulled Pseudocreodi. *Ann. N.Y. Acad. Sci.* **37**, 163–257.

de Robertis, E., Gerschenfeld, H. M., and Wald, F. (1960). Ultrastructure and function of glial cells. *In* "Structure and Function of the Cerebral Cortex" (D. B. Tower and J. P. Schade, eds.), pp. 69–80. Elsevier, Amsterdam.

DeVore, B. I. and Washburn, S. L. (1963). Baboon ecology and human evolution. *In* "African Ecology and Human Evolution" (F. C. Howell and F. Bourlière, eds.), pp. 335–367. Aldine, Chicago, Illinois.

Diamond, I. T. (1967). The sensory neocortex. *In* "Contributions to Sensory Physiology" (W. D. Neff, ed.), Vol. 2, pp. 51–100. Academic Press, New York.

Diamond, I. T., and Hall, W. C. (1969). Evolution of neocortex. *Science* **164**, 251–262.

Dixon, W. J., and Massey, F. J., Jr. (1969). "Introduction to Statistical Analysis," 3rd ed. McGraw-Hill, New York.

Dobzhansky, T. (1955). "Evolution, Genetics, and Man." Wiley, New York.

Dodgson, M. C. H. (1962). "The Growing Brain." Wright, Bristol.

Donaldson, H. H. (1898). Observations on the weight and length of the central nervous system and of the legs, in bull-frogs of different sizes. *J. Comp. Neurol.* **8**, 314–335.

Donaldson, H. H. (1924). "The Rat," Mem. No. 6. Wistar Inst. Anat. Biol., Philadelphia, Pennsylvania.

Dow, R. S., and Moruzzi, G. (1958). "The Physiology and Pathology of the Cerebellum." Univ. of Minnesota Press, Minneapolis.

Dubois, E. (1897). Sur le rapport du poids de l'encéphale avec la grandeur du corps chez mammifères. *Bull. Soc. Anthropol. Paris* [4] **8**, 337–376.

Dubois, E. (1898). The brain-cast of pithecanthropus erectus. *Proc. Int. Congr. Zool., 4th, 1898,* pp. 79–96.

Dubois, E. (1920). The quantitative relations of the nervous system determined by the mechanism of the neurone. *Proc., Kon. Ned. Akad. Wetensch.* **22**, 665–680.

DuBrul, E. L. (1967). Pattern of genetic control of structure in the evolution of behavior. *Perspect. Biol. Med.* **10**, 524–539.

Ebbesson, S. O. E., and Northcutt, R. G., (in press). Neurology of anamniotic vertebrates. *Proc. Conf. Evol. Nerv. Syst. Behav.,* 1973.

Eccles, J. C., ed. (1966). "Brain and Conscious Experience." Springer-Verlag, Berlin and New York.

Eccles, J. C. (1973). The cerebellum as a computer: patterns in space and time. *J. Physiol.* **229**, 1–32.

Edinger, L. (1885). "Zehn Vorlesungen über den Bau der nervösen Zentralorgane." Vogel, Leipzig (cited by Magoun, 1960).

Edinger, T. (1926). The brain of *Archaeopteryx*. *Ann. Mag. Natur. Hist.* [9] **18**, 151–156.

Edinger, T. (1927). Das Gehirn der Pterosaurier. *Z. Anat. Entwicklungsgesch.* **83**, 105–112.

Edinger, T. (1929). Die fossilen Gehirne. *Ergeb. Anat. Entwicklungsgesch.* **28**, 1–249.

Edinger, T. (1938). Das Gehrin des *Libypithecus*. *Zentralbl. Mineral., Geol. Palaeontol., Abt. B* No. 4, 122–128.

Edinger, T. (1939). Two notes on the central nervous system of fossil Sirenia. *Bull. Fac. Sci., Fouad I Univ.* **19**, 41–57.

Edinger, T. (1941). The brain of *Pterodactylus*. *Amer. J. Sci.* **239**, 665–682.

Edinger, T. (1942). The pituitary body in giant animals fossil and living: A survey and a suggestion. *Quart. Rev. Biol.* **17**, 31–45.

Edinger, T. (1948). Evolution of the horse brain, Baltimore, Maryland. *Geol. Soc. Amer., Mem.* **25**.

Edinger, T. (1949). Paleoneurology versus comparative brain anatomy. *Confin. Neurol.* **11**, 5–24.

Edinger, T. (1951). The brain of the Odontognathae. *Evolution* **5**, 6–24.

Edinger, T. (1955a). Hearing and smell in cetacean history. *Monatsschr. Psychiat. Neurol.* **129**, 37–58.

Edinger, T. (1955b). The size of parietal foramen and organ in reptiles. A rectification. *Bull. Mus. Comp. Zool., Harvard Univ.* **114**, 1–34.

Edinger, T. (1956a). Objets et résultats de la paléoneurologie. *Ann. Paleontol.* **62**, 97–116.

Edinger, T. 1956b). Paired pineal organs. *In* "Progress in Neurobiology" (J. Ariëns Kappers (ed.), pp. 121–129. Elsevier, Amsterdam.

Edinger, T. (1962). Anthropocentric misconceptions in paleoneurology. *Proc. Rudolf Virchow Med. Soc. City N.Y.* **19**, 56–107.

Edinger, T. (1964a). Midbrain exposure and overlap in mammals. *Amer. Zool.* **4**, 5–19.

Edinger, T. (1964b). Recent advances in paleoneurology. *Progr. Brain Res.* **6**, 147–160.

Edinger, T. (1966). Brains from 40 million years of Camelid history. *In* "Evolution of the Forebrain" (R. Hassler and H. Stephan, eds.), pp. 153–161. Thieme, Stuttgart.

Elias, H., and Schwartz, D. (1969). Surface areas of the cerebral cortex of mammals determined by stereological methods. *Science* **166**, 111–113.

Elliot Smith, G. (1903). On the morphology of the brain in the mammalia, with special reference to that of the lemurs, recent and extinct. *Trans. Linn. Soc. London (Zool.)* **8**, 319–431.

Elliot Smith, G. (1908). On the form of the brain in extinct lemurs of Madagascar with some remarks on the affinities of the Indrisinae. *Trans. Zool. Soc. London* **18**, 163–178.

Estes, R. (1961). Cranial anatomy of the cynodont reptile *Thrinaxodon liorhinus*. *Bull. Mus. Comp. Zool., Harvard Univ.* **125**, 163–180.

Evans, F. G. (1944). The morphological status of the modern Amphibia among the Tetrapoda. *J. Morphol.* **74**, 43–100.

Evans, H. M. (1940). "Brain and Body of Fish." McGraw-Hill (Blakiston), New York.

Evernden, J. F., Savage, D. E., Curtis, G. H., and James, G. T. (1964). Potassium-argon dates and the Cenozoic Mammalian chronology of North America. *Amer. J. Sci.* **262**, 145–198.

Every, R. G. (1965). The teeth as weapons: Their influence on behavior. *Lancet* **1**, 685–688.
Ewer, R. F. (1968). "Ethology of Mammals." Plenum, New York.
Filhol, H. (1883). Observations relatives au mémoire de M. Cope intitulé: Relations des horizons renferment des débris d'animaux vertébrés fossiles en Europe et en Amerique. *Annales des Sci. Geol.* **14(5)**, 1–51.
Flourens, P. (1842). "Recherches expérimentales sur les propriétés et les fonctions du système nerveux dans les animaux vertébrés," 2nd ed. Crevot, Paris.
Fraenkel, G. S., and Gunn, D. L. (1961). "The Orientation of Animals: Kineses, Taxes and Compass Reactions." Dover, New York [reprint of original edition, Oxford Univ. Press (Clarendon), London and New York, 1940].
Franz, V. (1911). Das Mormyridenhirn. *Zool. Jahrb. abt. Anat. u. Ontog. der Tiere* **32**, 465–492.
Freedman, L. (1960). Some new cercopithecoid specimens from Makapansget, South Africa. *Palaeontol. Afr.* **7**, 7–45.
Freedman, L. (1961). New cercopithecoid fossils, including a new species, from Taung, Cape Province, South Africa. *Ann. S. Afr. Mus.* **46**, 1–14.
Friede, R. L., and van Houten, W. H. (1962). Neuronal extension and glial supply: Functional significance of glia. *Proc. Nat. Acad. Sci. U.S.* **48**, 817–821.
Gaudry, A. (1862). "Animaux Fossiles et Géologie de l'Attique." Soc. Geol. Fr. (Savy), Paris.
Gazin, C. L. (1958). A review of the Middle and Upper Eocene primates of North America. *Smithson. Misc. Collect.* **136**, No. 1, 1–112.
Gazin, C. L. (1965a). A study of the early Tertiary condylarthran mammal *Meniscotherium*. *Smithson. Misc. Collect.* **149**, No. 2, 1–98.
Gazin, C. L. (1965b). An endocranial cast of the Bridger middle Eocene primate, *Smilodectes gracilis*. *Smithson. Misc. Collect.* **149**, No. 4, 1–14.
Gazin, C. L. (1968). A study of the Eocene condylarthran mammal *Hyopsodus*. *Smithson. Misc. Collect.* **153**, No. 4, 1–90.
Gervais, P. (1872). Forme cérébrale du *Cephalogale Geoffroyi*. *J. Zool.* **1**, 131–133.
Geschwind, N. (1965a). Disconnexion syndromes in animals and man. Parts I & II. *Brain* **88**, 237–294.
Geschwind, N. (1965b). Disconnexion syndromes in animals and man. Part III. *Brain* **88**, 585–644.
Gihr, M., and Pilleri, G. (1969). Hirn-Körpergewichts-Beziehungen bei Cetaceen. *Investigations on Cetacea* **1**, 109–126.
Gilmore, C. W. (1909). Osteology of the Jurassic reptile *Camptosaurus*. *Proc. U.S. Nat. Mus.* **36**, 197–332.
Gilmore, C. W. (1920). Osteology of the carnivorous Dinosauria in the United States National Museum, with special reference to the genera *Antrodemus* (*Allosaurus*) and *Ceratosaurus*. *U.S., Nat. Mus., Bull.* **110**, 1–159.
Gjukic, M. (1955). Ein Beitrag zum Problem der Korrelation zwischen Hirngewicht und Körpergewicht. *Z. Morphol. Anthropol.* **47**, 43–57. [Cited by Bauchot and Stephan (1969).]
Goldberg, J. M., and Greenwood, D. D. (1966). Response of neurons of the dorsal and posteroventral cochlear nuclei of the cat to acoustic stimuli of long duration. *J. Neurophysiol.* **29**, 72–93.
Gould, S. J. (1966). Allometry and size in ontogeny and phylogeny. *Biol. Rev. Cambridge Phil. Soc.* **41**, 587–640.

Gould, S. J. (1971). Geometric similarity in allometric growth: A contribution to the problem of scaling in the evolution of size. *Amer. Natur.* **105**, 113–136.

Granit, R. (1968). The development of retinal neurophysiology. *Science* **160**, 1192–1196.

Gray, J. (1953). "How Animals Move." Cambridge Univ. Press, London and New York.

Gray, J. (1968). "Animal Locomotion." Weidenfeld & Nicholson, London.

Gregory, W. K. (1920). On the structure and relations of *Notharctus,* an American Eocene primate. *Mem. Amer. Mus. Natur. Hist.* [N. S.] **3**, Part 2, 49–243.

Griffin, D. R. (1958). "Listening in the Dark." Yale Univ. Press, New Haven, Connecticut.

Grinnell, A. D. (1970). Comparative auditory neurophysiology of neotropical bats employing different echolocation signals. *Z. Vergl. Physiol.* **68**, 117–153.

Gross, W. (1937). Das Kopfskelett von *Cladodus wildungensis* Jaekel. 1. Endocranium und palatoquadratum. *Senckenbergiana* **19**, 80–107.

Gyldenstolpe, N. (1928). Zoological results of the Swedish expedition to Central Africa, 1921. Vertebrata 5. *Ark. Zool.* **20A**, 1–76.

Hailman, J. P. (1970). Comments on the coding of releasing stimuli. *In* "Development and Evolution of Behavior: Essays in Memory of T. C. Schneirla" (L. R. Aronson *et al.,* eds.), pp. 138–157. Freeman, San Francisco, California.

Hall, E. R., and Kelson, K. R. (1959). "The Mammals of North America," 2 vols. Ronald Press, New York.

Hall, K. R. L. (1960). Social vigilance behaviour of the chacma baboon, *Papio ursinus. Behaviour* **16**, 261–294.

Hall, K. R. L. (1963). Tool-using performances as indicators of behavioral adaptability. *Curr. Anthropol.* **4**, 479–494.

Halstead, L. B. (1969). "The Pattern of Vertebrate Evolution." Oliver & Boyd, Edinburgh.

Halstead Tarlo, L. B. (1965). Psammosteiformes (Agnatha): A review with descriptions of new material from the lower Devonian of Poland. I. General part. *Paleontol. Polon.* No. 13.

Harlow, H. F., and Harlow, M. K. (1962). Social deprivation in monkeys. *Sci. Amer.* **207(5)**, 136–146.

Harman, P. J. (1957). Paleoneurologic, neoneurologic, and ontogenetic aspects of brain phylogeny. "James Arthur Lecture on the Evolution of the Human Brain." Amer. Mus. Natur. Hist., New York.

Hartman, F. A. (1961). Locomotor mechanisms of birds. *Smithson. Misc. Collect.* **143**, No. 1, 1–91.

Hassler, R., and Stephan, H., eds. (1966). "Evolution of the Forebrain." Thieme, Stuttgart.

Haug, H. (1956). Remarks on the determination and significance of the gray/cell coefficient. *J. Comp. Neurol.* **104**, 473–492.

Hawkins, A., and Olszewski, J. (1957). Glia/nerve cell index for cortex of the whale. *Science* **126**, 76–77.

Hay, O. P. (1909). On the skull and the brain of *Triceratops,* with notes on braincases of *Iguanodon and Megalosaurus. Proc. U.S. Nat. Mus.* **36**, 95–108.

Heath, J. E. (1968). The origins of thermoregulation. *In* "Evolution and Environment" (E. T. Drake, ed.), pp. 259–278. Yale Univ. Press, New Haven, Connecticut.

Hebb, D. O. (1949). "The Organization of Behavior: A Neuropsychological Theory." Wiley, New York.

Heilmann, G. (1927). "The Origin of Birds." Appleton, New York.

Heimer, L. (1968). Synaptic distribution of centripetal and centrifugal nerve fibres in the olfactory system of the rat. An experimental anatomical study. *J. Anat.* **103**, 413–432.

Heller, I. H. and Elliott, K. A. C. (1954). Desoxyribonucleic acid content and cell density in brain and human brain tumors. *Canad. J. Biochem. Physiol.* **32**, 584–592.

Henry, J. P. (1966). "Biomedical Aspects of Space Flight." Holt, New York.

Heptonstall, W. B. (1970). Quantitative assessment of the flight of *Archaeopteryx*. *Nature (London)* **228**, 185–186.

Heptonstall, W. B. (1971a). An analysis of the flight of the Cretaceous pterodactyl *Pteranodon ingens* (Marsh). *Scot. J. Geol.* **7**, 61–78.

Heptonstall, W. B. (1971b). The flying ability of *Archaeopteryx*. *Nature (London)* **231**, 128.

Heptonstall, W. B. (1971c). *Archaeopteryx* again. *Nature (London)* **234**, 479.

Herrick, C. J. (1905). The central gustatory paths in the brains of bony fishes. *J. Comp. Neurol.* **15**, 375–456.

Herrick, C. J. (1948). "The Brain of the Tiger Salamander *Ambystoma Tigrinum*." Univ. of Chicago Press, Chicago, Illinois.

Herrick, C. J., and Obenchain, J. B. (1913). Notes on the anatomy of a cyclostome brain: *Ichthyomyzon concolor*. *J. Comp. Neurol.* **23**, 635–665.

Hersh, A. H. (1934). Evolutionary relative growth in the titanotheres. *Amer. Natur.* **68**, 537–561.

Hinde, R. A. (1969a). "Animal Behaviour," 2nd ed. McGraw-Hill, New York.

Hinde, R. A., ed. (1969b). "Bird Vocalizations: Essays Presented to W. H. Thorpe." Cambridge Univ. Press, London and New York.

Hodges, J. L., Jr., and Lehmann, E. L. (1968). A compact table for power of the t-test. *Ann. Math. Statist.* **39**, 1629–1637.

Hofer, H. (1969). In memoriam Tilly Edinger (Mit einem Schriftenverzeichnis von B. Kummel und H. Tobien). *Morphol. Jahrb.* **113**, 303–317.

Hofer, H. O., and Wilson, J. A. (1967). An endocranial cast of an early Oligocene primate. *Folia Primatol.* **5**, 148–152.

Holloway, R. L. (1966). Cranial capacity, neural reorganization, and hominid evolution: A search for more suitable parameters. *Amer. Anthropol.* **68**, 103–121.

Holloway, R. L. (1968). The evolution of the primate brain: Some aspects of quantitative relations. *Brain Res.* **7**, 121–172.

Holloway, R. L. (1970). New endocranial values for the australopithecines. *Nature (London)* **227**, 199–200.

Hopson, J. A. (1966). The origin of the mammalian middle ear. *Amer. Zool.* **6**, 437–450.

Hopson, J. A. (1967). Comments on the competitive inferiority of the multituberculates. *Syst. Zool.* **16**, 352–355.

Hopson, J. A., and Crompton, A. W. (1969). Origin of mammals. *Evol. Biol.* **3**, 15–72.

Hotton, N., III. (1959). The pelycosaur tympanum and early evolution of the middle ear. *Evolution* **13**, 99–121.

Hotton, N., III. (1960). The chorda tympani and middle ear as guides to origin and divergence of reptiles. *Evolution* **14**, 194–211.

Howell, F. C. (1967). Recent advances in human evolutionary studies. *Quart. Rev. Biol.* **42**, 471–513.

Howell, F. C. (1969). Remains of Hominidae from Pliocene/Pleistocene formations in the lower Omo basin, Ethiopia. *Nature (London)* **223**, 1234–1239.

Hrdlička, A. (1906). Brains and brain preservatives. *Proc. U.S. Nat. Mus.* **30**, 245–320.

Huxley, J. S. (1932). "Problems of Relative Growth." Allen & Unwin, London.

Huxley, J. (1942). "Evolution The Modern Synthesis." Harper, New York and London.

Jacobs, M. S., Morgane, P. J., and McFarland, W. L. (1971). The anatomy of the brain of the bottlenose dolphin (*Tursiops truncatus*). Rhinic lobe (Rhinencephalon). I. The paleocortex. *J. Comp. Neurol.* **141**, 205–272.

Janensch, W. (1935). Die Schädel der Sauropoden *Brachiosaurus, Barosaurus* und *Dicraeosaurus* aus den Tendaguruschichten Deutsch-Ostafrikas. *Palaeontographica, Suppl.* **7**, 147–248.

Jarvik, E. (1955). The oldest tetrapods and their forerunners. *Sci. Mon.* **80**, 141–154.

Jerison, H. J. (1955). Brain to body ratios and the evolution of intelligence. *Science* **121**, 447–449.

Jerison, H. J. (1961). Quantitative analysis of evolution of the brain in mammals. *Science* **133**, 1012–1014.

Jerison, H. J. (1963). Interpreting the evolution of the brain. *Hum. Biol.* **35**, 263–291.

Jerison, H. J. (1968). Brain evolution and *Archaeopteryx*. *Nature (London)* **219**, 1381–1382.

Jerison, H. J. (1969). Brain evolution and dinosaur brains. *Amer. Natur.* **103**, 575–588.

Jerison, H. J. (1970a). Gross brain indices and the analysis of fossil endocasts. *In* "The Primate Brain" (C. R. Noback and W. Montagna, eds.), pp. 225–244. Appleton, New York.

Jerison, H. J. (1970b). Brain evolution: New light on old principles. *Science* **170**, 1224–1225.

Jerison, H. J. (1971a). More on why birds and mammals have big brains. *Amer. Natur.* **105**, 185–189.

Jerison, H. J. (1971b). Quantitative analysis of the evolution of the camelid brain. *Amer. Natur.* **105**, 227–239.

Kandel, E. R. (1967). Cellular studies of learning. *In* "The Neurosciences" (G. C. Quarton *et al.*, eds.), pp. 666–689. Rockefeller Univ. Press, New York.

Karten, H. J. (1969). The organization of the avian telencephalon and some speculations on the phylogeny of the amniote telencephalon, *Ann. N. Y. Acad. Sci.* **167**, 164–179.

Keith, A. (1948). "A New Theory of Human Evolution." Watts, London.

Kellogg, R. (1936). A review of the Archaeoceti. *Carnegie Inst. Wash. Publ.* **482**, 1–366.

Kellogg, W. N. (1961). "Porpoises and Sonar." Univ. of Chicago Press, Chicago, Illinois.

Kemeny, J. G., Mirkil, H., Snell, J. L., and Thompson, G. L. (1959). "Finite Mathematical Structures." Prentice-Hall, Englewood Cliffs, New Jersey.

Kennard, M. A., and Willner, M. D. (1941). Weights of brains and organs of 132 new and old world monkeys. *Endocrinology* **28**, 977–984.

Kennedy, D. (1967). Small systems of nerve cells. *Sci. Amer.* **216(5)**, 44–52.

Kermack, K. A. (1963). The cranial structure of triconodonts. *Philo. Trans. Roy. Soc. London, Ser. B* **246**, 83–103.
Killackey, H., and Diamond, I. T. (1971). Visual attention in the tree shrew: An ablation study of the striate and extrastriate visual cortex. *Science* **171**, 696–699.
Klüver, H. (1951). Functional differences between the occipital and temporal lobes. *In* "Cerebral Mechanisms in Behavior" (L. A. Jeffress, ed.), pp. 147–199. Wiley, New York.
Knight, C. R. (1947). "Animal Anatomy and Psychology for Artists and Laymen." McGraw-Hill, New York (re-issued: "Animal Drawing; Anatomy and Action for Artists." Dover, New York, 1959).
Knoll, R. L., and Stenson, H. H. (1968). A computer program to generate and measure random forms. *Percept. Psychophys.* **3**, 311–316.
Kochetkova, V. I. (1960). L'évolution des régions spécifiquement humaines de l'écorce cérébrale chez les hominides. *Actes Congr. Int. Sci. Anthropolo. Ethnolo., 6th, 1960* Vol. 1, pp. 623–630.
Koffka, K. (1935). "Principles of Gestalt Psychology." Harcourt, New York.
Kohler, I. (1964). The formation and transformation of the visual world. *Psychol. Issues* **3**, 28–46.
Kulp, J. (1961). Geologic time scale. *Science* **133**, 1105–1114.
Lapique, L. (1907). Tableau générale des poids somatique et encéphalique dans les espèces animales. *Bull. Soc. Anthropol. Paris* [5] **8**, 248–262.
Lartet, E. (1868). De quelques cas de progression organique verifiables dans la succession des temps géologiques sur des mammifères de même famille et de même genre. *C.R. Acad. Sci.* **66**, 1119–1122.
Lashley, K. S. (1929). "Brain Mechanisms and Intelligence." Univ. of Chicago Press, Chicago, Illinois.
Lashley, K. S. (1949). Persistent problems in the evolution of mind. *Quart. Rev. Biol.* **24**, 28–42.
Lashley, K. S. (1950). In search of the engram. *Symp. Soc. Exp. Biol.* **4**, 454–482.
Latimer, H. B. (1956). The weights of the parts of the brain in several species of animals. *Trans. Kans. Acad. Sci.* **59**, 432–441.
Leakey, M. D., Clarke, R. J., and Leakey, L. S. B. (1971). New hominid skull from Bed I, Olduvai Gorge, Tanzania. *Nature (London)* **232**, 308–312.
Leakey, R. E. F. (1971). Further evidence of lower Pleistocene hominids from East Rudolf, North Kenya. *Nature (London)* **231**, 241–245.
Leakey, R. E. F. (1973). Evidence for an advanced Plio-Pleistocene hominid from East Rudolf, Kenya. *Nature (London)* **242**, 447–450.
LeCren, E. D. (1951). The length-weight relationship and seasonal cycle in gonad weight and condition in the perch (*Perca fluviatilis*). *J. Anim. Ecol.* **20**, 201–219.
LeGros Clark, W. E. (1945). Deformation patterns in the cerebral cortex. *In* "Essays on Growth and Form Presented to D'Arcy Wentworth Thompson" (W. E. LeGros Clark and P. B. Medawar, eds.), pp. 1–22. Oxford Univ. Press (Clarendon), London and New York.
LeGros Clark, W. E. (1962). "The Antecedents of Man," 2nd ed. Quadrangle Books, Chicago, Illinois.
LeGros Clark, W. E. (1967). "Man-Apes or Ape-Men? The Stories of Discoveries in Africa." Holt, New York.
LeGros Clark, W. E., and Thomas, D. P. (1952). "Fossil Mammals of Africa," No. 5. The Miocene Lemuroids of East Africa. Brit. Mus. (Natur. Hist.), London.

Lende, R. A. (1963). Cerebral cortex: A sensory motor amalgam in Marsupialia. *Science* **141**, 730–732.
Lende, R. A. (1964). Representation in the cerebral cortex of a primitive mammal: Sensorimotor, visual, and auditory fields in the Echidna (*Tachyglossus aculeatus*). *J. Neurophysiol.* **27**, 37–48.
Lende, R. A., and Sadler, K. M. (1967). Sensory and motor areas in neocortex of hedgehog. *Brain Res.* **5**, 390–405.
Lenneberg, E. H. (1967). "Biological Foundations of Language." Wiley, New York.
Lettvin, J. Y., Maturana, H. R., McCulloch, W. S., and Pitts, W. H. (1959). What the frog's eye tells the frog's brain. *Proc. IRE* **47**, 1940–1951.
Lieberman, P., and Crelin, E. S. (1971). On the speech of neanderthal man. *Linguistic Inquiry* **2**, 203–222.
Lilly, J. C. (1961). "Man and Dolphin." Doubleday, Garden City, New York.
Lissmann, H. W. (1958). On the function and evolution of electric organs in fish. *J. Exp. Biol.* **35**, 156–191.
Lloyd, D. P. C. (1960). Spinal mechanisms involved in somatic activities. In "Handbook of Physiology" (Amer. Physiol. Soc., J. Field, ed.), Sect. 1, Vol. 2, pp. 929–950. Williams & Wilkins, Baltimore, Maryland.
Lorenz, K. (1935). Der Kumpan in der Umwelt des Vogels; die Artgenosse als auslösendes Moment sozialer Verhaltungsweisen. *J. Ornithol.* **83**, 137–215 and 289–413, abridged as: The companion in the bird's world. *Auk* **54**, 245–273 (1937).
Lull, R. S. (1920). New tertiary Artiodactyla. *Amer. J. Sci.* **200**, 82–130.
Lull, R. S., and Wright, N. E. (1942). Hadrosaurian dinosaurs of North America. *Geol. Soc. Amer., Spec. Pap.* **40**, 1–242.
McKenna, M. C. (1963). Primitive Paleocene and Eocene Apatemyidea (Mammalia, Insectivora) and the primate-insectivore boundary. *Amer. Mus. Nov.* **2160**, 1–39.
McKenna, M. C. (1966). Paleontology and the origin of the primates. *Folia Primatol.* **4**, 1–25.
McKibben, P. S. (1913). The eye-muscle nerves in *Necturus*. *J. Comp. Neurol.* **23**, 153–172.
Magoun, H. W. (1960). Evolutionary concepts of brain function following Darwin and Spencer. In "Evolution After Darwin" (S. Tax, ed.), Vol. 2, p. 199. Univ. of Chicago Press, Chicago, Illinois.
Magoun, H. W. (1963). "The Waking Brain," 2nd ed. Thomas, Springfield, Illinois.
Mandel, P., Rein, H., Harth-Edel, S., and Mardell, R. (1964). Distribution and metabolism of ribonucleic acid in the vertebrate central nervous system. In "Comparative Neurochemistry" (D. Richter, ed.), pp. 149–163. MacMillan, New York.
Manley, G. A. (1971). Some aspects of the evolution of hearing in vertebrates. *Nature (London)* **230**, 506–509.
Manouvrier, L. (1885). Sur l'interprétation de la quantité dans l'encéphale et dans le cerveau en particulier. *Bull. Soc. Anthropol. Paris* **3**, 137–323.
Marler, P. (1965). Communication in monkeys and apes. In "Primate Behavior: Field Studies of Monkeys and Apes" (I. DeVore, ed.), pp. 544–584. Holt, New York.
Marler, P., and Hamilton, W. J., III. (1966). "Mechanisms of Animal Behavior." Wiley, New York.
Marquis, D. G. (1934). Effects of removal of the visual cortex in mammals, with

observations on the retention of light discrimination in dogs. *Res. Publ., Ass. Res. Nerv. Ment. Dis.* **13**, 558–592.
Marsh, O. C. (1874). Small size of the brain in Tertiary mammals. *Amer. J. Sci. Arts* **8**, 66–67.
Marsh, O. C. (1876). On some characters of the genus *Coryphodon* (Owen). *Amer. J. Sci. Arts* **11**, 425–428.
Marsh, O. C. (1880). "Odontornithes: A monograph on the extinct toothed birds of North America," U.S. Geol. Exploration 40th Parallel, Vol. 7.
Marsh, O. C. (1886). Dinocerata. *U.S., Geol. Surv., Monogr.* **10** (Whole No.), 1–243.
Marsh, O. C. (1893). Restoration of *Coryphodon*. *Amer. J. Sci.* **46**, 321–326.
Marshall, A. J., ed. (1961). "Biology and Comparative Physiology of Birds," Vol. 2. Academic Press, New York.
Marshall, J. (1892). On the relations between the weight of the brain and its parts and the stature and mass of the body in man. *J. Anat. Physiol. Norm. Pathol. (London)* **26**, 445–500.
Marshall, N. B. (1966). "The Life of Fishes." World Publ., Cleveland, Ohio.
Martin, C. J. (1903). Thermal adjustment and respiratory exchange in monotremes and marsupials. A study in the development of homoeothermism. *Phil. Trans. Roy. Soc. London, Ser. B* **195**, 1–37.
Masterton, R. B., Jane, J. A., and Diamond, I. T. (1968). Role of brain-stem auditory structures in sound localization. II. Inferior colliculus and its brachium. *J. Neurophysiol.* **31**, 96–108.
Masterton, R. B., Heffner, H., and Ravizza, R. (1969). Evolution of human hearing. *J. Acoust. Soc. Amer.* **45**, 966–985.
Matthew, W. D. (1909). The Carnivora and Insectivora of the Bridger Basin, middle Eocene. *Mem. Amer. Mus. Natur. Hist.* **9**, Part 6, 289–567.
Maynard Smith, J. (1952). The importance of the nervous system in the evolution of animal flight. *Evolution* **6**, 127–129.
Mayr, E. (1963). "Animal Species and Evolution." Cambridge Univ. Press, London and New York.
Merriam, J. C., and Stock, C. (1932). The Felidae of Rancho La Brea, Washington, D.C. *Carnegie Inst. Wash. Publ.* **422**.
Mettler, F. A. (1956). "Culture and the Structural Evolution of the Neural System. James Arthur Lecture." Amer. Mus. Natur. Hist., New York.
Meyer, D. R., and Harlow, H. F. (1949). The development of transfer of response to patterning by monkeys. *J. Comp. Physiol. Psychol.* **42**, 454–462.
Millot, J., and Anthony, J. (1956). Considérations préliminaires sur le squelette axial et le système nerveux central de *Latimeria chalumnae* Smith. *Mém. Inst. Sci. Madagascar, Ser. A*. **11**, 167–188.
Millot, J., and Anthony, J. (1966). "Anatomie de *Latimeria chalumnae*," Tome II. CNRS, Paris.
Moodie, R. L. (1915). A new fish brain from the Coal Measures of Kansas, with a review of other fossil brains. *J. Comp. Neurol.* **25**, 135–181.
Moodie, R. L. (1922). On the endocranial anatomy of some Oligocene and Pleistocene mammals. *J. Comp. Neurol.* **34**, No. 4, 343–370.
Moulton, D. G., and Beidler, L. M. (1967). Structure and function in the peripheral olfactory system. *Physiol. Rev.* **47**, 1–52.
Myers, R. D., and Veale, W. L. (1970). Body temperature—possible ionic mechanism in the hypothalamus controlling the set point. *Science* **170**, 95–97.

Newell, N. D. (1949). Phyletic size increase, an important trend illustrated by fossil invertebrates. *Evolution* **3**, 103–124.
Newton, E. T. (1888). On the skull, brain, and auditory organ of a new species of pterosaurian (*Scaphognathus Purdoni*), from the upper Lias near Whitby, Yorkshire. *Phil. Trans. Roy. Soc. London, Ser. B* **179**, 503–537.
Nieuwenhuys, R. (1962). Trends in the evolution of the actinopterygian forebrain. *J. Morphol.* **111**, 69–88.
Nieuwenhuys, R. (1966). The interpretation of the cell masses in the teleostean forebrain. *In* "Evolution of the Forebrain" (R. Hassler and H. Stephan, eds.), p. 32–39. Thieme, Stuttgart.
Noback, C. R., and Moskowitz, N. (1962). Structural and functional correlates of "encephalization" in the primate brain. *Ann. N.Y. Acad. Sci.* **102**, 210–218.
Noble, G. K. (1931). "The Biology of the Amphibia." McGraw-Hill, New York.
Nyberg, D. (1971). An hypothesis concerning the larger brains of homoiotherms. *Amer. Natur.* **105**, 183–185.
Olson, E. C. (1944). Origin of mammals based on the cranial morphology of therapsid suborders, *Geol. Soc. Amer., Spec. Pap.* **55**, 1–136.
Olson, E. C. (1959). The evolution of mammalian characters. *Evolution* **13**, 344–353.
Olson, E. C. (1961). The food chain and the origin of mammals. *In* "International Colloquium on the Evolution of Lower and Nonspecialized Mammals" (G. Vandebroek, ed.), pp. 97–116. Kon. Vlaamse Acad. Wetensch. Lett. Sch. Kunsten België, Brussels.
Olson, E. C. (1965). "The Evolution of Life." Mentor, New York.
Olson, E. C. (1970). "Vertebrate Paleozoology." Wiley, New York.
Orlov, J. A. (1948). *Perunium ursogulo* Orlov, a new gigantic extinct mustelid (a contribution to the morphology of the skull and brain and to the phylogeny of mustelidae). *Acta Zool.* **29**, 63–105.
Osborn, H. F. (1898a). Article IV. A complete skeleton of *Teleoceras fossiger*. Notes upon the growth and sexual characters of this species. *Bull. Amer. Mus. Natur. Hist.* **10**, 51–60.
Osborn, H. F. (1898b). A complete skeleton of *Coryphodon radians*. Notes upon the locomotion of this animal. *Bull. Amer. Mus. Natur. Hist.* **10**, 81–91.
Osborn, H. F. (1898c). Remounted skeleton of *Phenacodus primaevus*. Comparison with *Euprotogonia*. *Bull. Amer. Mus. Natur. Hist.* **10**, 159–164.
Osborn, H. F. (1910). "The Age of Mammals in Europe, Asia and North America." Macmillan, New York.
Osborn, H. F. (1912). Crania of *Tyrannosaurus* and *Allosaurus*. *Mem. Amer. Mus. Natur. Hist.* [N.S.] **1**, 1–30.
Osborn, H. F. (1929). The Titanotheres of Ancient Wyoming, Dakota and Nebraska. 2 vols. *U.S., Geol. Surv., Monogr.* **55**, 1–953.
Osborn, H. F. (1936–1942). "Proboscidea," 2 vols. Amer. Mus. Natur. Hist., New York.
Ostrom, J. H. (1970). *Archaeopteryx*: Notice of a "new" specimen. *Science* **170**, 537–538.
Owen, R. (1863). On the *Archaeopteryx* of von Meyer, with a description of the fossil remains of a long-tailed species, from the lithographic stone of Solenhofen. *Phil. Trans. Roy. Soc. London* **153**, 33–47.
Papez, J. W. (1929). "Comparative Neurology." Crowell-Collier, New York.

Parrington, F. R. (1936). On the tooth-replacement in theriodont reptiles. *Phil. Trans. Roy. Soc. London, Ser. B* **226**, 121–142.
Parrington, F. R. (1967). The origins of mammals. *Advan. Sci.* **24**, 165–173.
Patterson, B. (1937). Some notoungulate braincasts. *Field Mus. Natur. Hist., Geol. Ser.* **6**, 273–301.
Patterson, B., and Olson, E. C. (1961). A triconodontid mammal from the Triassic of Yunnan. *In* "International Colloquium on the Evolution of Lower and Nonspecialized Mammals" (G. Vandebroek, ed.), pp. 129–191. Kon. Vlaamse Acad. Wetensch. Lett. Sch. Kunsten België, Brussels.
Penfield, W., and Roberts, L. (1959). "Speech and Brain-mechanisms." Atheneum, New York.
Petras, J. M., and Noback, C. R., eds. (1969). Comparative and evolutionary aspects of the vertebrate central nervous system. *Ann. N.Y. Acad. Sci.* **167**, 1–513.
Pilbeam, D. R., and Simons, E. L. (1965). Some problems of hominid classification. *Amer. Sci.* **53**, 237–259.
Pilleri, G. and Gihr, M. (1969). Uber adriatische *Tursiops truncatus* (Montagu 1821) und vergleichende Untersuchungen über mediterrane und atlantischer Tümmler. *Investigations on Cetacea* **1**, 66–73.
Pilleri, G. and Gihr, M. (1969). On the anatomy and behavior of Risso's dolphin (*Grampus griseus* Cuvier). *Investigations on Cetacea* **1**, 74–93.
Piveteau, J. (1951). Recherches sur l'évolution de l'encéphale chez les carnivores fossiles. *Ann. Paleontol.* **37**, 133–152.
Piveteau, J. (1955). "Traité de Paléontologie," Tome V. Masson, Paris.
Piveteau, J. (1957). "Traité de Paléontologie," Tome VII. Masson, Paris.
Piveteau, J., ed. (1958). "Traité de Paléontologie," Tome VI, Vol. 2. Masson, Paris.
Piveteau, J., ed. (1961). "Traité de Paléontologie," Tome VI, Vol. 1. Masson, Paris.
Piveteau, J., ed. (1966). "Traité de Paléontologie," Tome IV, Vol. 3. Masson, Paris.
Polyak, S. (1957). "The Vertebrate Visual System" (H. Kluver, ed.). Univ. of Chicago Press, Chicago, Illinois.
Portmann, A. (1946). Etudes sur la cérébralisation chez les oiseaux. I. *Alauda* **14**, 2–20.
Portmann, A. (1947). Etudes sur la cérébralisation chez les oinseaux. II. *Alauda* **15**, 1–15.
Portmann, A., and Stingelin, W. (1961). The central nervous system. *In* "Biology and Comparative Physiology of Birds" (A. J. Marshall, ed.), Vol. 2, pp. 1–36. Academic Press, New York.
Poulson, T. L. (1963). Cave adaptation in amblyopsid fishes. *Amer. Midl. Natur.* **70**, 257–290.
Premack, D. (1971). Language in chimpanzee? *Science* **172**, 808–822.
Pringle, J. W. S. (1957). "Insect Flight." Cambridge Univ. Press, London and New York.
Prosser, C. L., and Brown, F. A., Jr., eds. (1961). "Comparative Animal Physiology," 2nd ed. Saunders, Philadelphia, Pennsylvania.
Quiring, D. P. (1941). The scale of being according to the power formula. *Growth* **5**, 301–327.
Quiring, D. P. (1950). "Functional Anatomy of the Vertebrates." McGraw-Hill, New York.
Radinsky, L. B. (1966). The adaptive radiation of the phenacodontid condylarths and the origin of the Perissodactyla. *Evolution* **20**, 408–417.

Radinsky, L. B. (1967a). Relative brain size: A new measure. *Science* **155**, 836–838.
Radinsky, L. B. (1967b). The oldest primate endocast. *Amer. J. Phys. Anthropol.* **27**, 385–388.
Radinsky, L. B. (1967c). *Hyrachyus, Chasmotherium,* and the early evolution of helaletid tapiroids. *Amer. Mus. Nov.* **2313**, 1–23.
Radinsky, L. B. (1968a). A new approach to mammalian cranial analysis, illustrated by examples of prosimian primates. *J. Morphol.* **124**, 167–180.
Radinsky, L. B. (1968b). Evolution of somatic sensory specialization in otter brains. *J. Comp. Neurol.* **134**, 495–505.
Radinsky, L. B. (1969). Outlines of canid and felid brain evolution. *Ann. N.Y. Acad. Sci.* **167**, 277–288.
Radinsky, L. B. (1970). The fossil evidence of prosimian brain evolution. In "The Primate Brain" (C. R. Noback and W. Montagna, eds.), pp. 209–224. Appleton, New York.
Reiss, R. F., ed. (1964). "Neural Theory and Modeling." Stanford Univ. Press, Stanford, California.
Rensch, B. (1956). Increase of learning capability with increase of brain size. *Amer. Natur.* **90**, 81–95.
Rensch, B. (1959). "Evolution Above the Species Level." Methuen, London.
Rensch, B. (1967). The evolution of brain achievements. *Evol. Biol.* **1**, 26–68.
Repérant, J. (1970). Moulages endocraniens de tylopodes fossiles. *Ann. Paleontol.* **56**, 111–145.
Repérant, J. (1971). Un nouveau moulage endocrânien de *Procamelus* du Pliocène inférieur. *Bull. Mus. Hist. Natur. Paris* [3] **17**, 1–9.
Riggs, E. S. (1935). A skeleton of *Astrapotherium. Field Mus. Natur. Hist., Geol. Ser.* **6**, 167–177.
Riggs, E. S. (1937). Mounted skeleton of *Homalodotherium. Field Mus. Natur. Hist., Geol. Ser.* **6**, 233–243.
Robb, R. C. (1935a). A study of mutations in evolution. I. Evolution in the equine skull. *J. Genet.* **31**, 39–46.
Robb, R. C. (1935b). A study of mutations in evolution. II. Ontogeny in the equine skull. *J. Genet.* **31**, 47–52.
Robinson, J. T. (1967). Variation and the taxonomy of the early hominids. *Evol. Biol.* **1**, 69–100.
Roe, A., and Simpson, G. G., eds. (1958). "Behavior and Evolution." Yale Univ. Press, New Haven, Connecticut.
Romer, A. S. (1937). The braincase of the Carboniferous crossopterygian *Megalichthys nitidus. Bull. Mus. Comp. Zool., Harvard Univ.* **82**, 1–73.
Romer, A. S. (1959). "The Vertebrate Story." Univ. of Chicago Press, Chicago, Illinois.
Romer, A. S. (1961). Synapsid evolution and dentition. In "International Colloquium on the Evolution of Lower and Non-specialized Mammals" (G. Vandebroek, ed.), pp. 9–56. Kon. Vlaamse. Acad. Wetensch. Lett. Sch. Kunsten België, Brussels.
Romer, A. S. (1964). The braincase of the paleozoic elasmobranch Tamiobatis. *Bull. Mus. Comp. Zool., Harvard Univ.* **131**, 87–105.
Romer, A. S. (1966). "Vertebrate Paleontology," 3rd ed., Univ. of Chicago Press, Press, Chicago, Illinois.
Romer, A. S. (1968). "Notes and Comments on Vertebrate Paleontology." Univ. of Chicago, Illinois.

Romer, A. S. (1969). Vertebrate history with special reference to factors related to cerebellar evolution. *In* "Neurobiology of Cerebellar Evolution and Development" (R. Llinas, ed.), pp. 1–18. Amer. Med. Ass., Educ. Res. Found., Chicago, Illinois.

Romer, A. S., and Edinger, T. (1942). Endocranial casts and brains of living and fossil Amphibia. *J. Comp. Neurol.* **77**, 355–389.

Rovainen, C. M. (1967). Physiological and anatomical studies on large neurons of central nervous system of the sea lamprey (*Petromyzon marinus*). I. Müller and Mauthner cells. *J. Neurophysiol.* **30**, 1000–1023.

Ruch, T. C. (1935). Cortical localization of somatic sensibility. The effect of precentral, postcentral and posterior parietal lesions upon the performance of monkeys trained to discriminate weights. *Res. Publ., Ass. Res. Nerv. Ment. Dis.* **15**, 289–330.

Russell, D. E. (1964). Les mammifères Paléocènes d'Europe. *Mem. Mus. Hist. Natur. Paris, Ser. C* [N.S.] **13**, 1–324.

Russell, D. E., and Sigogneau, D. (1965). Etude de moulages endocraniens de mammifères Paléocènes. *Mem. Mus. Hist. Natur. Paris, Ser. C* [N.S.] **14**, 1–36.

Sacher, G. A. (1970). Allometric and factorial analysis of brain structure in insectivores and primates. *In* "The Primate Brain" (C. R. Noback and W. Montagna, eds.), pp. 245–287. Appleton, New York.

Savage, R. J. G. (1957). The Anatomy of *Potamotherium*, an Oligocene lutrine. *Proc. Zool. Soc. London* **129**, 151–242.

Säve-Söderbergh, C. (1952). On the skull of *Chirodipterus wildungensis* Gross, an Upper Devonian Dipnoan from Wildungen, Stockholm. *Kgl. Sv. Vetenskapsakad., Handl.* [4] **3 (4)**.

Sawin, H. J. (1941). The cranial anatomy of *Eryops megacephalus*. *Bull. Mus. Comp. Zool., Harvard Mus.* **88**, 407–463.

Schaeffer, B. (1967). Comments on elasmobranch evolution. *In* "Sharks, Rays, and Skates" (P. W. Gilbert, R. F. Mathewson, and D. P. Rall, eds.). pp. 3–35. Johns Hopkins Press, Baltimore, Maryland.

Schaeffer, B. (1969). Adaptive radiation of the fishes and the fish-amphibian transition. *Ann. N.Y. Acad. Sci.* **167**, 5–17.

Schaller, G. B. (1963). "The Mountain Gorilla: Ecology and Behavior." Univ. of Chicago Press, Chicago, Illinois.

Schaller, G. B. (1967). "The Deer and the Tiger. A Study of Wildlife in India." Univ. of Chicago Press, Chicago, Illinois.

Schepers, G. W. H. (1946). The endocranial casts of the South African ape-men. *Transvaal Mus. Mem.* No. 2, pp. 155–272.

Schuchert, C., and LeVene, C. M. (1940). "O. C. Marsh, Pioneer in Paleontology." Yale Univ. Press, New Haven, Connecticut.

Schultz, A. H. (1950). Morphological observations on gorillas. *In* "The Anatomy of the Gorilla" (W. K. Gregory, ed.), pp. 228–253. Columbia Univ. Press, New York.

Schwarzbach, M. (1963). "Climates of the Past." Van Nostrand-Reinhold, Princeton, New Jersey.

Scott, W. B. (1886). On some new and little known creodonts. *J. Acad. Natur. Sci. Philadelphia* [2] **9**, 155–185.

Scott, W. B. (1937). "A History of Land Mammals in the Western Hemisphere" (rev. ed.). Macmillan, New York.

Scott, W. B., and Jepsen, G. L. (1936). The mammalian fauna of the White River

Oligocene. Part 1. Insectivora and Carnivora. *Trans. Amer. Phil. Soc.* [N.S.] **28**, 1–270.

Scott, W. B., and Osborn, H. F. (1890). Preliminary account of the fossil mammals from the White River and Loup Fork Formations, contained in the Museum of Comparative Zoology. Part II. Carnivora and Artiodactyla. *Bull. Mus. Comp. Zool., Harvard Univ.* **20**, 65–100.

Sebeok, T. A. (1965). Animal communication. *Science* **147**, 1006–1014.

Seeley, H. G. (1901). "Dragons of the Air: An Account of Extinct Flying Reptiles." Methuen, London.

Shariff, G. A. (1953). Cell counts in the primate cerebral cortex. *J. Comp. Neurol.* **98**, 381–400.

Sherrington, C. S. (1950). "Man On His Nature," 2nd ed. Yale Univ. Press, New Haven, Connecticut.

Shevchenko, Yu. G. (1971). "Evolution of the Cerebral Cortex in Primates and Man." University Press, Moscow (in Russian).

Sholl, D. A. (1948). The quantitative investigation of the vertebrate brain and the applicability of allometric formulae to its study. *Proc. Roy. Soc. London, Ser. B* **135**, 243–258.

Sholl, D. A. 1956. "The Organization of the Cerebral Cortex." Methuen, London.

Sigogneau, D. (1968). Le genre *Dremotherium* (Cervoidea). Anatomie du crâne, denture et moulage endocranien. *Ann. Paleontol.* **14**, 39–115.

Simons, E. L. (1960). The Paleocene Pantodonta. *Trans. Amer. Phil. Soc.*, **50**, Part 6, 3–81.

Simons, E. L. (1964). The early relatives of man. *Sci. Amer.* **211(1)**, 50–62.

Simons, E. L. (1967a). New evidence on the anatomy of the earliest catarrhine primates. *In* "Neue Ergebnisse der Primatologie: Progress in Primatology" (D. Starck, R. Scheider, and H. J. Kuhn, eds.), pp. 15–18.

Simons, E. L. (1967b). Fossil primates and the evolution of some primate locomotor systems. *Amer. J. Phys. Anthropol.* **26**, 241–253.

Simons, E. L., and Pilbeam, D. R. (1965). A preliminary revision of Dryopithecinae (Pongidae, Anthropoidea). *Folia Primatol.* **3**, 81–152.

Simpson, G. G. (1927). Mesozoic mammalia. IX. The brain of Jurassic mammals. *Amer. J. Sci.* **214**, 259–268.

Simpson, G. G. (1928). "A Catalogue of the Mesozoic Mammalia in the Geological Department of the British Museum," British Museum, London.

Simpson, G. G. (1932). Skulls and brains of some mammals from the Notostylops Beds of Patagonia. *Amer. Mus. Nov.* **578**, 1–11.

Simpson, G. G. (1933a). Braincasts of *Phenacodus, Notostylops,* and *Rhyphodon*. *Amer. Mus. Nov.* **622**, 1–19.

Simpson, G. G. (1933b). Braincasts of two typotheres and a litoptern. *Amer. Mus. Nov.* **629**, 1–18.

Simpson, G. G. (1937). Skull structure of the Multituberculata. *Bull. Amer. Mus. Natur. Hist.* **73**, 727–763.

Simpson, G. G. (1945). The principles of classification and a classification of mammals. *Bull. Amer. Mus. Natur. Hist.* **85**, 1–350.

Simpson, G. G. (1948). The beginning of the age of mammals in South America. *Bull. Amer. Mus. Natur. Hist.* **91**, 1–232.

Simpson, G. G. (1951). "Horses." Oxford Univ. Press, London and New York.

Simpson, G. G. (1953). "The Major Features of Evolution." Columbia Univ. Press, New York.

Simpson, G. G. (1959). Mesozoic mammals and the polyphyletic origin of mammals. *Evolution* **13**, 405–414.
Simpson, G. G. (1960a). The history of life. *In* "Evolution After Darwin" (S. Tax, ed.), Vol. 1, pp. 117–180. Univ. of Chicago Press, Chicago, Illinois.
Simpson, G. G. (1960b). Diagnosis of the classes Reptilia and Mammalia. *Evolution* **14**, 388–392.
Simpson, G. G. (1961). Evolution of mesozoic mammals. *In* "International Colloquium on the Evolution of Lower and Non-specialized Mammals" (G. Vandebroek, ed.), pp. 57–95. Kon. Vlaanse Acad. Wetensch. Lett. Sch. Kunstein België, Brussels.
Simpson, G. G. (1964). "This View of Life." Harcourt, New York.
Simpson, G. G. (1965). "The Geography of Evolution." Capricorn, New York.
Simpson, G. G. (1967). "The Meaning of Evolution" (rev. ed.). Yale Univ. Press, New Haven, Connecticut.
Sloan, R. E., and Van Valen, L. (1965). Cretaceous mammals from Montana. *Science* **148**, 220–227.
Snell, O. (1891). Die Abhängigkeit des Hirngewichtes von dem Körpergewicht und den geistigen Fähigkeiten. *Arch. Psychiat. Nervenkr.* **23**, 436–446.
Spector, W. S., ed. (1956). "Handbook of Biological Data." Saunders, Philadelphia, Pennsylvania.
Sperry, R. W. (1951). Mechanisms of neural maturation. *In* "Handbook of Experimental Psychology" (S. S. Stevens, ed.), pp. 236–280. Wiley, New York.
Stager, K. E. (1964). The role of olfaction in food location by the turkey vulture (*Cathartes aura*). *Los Angeles County Mus. Contrib. Sci.* No. 81, pp. 1–63.
Stebbins, G. L. (1969). "The Basis of Progressive Evolution." Univ. of North Carolina Press, Chapel Hill.
Stensiö, E. (1925). On the head of the macropetalichthyids, etc. *Field Mus. Natur. Hist., Geol. Ser.* **4**, 94.
Stensiö, E. (1963). The brain and the cranial nerves in fossil, lower craniate vertebrates. *Skr. Nor. Videnskaps-Akad. Oslo, Mat. Naturv. Kl.* [N.S.] No. 13.
Stenson, H. H. (1966). The physical factor structure of random forms and their judged complexity. *Percept. Psychophys.* **1**, 303–310.
Stephan, H. (1960). Methodische Studien über den quantitativen Vergleich architektonischer Struktureinheiten des Gehirns. *Z. Wiss. Zool.* **164**, 143–172.
Stephan, H. (1966). Grössenänderungen im olfaktorischen und limbischen System während der phylogenetischen Entwicklung der Primaten. *In* "Evolution of the Forebrain" (R. Hassler and H. Stephan, eds.), pp. 377–388. Thieme, Stuttgart.
Stephan, H., and Andy, O. J. (1969). Quantitative comparative neuroanatomy of primates: An attempt at a phylogenetic interpretation. *Ann. N.Y. Acad. Sci.* **167**, 370–387.
Stephan, H. and Spatz, H. (1962). Vergleichend-anatomische Untersuchungen an Insektivorengehirnen IV. Gehirne afrikanischer Insektivoren Versuch einer Zuordnung von Hirnbau und Lebensweise. *Morphol. Jahrb.* **103**, 108–174.
Stephan, H., Bauchot, R., and Andy, O. J. (1970). Data on size of the brain and of various parts in insectivores and primates. *In* "The Primate Brain" (C. R. Noback and W. Montagna, eds.), pp. 289–297. Appleton, New York.
Stettner, L. J. and Matyniak, K. A. (1968). The brain of birds. *Sci. Amer.* **218(6)**, 64–76.
Stevens, C. F. (1966). "Neurophysiology: A Primer." Wiley, New York.

Stevens, S. S. (1946). On the theory of scales and measurement. *Science* **103**, 677–680.
Stingelin, W. (1958). "Vergleichend morphologische Untersuchungen am Vorderhirn der Vögel auf cytologischer und cytoarchitektonischer Grundlage." Verlag Helbing & Lichtenhahn, Basel.
Stirton, R. A. (1953). A new genus of interatheres from the Miocene of Columbia. *Univ. Calif., Berkeley, Publ. Geol. Sci.* **29**, 265–348.
Stock, C. (1956). "Rancho La Brea: A Record of Pleistocene Life in California," 6th ed. Los Angeles County Museum, Los Angeles.
Streeter, G. L. (1904). The structure of the spinal cord of the Ostrich. *Amer. J. Anat.* **3**, 1–27.
Swinton, W. E. (1958). Dinosaur brains. *New Sci.* **4**, 707–709.
Swinton, W. E. (1965). "Fossil Birds," 2nd ed. British Museum (Natural History), London.
Szalay, F. S. (1971). Cranium of the late Paleocene primate *Plesiadapis Tricuspidens*. *Nature (London)* **230**, 324–325.
Szalay, F. S., and Gould, S. J. (1966). Asiatic Mesonychidae (Mammalia, Condylarthra). *Bull. Amer. Mus. Natur. Hist.* **132**, 127–174.
Tarling, D. H. (1971). Gondwanaland, palaeomagnetism and continental drift. *Nature (London)* **229**, 17–21.
Thompson, D'Arcy W. (1942). "On Growth and Form," 2nd ed. Cambridge Univ. Press, London and New York.
Thorpe, M. C. (1937). The Merycoidodontidae: An extinct group of ruminant mammals. *Mem. Peabody Mus. Yale* **3**, 1–428.
Tilney, F. (1928). "The Brain from Ape to Man," 2 vols. Harper (Hoeber), New York.
Tilney, F. (1931). Fossil brains of some early Tertiary mammals of North America. *Bull. Neurol. Inst. New York* **1**, 430–505.
Tinbergen, N. (1951). "The Study of Instinct." Oxford Univ. Press, London and New York.
Tinbergen, N. (1961). "The Herring Gull's World." Basic Books, New York.
Tobias, P. V. (1965). Early man in East Africa. *Science* **149**, 22–33.
Tobias, P. V. (1966). The distinctiveness of *Homo habilis*. *Nature (London)* **209**, 953–957.
Tobias, P. V. (1967). "The Cranium and Maxillary Dentition of *Australopithecus* (*Zinjanthropus*) *boisei*. Olduvai Gorge," Vol. 2. Cambridge Univ. Press, London and New York.
Tobias, P. V. (1970). Brain-size, grey matter and race—fact or fiction? *Amer. J. Phys. Anthropol.* **32**, 3–26.
Tobias, P. V. (1971). "The Brain in Hominid Evolution." Columbia Univ. Press, New York and London.
Tolman, E. C. (1948). Cognitive maps in rats and men. *Psychol. Rev.* **55**, 189–208.
Tower, D. B. (1954). Structural and functional organization of mammalian cerebral cortex: The correlation of neurone density with brain size. *J. Comp. Neurol.* **101**, 19–51.
Tower, D. B. and Young, O. M. (1973). The activities of butyrylcholinesterase and carbonic anhydrase, the rate of anaerobic glycolysis, and the question of a constant density of glial cells in cerebral cortices of mammalian species from mouse to whale. *J. Neurochem.* **20**, 269–278.
Valenstein, E. S., Cox, V. C., and Kakalewski, J. W. (1970). Reexamination of the role of the hypothalamus in motivation. *Psychol. Rev.* **77**, 16–31.

Van Valen, L. (1960). Therapsids as mammals. *Evolution* **14**, 304–313.
Van Valen, L. (1965). Treeshrews, primates, and fossils. *Evolution* **19**, 137–151.
Van Valen, L., and Sloan, R. E. (1965). The earliest primate. *Science* **150**, 743–745.
Van Valen, L., and Sloan, R. E. (1966). The extinction of the multituberculates. *Syst. Zool.* **15**, 261–278.
von Békésy, G., and Rosenblith, W. A. (1951). The mechanical properties of the ear. *In* "Handbook of Experimental Psychology" (S. S. Stevens, ed.), pp. 1075–1115. Wiley, New York.
von Bonin, G. (1937). Brain weight and body weight in mammals. *J. Gen. Psychol.* **16**, 379–389.
von Bonin, G. (1963). "The Evolution of the Human Brain." Univ. of Chicago Press, Chicago, Illinois.
von Bonin, G., and Bailey, P. (1947). "The Neocortex of Macaca Mulatta." Univ. of Illinois Press, Urbana.
von Haller, A. (1762). "Elementa Physiologiae Corporis Humanis," Tome 4. Lausanne.
von Neumann, J. (1951). The general and logical theory of automata. *In* "Cerebral Mechanisms in Behavior" (L. A. Jeffress, ed.), pp. 1–41. Wiley, New York.
von Uexküll, J. (1934). "Streifzüge durch die Umwelten von Tieren und Menschen." Springer-Verlag, Berlin and New York (transl. in C. H. Schiller, ed., "Instinctive Behavior: The Development of a Modern Concept," pp. 5–80. International Universities Press, New York, 1957).
Walker, E. P. (1964). "Mammals of the World," 3 vols. Johns Hopkins Press, Baltimore, Maryland.
Walls, G. (1942). "The Vertebrate Eye and its Adaptive Radiation." Cranbrook Press, Bloomfield Hills, Michigan.
Warnke, P. (1908). Mitteilung neuer Gehirn und Körpergewichtsbestimmungen bei Säugern, nebst Zusammenstellung der gesamten bisher beobachteten absoluten und relativen Gehirngewichte bei den verschiedenen Spezies. *J. Psychol. Neurol.* **13**, 355–403.
Warren, J. M. (1965). Primate learning in comparative perspective. *In* "Behavior of Nonhuman Primates" (A. M. Schrier, H. F. Harlow, and F. Stollnitz, eds.), Vol. 1, pp. 249–281. Academic Press, New York.
Washburn, S. L. (1951). The analysis of primate evolution with particular reference to the origin of man. *Cold Spring Harbor Symp. Quant. Biol.* **15**, 67–77.
Washburn, S. L. (1965). An ape's eye view of human evolution. In "The Origin of Man" (P. L. DeVore, ed.), pp. 89–107. Wenner-Gren Found., New York.
Washburn, S. L., and Howell, F. C. (1960). Human evolution and culture. *In* "Evolution After Darwin" (S. Tax, ed.), Vol. 2, pp. 33–56. Univ. of Chicago Press, Chicago, Illinois.
Watson, D. M. S. (1913). Further notes on the skull, brain and organs of special sense of *Diademodon*. *Ann. Mag. Natur. Hist.* [8] **12**, 217–228.
Watson, D. M. S. (1951). "Paleontology and Modern Biology." Yale Univ. Press, New Haven, Connecticut.
Webb, P., ed. (1964). "Bioastronautics Data Book," NASA SP-3006. National Aeronautics and Space Administration. Washington, D.C.
Weber, M. (1896). Vorstudien über das Hirngewicht der Säugethiere. *Festschr. f. Carl Gegenbaur* **3**, 105–123.
Weiskrantz, L. (1961). Encephalisation and the scotoma. *In* "Current Problems in Animal Behaviour" (W. H. Thorpe and O. L. Zangwill, eds.), pp. 30–58. Cambridge Univ. Press, London and New York.

Welker, W. I., and Campos, G. B. (1963). Physiological significance of sulci in somatic sensory cerebral cortex in mammals of the family Procyonidae. *J. Comp. Neurol.* **120**, 19–36.

Welker, W. I., and Seidenstein, S. (1959). Somatic sensory representation in the cerebral cortex of the raccoon (*Procyon lotor*). *J. Comp. Neurol.* **111**, 469–502.

Wever, E. G. (1965). Structure and function of the lizard ear. *J. Audit. Res.* **5**, 331–371.

Wheeler, W. H. (1961). Revision of the uintatheres. *Bull. Peabody Mus. Yale* **14**, 1–93.

White, J. F., and Gould, S. J. (1965). Interpretation of the coefficient in the allometric equation. *Amer. Natur.* **99**, 5–18.

Whiting, H. P., and Halstead Tarlo, L. B. (1965). The brain of the Heterostraci (Agnatha). *Nature (London)* 207, 829–831.

Wilkie, D. R. (1959). The work output of animals: Flight by birds and by manpower. *Nature (London)* **183**, 1515–1516.

Wilson, D. M. (1964). The origin of the flight-motor command in grasshoppers. *In* "Neural Theory and Modeling" (R. F. Reiss, ed.), pp. 331–345. Stanford Univ. Press, Stanford, California.

Wilson, D. M. (1968). The flight-control system of the locust. *Sci. Amer.* **218(5)**, 83–90.

Wilson, J. A. (1966). A new primate from the earliest Oligocene, west Texas, preliminary report. *Folia Primatol.* **4**, 227–248.

Wirz, K. (1950). Zur quantitativen Bestimmung der Rangordung bei Säugetieren. *Acta Anat.* **9**, 134–196.

Wolin, L. R., and Massopust, L. C., Jr. (1970). Morphology of the primate retina. *In* "The Primate Brain" (C. R. Noback and W. Montagna, eds.), pp. 1–27. Appleton, New York.

Woodward, A. S. (1900). On some remains of *Grypotherium* (*Neomylodon*) *listai* and associated mammals from a cavern near Consuelo Cove, Last Hope, Patagonia. *Proc. Zool. Soc. London* **5**, 64–79.

Woolsey, C. N. (1958). Organization of somatic sensory and motor areas of the cerebral cortex. *In* "Biological and Biochemical Bases of Behavior" (H. F. Harlow and C. N. Woolsey, eds.), pp. 63–81. Univ. of Wisconsin Press, Madison.

Wright, R. D. (1934). Some mechanisms in the evolution of the central nervous system. *J. Anat.* **69**, 86–88.

Yalden, D. W. (1971a). Flying ability of *Archaeopteryx*. *Nature (London)* **231**, 127.

Yalden, D. W. (1971b). *Archaeopteryx* again. *Nature (London)* **234**, 478–479.

Young, J. Z. (1962). "The Life of Vertebrates," 2nd ed. Oxford Univ. Press, London and New York.

Young, J. Z. (1964). "A Model of the Brain." Oxford Univ. Press (Clarendon), London and New York.

Zamenhof, W. S., Van Marthens, E., and Bursztyn, H. (1971). The effect of hormones on DNA synthesis and cell number in the developing chick and rat brain. *In* "Hormones in Development" (M. Hamburgh and E. J. W. Barrington, eds.), pp. 101–119. Appleton, New York.

Zangerl, R. (1960). The vertebrate fauna of the Selma Formation of Alabama, Part V. An advanced chelonid sea turtle. *Fieldiana, Geol. Mem.* **3**, No. 5, 277–312.

Zapfe, H. (1963). Lebensbild von *Megaladapis edwardsi* (Grandidier): Ein Reconstruktionversuch. *Folia Primatol.* **1**, 178–187.

Appendix I

Wirz's Analysis of Relative Size of Parts of the Brain

It is unfortunate that Wirz (1950) published only indices and omitted data on the actual weights of the parts of the brain. The indices were presented only for families of mammals rather than for individuals or species. In order to use her data as the background against which to compare the fossil sample, it was necessary to derive measurements of specific weights of brain and body by transformations of her indices, and we must, therefore, review and understand her method of determining her indices. This appendix should be read as a continuation of the exposition introducing the data of Table 11.2, p. 247, where some of the basic terms are defined.

Wirz assumed that the "stem complex" in insectivores, the most "primitive" of living placentals, could be used as the base against which the development of the brain and its parts could be compared. Like Stingelin (1958), she was a student of Portmann's and she followed the same basic approach (Portmann, 1946, 1947). The allometry of organ and body size is taken into account, in their approach, through the use of an allometric equation relating the weight of the stem complex of insectivores to their body weights. Let us call this the basal stem complex. She reported this equation as

$$\text{basal stem complex} = x \; P^{0.48} \qquad (I.1)$$

presumably in cgs units but did not give a value for x. She then defined her brain indices for a mammal of body weight P (grams) as

$$\text{index} = \frac{\text{brain fraction weight}}{x \; P^{0.48}} \qquad (I.2)$$

She reported her indices for many families of mammals, presumably averaging the values for species within a family. Of her indices the following are used here: total index (T.I.), in which the numerator in Eq. (I.2) is the total brain weight; neopallial index (N.I.), with neopallial weight as numerator; cerebellar index (C.I.), with cerebellar weight as numerator; and stem index (St.I.), with the stem complex weight as numerator.

It will be helpful for understanding what is actually being done when these indices are derived and used if we look more closely at the total index (T.I.). This is very similar to Dubois' (1897) index of cephalization, that is, k in Eq. (2.1), calculated for a particular specimen's brain and body weights. (Dubois used an exponent of $\frac{5}{9}$ rather than $\frac{2}{3}$, for reasons discussed in Chapter 3.) The equation of the transformation of k to T.I. can be derived from Eqs. (2.1) and I.2). We note, first, that from Eq. (I.2)

$$(\text{T.I.}) = \frac{E}{x\, P^{\frac{1}{2}}} \tag{I.2a}$$

(the exponent, 0.48, has been approximated as $\frac{1}{2}$). The brain weight E is given in terms of k in Eq. (2.1), which may be combined with Eq. (I.2a) to yield

$$(\text{T.I.}) = (k\, P^{\frac{2}{3}})/(x\, P^{\frac{1}{2}})$$
$$= (1/x)\, (k\, P^{\frac{1}{6}})$$

Shortly it is shown that x is approximately 0.03. The equation relating Wirz's total index T.I. to the index of cephalization k, as used here is, therefore,

$$(\text{T.I.}) = 33.3\, k\, P^{1/6} \tag{I.3}$$

In other words the difference between Dubois' index of cephalization as modified by von Bonin (1937) and Wirz's total index is in the way that body weight is taken into account.

It is possible to solve for x in Eqs. (I.1) and (I.2) by using published data on living hedgehogs. This enables us to reconstruct the various brain fractions in a number of living mammals and to have that as a framework against which to measure relative brain development in the archaic Tertiary series of mammals. We proceed, first, by noting that the stem index for the hedgehog is, by definition, 1.0 because it is the quotient of the stem complex of this insectivore divided by itself. Next, Wirz informs us in her tables that the total index was (T.I.) $= 4.50$ for the family Erinaceidae. Now we must assume that the index for the family was appropriate for the European hedgehog *Erinaceus europaeus*. We then seek data on total brain and body weights in this species. Bauchot and Stephan (1966) recently reviewed such data and indicated a typical (or average) pair of weights of $E = 3.35$ g and $P = 860$ g. This is sufficient information to derive the weight of the hedgehog's stem complex because according to Eq. (I.2) applied to the total index

Appendix I

$$(\text{T.I.}) = \text{brain weight/stem complex weight (in insectivores)}$$

or

$$4.50 = 3.35 \text{ g/stem complex weight (in hedgehog)}$$

and

$$\text{stem complex weight} = 3.35/4.50 \text{ g} = 0.744 \text{ g}$$

This weight is presented to three significant figures at this time; rounding is performed later, and tabled data are given to one or two decimal places.

One may now solve Eq. (I.1) for x in the case of the hedgehog

$$0.744 \text{ g} = x \ (860 \text{ g})^{0.48}$$

To three decimal places this results in $x = 0.029$. The result may be approximated as indicating that the basal or insectivore level for the weight of the stem complex in mammals, used as the divisor in all Wirz's indices, is 3% of the square root of the body weight, or

$$(\text{basal St.I.}) = 0.03 \ P^{\frac{1}{2}} \tag{I.1a}$$

In the computations of indices or derivation of weights from Wirz's indices (Table 11.2) the following solution of Eq. (I.1) is used

$$(\text{basal St.I.}) = 0.029 \ P^{0.48} \tag{I.1b}$$

In more general analyses, to show the relationships among indices, the simpler form, Eq. (I.1a) has been used.

We can now calculate particular brain or brain part weights from Wirz's indices, simply by multiplying the index by the value of the basal St.I. for a particular animal of known weight. The computation can be illustrated with selected data on man. Wirz gives T.I. for man as 214. If one assumes a typical human (male) weight of 70 kg, then the basal St.I. for a 70 kg mammal, from Eq. (I.1b), is 6.14 g. The total brain weight should, therefore, be (214) (6.14) g or 1314 g, the figure presented in the text (p. 246). The cerebellar index C.I. for man, according to Wirz, is 25.7. Multiplied by the basal St.I. of 6.14 g, this results in an estimation of a cerebellar weight of 158 g, which is within the weight range for the human cerebellum of 136–169 g reported by Blinkov and Glezer (1968).

We are especially interested in using Wirz's neopallial index N.I. to estimate the size of the forebrain, and Marshall's (1892) data were cited in the text. One of Marshall's groups, an unspecified group of normal men of middle height (167–173 cm), between 50–60 years old when they died, averaged a total brain weight of 1327 g; their cerebral hemispheres were reported to average 1157 g and their cerebellums 145 g. Assuming again a value of $P = 70$ kg, the neopallial index N.I. for man, according to Wirz, was 170, and this leads to a neopallial weight of (6.14 g) (170) = 1045 g, a weight only 10% less than the cerebral hemisphere weight reported by Marshall. Although neither the cerebral hemispheres nor Wirz's neopallial structures correspond exactly to the "forebrain" of archaic mammals as measured here, they seem reasonable approximations; at worst the neopallial weight of Wirz should underestimate the absolute mass of the region measured as forebrain in the fossils.

The method applied to human data in the previous paragraph was applied to seven "progressive" mammalian species in families for which indices were reported by Wirz, and the data were summarized in Table 11.2 in Chapter 11. This table, thus, has information for an analysis of the relative size of the brain and its parts as functions of body size in living progressive mammals of about the same range of body sizes as the fossil sample. The living mammals are probably also similar in niche to the fossils. The data on brain and body weights in Table 11.2 were from Weber (1896). Wirz's indices are for the families to which these species belong. The "basal St.I." in Table 11.2 was computed from Eq. (I.1b) for the body weights reported by Weber. The estimates of the total brain weight (E_T) and the weights of the forebrain (E_F) and hindbrain (E_H) followed formulas indicated in the notes to Table 11.2.

It is instructive, before returning to the analysis of the fossil evidence, to examine the brain:body relationships suggested by Table 11.2. One may ask, first, how accurate is the use of Wirz's indices to reconstruct whole brain weights. This can be determined immediately by comparing the reported actual weights, from Weber, shown under E and the independently calculated weights shown under E_T, using Wirz's index, which involves only the information on body weight. The correspondence is remarkable when one realizes that it is not even certain that Wirz's sample for deriving her indices included the species of Table 11.2.

A final word about Wirz's index is in order because of its apparent complexity and a certain perverse appeal due to that complexity. When indices have been discussed in recent years, it has usually been asserted that gross brain size is a priori inappropriate for valid indices because it obscures the specialized functions of particular nuclei or complexes in the brain (Sholl, 1948, 1956; Anthony, 1951a,b; Haug, 1956). The complexity of Wirz's index has made it more acceptable to some. This may be a result of insufficient awareness of how simple Wirz's indices really are and a commitment to complexity because of insufficient awareness of how excellent a biological statistic the gross brain size is. Wirz's index is essentially the same as Dubois', and her use of it is almost identical with Brummelkamp's (1940) use of an extension of Dubois' work. As such

Appendix I

it is subject to some of the same criticisms, such as Sholl's (1948), although these criticisms have probably been overdrawn.

Let us now examine her method. Because of the high correlation between the neopallial weight and the gross brain weight, Wirz obtained results with her neopallial index that were essentially the same as she would have gotten had she used the total index. Effectively, she took brain:body data of the type presented elsewhere in this book, a set of points representing brain and body weights on log–log coordinates, and superimposed upon them a set of parallel lines with a slope of 0.48. One of her analyses involved two of those lines, which were chosen as dividing lines to differentiate groups of points into sets of low, middle, and high brain development groups. The procedure is formally the same as Brummelkamp's, although it does not suggest that the distance between the lines is particularly significant. (Brummelkamp thought they were divided by steps of $\sqrt{2}$ in k of Dubois' version of Eq. (2.1) or (3.1) with an exponent of 0.56. Wirz's procedure produces somewhat different divisions because it is based on a line of slope 0.48 rather than 0.56.)

The derivation of indices from this kind of procedure involves assumptions about the relationship between the base level, such as the basal stem complex of Wirz, and additional neural tissue. Wirz's indices all involve the multiple required to reach the weight of the additional tissue from the basal level. This is identical with all uses of an index of cephalization, as by Dubois, Brummelkamp, von Bonin, and others including myself. In recognizing the difficulty in this approach to extra tissue, I have tried an alternate approach (see Chapter 3 and Jerison, 1963) in which a basal amount of brain is determined for primitive mammals as a minimum for total (not simply vegetative) information processing by a mammalian brain. This is essentially a line fitted to the archaic mammalian polygon, Eq. (3.17). Then, additional neural tissue is considered as if it were a statistic to estimate an added (not multiplied) number of neurons for processing information beyond the primitive mammalian level. The approach produces yet another index, the number of "extra neurons," which is directly related to hypotheses about information-processing capacity. The index is applied to data on higher primates (Chapter 16, Table 16.2 and 16.3, and pp. 395–399.

Appendix II

Statistical Tests on Mammalian Data

One of the main results in the analysis of the evolution of the mammalian brain was the discovery that diversity evolved just as encephalization evolved. A number of approaches to the statistical analysis of these results were possible. As indicated in an early summary of this result (Jerison, 1970b), a nonparametric test was first applied to the cumulative frequency distributions of EQ of each of several groups. A second look at the data indicated that a parametric analysis, a conventional simple analysis of variance, was possible if the EQ data were treated with a logarithmic transformation. The success of the transformation for this purpose is, itself, of some interest because it suggests the relationship between evolving encephalization and diversity.

The effect of the transformation may be demonstrated by regraphing the cumulative frequency distributions of figures in the text on "probability $\times$ log" graph paper. It was possible to add new curves that could not easily be shown on the coordinate systems of the previous figures because of their extended ranges of EQ. The results of such graphing are shown in Fig. II.1. I did not bother to fit straight lines to the data of these figures; instead, I merely connected the successive points in the distribution. The figures are drawn to show that all the curves are similarly oriented in this new coordinate system, and this means that the diversity of EQ is proportional to the average value of EQ.

Until we reached the comparisons within the class Mammalia, we were able to perform all analyses in this book without the use of statistics because we could compare the nonoverlapping minimum convex polygons for the several classes of vertebrates. The analysis of the logarithmic transformations of EQ as presented in Fig. II.1 may also be visualized as representing such polygons, but these overlap one another. In the previous analyses, one of the remarkable features was the similarity of shape as well as orientation of the polygons—especially when we could compare groups that had similar ranges of body sizes. This was most notable in the comparison of reptiles and mammals (Fig. 11.5). The implication of

Appendix II

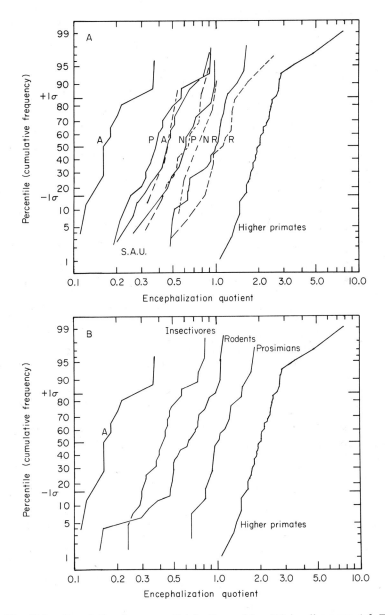

Fig. II.1. Cumulative frequency distributions of log *EQ* in all groups (cf. Table II.1). A, Distributions of fossil (S.A.U. = South American fossil ungulates) and living ungulates, carnivores, higher primates; abbreviations as in Fig. 13.10. B, Distribution of archaic ungulates (A), insectivores, rodents, and prosimians and higher primates. Data of archaic ungulates (A) and higher primates are shown in both graphs as anchoring distributions to indicate positions of other assemblages.

Fig. II.1 is that the mammalian assemblages considered in this book could also be represented by a set of (overlapping) polygons that were similar in shape and orientation. They would look like a set of polygons similar to the one for the archaic "ungulates" of Fig. 11.5.

For the purpose of the statistical analysis there are 14 assemblages—all the assemblages of mammals discussed in the last section of this book. A simple analysis of variance for log EQ in these assemblages indicated significant differences beyond the 0.001 level, as might be expected, with $F = 65.1$, $df = 13/251$. An instructive tabular analysis is presented in Table II.1, in which the means and standard deviations of EQ and the means and standard deviations of the logarithms of EQ are presented, and the result of a Duncan multiple range test is given in the last column. [These data differ from the data as analyzed in the text (e.g. Table 14.2) in that the computations did not exclude outlying points. The curve-fitting procedure used in the body of the text in which the points between $\pm 1\,\sigma$

Table II.1
Summary Statistics of Encephalization Quotients (EQ) in Mammalian Assemblages[a]

Assemblage (in rank order, log EQ)	N	EQ Mean	S.D.	Log EQ Mean	S.D.	Subsets[b] (range test)
Archaic ungulates	13	0.194	0.0815	−0.741	0.1574	]
Paleogene ungulates	22	0.415	0.1860	−0.417	0.1727	]
Recent insectivores	33	0.443	0.1555	−0.377	0.1418	
South American Paleogene ungulates	9	0.463	0.2020	−0.369	0.1868	
Archaic carnivores	4	0.448	0.0925	−0.357	0.0940	
South American Neogene ungulates	11	0.483	0.1576	−0.337	0.1397	]
Paleogene carnivores	11	0.618	0.1818	−0.228	0.1374	]
Neogene ungulates	13	0.635	0.2259	−0.225	0.1679	
Recent rodents	37	0.652	0.2509	−0.225	0.2016	
Neogene carnivores	6	0.765	0.1909	−0.128	0.1091	]
Recent ungulates	25	0.964	0.3153	−0.040	0.1498	]
Recent prosimians	20	1.105	0.3163	0.027	0.1195	
Recent carnivores	15	1.167	0.4816	0.037	0.1668	]
Recent higher primates	46	2.082	0.6268	0.302	0.1172	]

[a] Statistics are based on entire sample rather than on the central portion of the sample ($\pm 1\sigma$) used in text.

[b] Subsets indicated by brackets differed from one another by more than the shortest significant range ($\alpha = 0.05$) for a subset of that size. For example, the archaic ungulates form a unique subset that differs from any other subset of assemblages in this sample.

Appendix II 465

were used to fit a line on normal probability paper makes good biological sense but leads to difficulties in statistical analyses.]

The increasing diversity in *EQ* is shown in the increasing values of the standard deviations of Table II.1 for the nontransformed data. In examining the standard deviations of the log transforms of *EQ* for the various assemblages, we note much more homogeneous variance than in the nontransformed data. For a formal test of homogeneity we should combine the data for all the primates (prosimians and anthropoids), and in the log transforms one then obtains an overall mean of log $EQ = 0.219$ and a standard deviation for the sample of 66 primate species of 0.173. The smallest standard deviations for the assemblages, under the log transformation, were those for the archaic and Paleogene carnivores; these were probably accidents of sampling resulting from the small size of these groups.

One may use the simple analysis of variance as the basis of a test of significantly differentiable groups. This was performed using the Duncan multiple range test, with results as indicated in Table II.1 (Dixon and Masey, 1969). Seven subsets were distinguished with $\alpha = 0.05$. These included a subset consisting only of the archaic ungulates and another subset consisting only of the primate suborder Anthropoidea [suborders Platyrrhini and Catarrhini of Romer (1966)]. The remaining subsets are the ones identified in the body of the text. Thus, within the holarctic ungulates, the Paleogene, Neogene, and Recent had to be placed in different subsets under all groupings, indicating that these were significantly different from one another at "the 0.05 level." Although a similarly clear distinction could not be made among the carnivores, I would attribute that to the small samples available for all but the Recent carnivores. Thus, the archaic and Neogene carnivores had to be in different groups, but neither could be distinguished from the Paleogene carnivores at the 0.05 level.

It is also interesting to consider the other living orders of mammals in comparison with the fossil carnivores and ungulates to which I devoted much of the analysis in the text. The insectivores were significantly higher in *EQ* than the archaic ungulates; this is evidence that they cannot be considered as entirely unevolved with respect to enlarged brains. They could be grouped with the Paleogene ungulates and carnivores and the South American Paleogene and Neogene ungulates. The living rodents are significantly larger brained than the insectivores, according to this analysis, and smaller brained than the other Recent orders considered in the analysis of variance. Finally, the prosimians fell within the same subset as the Recent carnivores and ungulates, all clustered about $EQ = 1.0$ (log $EQ = 0$), and are, therefore, "average" living mammals.

Appendix III

Foramen Magnum, the Size Factor, and Brain Size

The data for this appendix are from Bauchot and Stephan (1967); raw data were measures of length and width of photographs of the foramen magnum and of the cross section of the medulla of four insectivores and five primates for which endocasts and brains are presented in the plates of that report. Data on brain weight (E) and body weight (P) were taken from the legends for the plates, and magnification factors as indicated in the photographs were accepted in converting all the measures to actual size. A summary of the data is presented in Table III.1.

As indicated in Fig. 16.5 of the text, log transforms of these data are approximately linearly related; a matrix of product–moment correlation coefficients among the logarithms of the four measures available on each of the nine specimens is presented in Table III.2, along with the slopes of the best-fitting lines (least squares, assuming errors in both axes). By interpreting these correlations as indicating the ratio of common variance to total variance, a number of issues are immediately clarified, although one or two special comments should first be made.

The simple brain:body correlation is unusually low; $r = 0.66$. In most analyses one may expect correlations of the order of 0.90 for these measures. The low value in the present data probably occurred for two reasons. First, there was a relatively narrow range of brain and body sizes in the species that were sampled; high correlations are the rule when one's sample is from a highly diverse group of mammals, such as the group indicated in Fig. 2.3. Second, the sample has a systematic bias in that four of the species are from the relatively small-brained order Insectivora and the remaining five are from the relatively large-brained order Primates. In a sense, the sample has been biased against the best brain:body analysis in order to determine how adequate the measures of the foramen magnum and the medulla were as independent estimators of body size.

The other correlations in Table III.2 are, therefore, of special interest. The cross section of the medulla shares relatively little variance with brain

Appendix III

Table III.1
Brain and Body Data for Analysis of Foramen Magnum Statistic

Species	Brain weight (g)	Body weight (g)	n^a (mm²)	m^b (mm²)
1. Tenrec ecaudatus	2.42	735	42	15
2. Solenodon paradoxus	4.45	640	62.4	14.5
3. Potamogale velox	4.7	800	60	31.4
4. Rhynchocyon stuhlmanni	6.2	500	70	17.5
5. Loris gracilis	6.3	195	36	8
6. Microcebus murinus	2.0	54	26.9	4.95
7. Lemur fulvus	23.9	1200	78	20.7
8. Propithecus verrauxi	26.9	3200	169	20.8
9. Callithrix jacchus	8.6	215	42.8	15.7

[a] n: maximum width × maximum height of foramen magnum as shown on skull or endocast.

[b] m: maximum width × maximum height of cross section of medulla at approximate level of foramen magnum or N.XII.

size ($r = 0.51$; 25% variance in common) but is highly correlated with body size, having ⅔ of its variance in common with the body size variable. The cross section of the foramen magnum, which is a measure of the skull rather than the brain, is even more highly correlated with body size, sharing over 80% common variance. This is Radinsky's preferred independent "size" variable, and, as a measure, it is also highly correlated with brain size. The source of that surprising correlation can be explored by an analysis of partial correlation coefficients. We note, finally, that among the simple correlations, the one between the foramen magnum and the cross-sectional area of the medulla is intermediate, indicating about 50% common variance.

Table III.2
Correlation Matrix for Data of Table III.1

Dependent variable	Independent variable[a]		
	Body (P)	n	m
Brain (E)	0.66	0.79	0.51
	(0.66)	(1.9)	(2.4)
Body (P)		0.90	0.82
		(2.4)	(2.5)
n			0.70
			(1.0)

[a] Main entries are product–moment correlations of the logarithms of the data; numbers in parentheses are slopes of least-squares fits, assuming errors in both variables.

In the partial correlations, each of the correlations of Table III.2 is considered with the contribution of one or both of the other variables removed from their common variance. For example, the correlation between gross brain weight and the cross-sectional area of the medulla can be associated entirely with the common variance that these two measures share with the area of the foramen magnum and the body weight. With the latter two variables controlled, we obtain the partial correlation $r_{Em \cdot nP} = 0.02$. (We may read this, "the correlation of E with m, given n and P.") The complete set of partial correlations is presented in Table III.3. This includes the partial correlations with one and with two variables controlled.

Table III.3
Partial Correlation Coefficients

$r_{EP \cdot m} = 0.49$	$r_{Pm \cdot E} = 0.75$	$r_{EP \cdot mn} = -0.16$
$r_{EP \cdot n} = -0.19$	$r_{Pm \cdot n} = 0.61$	$r_{Em \cdot nP} = 0.02$
$r_{Em \cdot n} = -0.10$	$r_{Pn \cdot E} = 0.82$	$r_{En \cdot mP} = 0.60$
$r_{Em \cdot P} = -0.07$	$r_{Pn \cdot m} = 0.80$	$r_{Pm \cdot nE} = 0.61$
$r_{En \cdot m} = 0.70$	$r_{mn \cdot E} = 0.56$	$r_{Pn \cdot mE} = 0.73$
$r_{En \cdot P} = 0.60$	$r_{mn \cdot P} = -0.15$	$r_{mn \cdot EP} = -0.14$

The analysis was performed in this way to emphasize underlying functional relationships rather than the predictive power of the variables. We may recognize that Radinsky's use of the measure n, the cross section of the foramen magnum, is justifiable according to the present analysis as the best predictive variable for estimating body size. The correlation between P and n in this diverse sample was 0.90, and even if common variances with brain size (E) and medulla (m) are taken into account in the partial correlation, there remains a common variance of about 50% ($r = 0.73$). From a functional point of view, however, there is a serious defect in the use of the foramen magnum as a measure because of its very strong correlation with brain size. Even partialling out both body size and the cross section of the medulla, we are left with $r_{En \cdot mP} = 0.60$, indicating that more than $\frac{1}{3}$ of the variance in E and n is common to these measures. As pointed out in the text (p. 376, *et seq.*), this means that the measure of the foramen magnum, though extremely useful when one wishes to estimate body size, is defective for a functional analysis of brain:body relations because it also estimates brain size. In statistical terms, the problem with the measure of the foramen magnum is that it reduces residuals in both the brain and body size measures by sharing variance with both measures. Its value as an estimator of body size becomes less important because its "errors" in estimating body size (the residuals for that dimension) will tend to be correlated with the direct measure of brain size.

Appendix III

The analysis in this appendix is based on a very small sample of species. Although its results make good biological sense, the analysis should be validated on a larger and more diverse sample. Had we used conventional "significance levels," such as $p < 0.05$, r_{Em} in Table III.2 would be described as "nonsignificant." (For a sample size $N = 9$, we reject the null hypothesis that $r = 0$ only if $r \geqslant 0.58$.) This is the same as the "functional" conclusion that we reached following the analysis with partial correlation coefficients (Table III.3). We noted there that the correlation between E and m should be attributed to spurious effects of their common variance with a "body size factor." More precise statements would be possible if we had larger samples.

I cannot develop the complete solution to the problems implied by the common variance of E, P, and n, but it is appropriate to make one suggestion. It may be possible to improve the validity of n in a functional analysis of the evolution of relative brain size (brain:body relations) by using multiple regression techniques in which the component of variance in n attributable to brain size (E) is systematically removed. Unless that is done, graphs of E as a function of n (such as Fig. 16.4) will tend to show too little variation in brain size. They will underestimate the deviation of actual brain size (or endocranial volume) from the "expected" brain size as represented by a line of best fit. They will, therefore, tend to permit type II errors, in which the judgment of "no difference" is made when actual differences exist among groups being compared with respect to relative brain size. I believe that Radinsky's analysis illustrated such an error, when he recognized only *Smilodectes* as a relatively small-brained Eocene prosimian genus. The error is not serious, as I pointed out in the body of the text. The evidence on prosimian brain evolution suggests that the modern size range was achieved relatively early in the Tertiary radiation, certainly before the end of the Paleogene, that is, 25 m.y. or so ago, and the available data are insufficient to make fine judgments of changes in relative brain size during the period of the major radiation of progressive mammals between 50–25 m.y. ago.

Index

Names are indexed to indicate discussions of concepts that are associated with a particular person. Authors cited are listed in the Bibliography.

Aardvark, 354
Abducens nerve (N.VI), 90
Adapis, 364, 366, 369–371, 374–376, 378–384
Adaptive radiations, 6–7, 37–41, 173–175, 256–258, 406–407
  birds, 178–181, 197–199, 275–281
  lower vertebrates, 82–84, 115–116, 122–123, 124–130, 138–141, 161–163
  mammals
    archaic, 200–204, 226–235, 253–255, 261–263, 273–275
    progressive, 289–293, 321–323, 341–355, 364–368, 418–424
Adaptive zones, *see* Niches
Additive hypothesis in theory of brain size, 361–362
Adinotherium, 326
Aegyptopithecus, 366–367, 403
Aerial niches, 39, 156–157, 167–172, 197–199
  mammalian, 352–354
Aggression, 363, 397, 409
Agnathans, 38, 40, 82–88
Aletomeryx, 301, 306
Allometric relationship, upper limit on, 344–345
Allometric analysis, 45–49, 57–60, 309–310, 327–329, 334–335
Allometry in related species, 359–362, 394
Allosaurus, 145
Alouatta, 392
Amblotherium, 203–205
Amblypods, 227, 230, 230–233, 236–237, 305, 308, 321
Ameiurus, 112
Ammocoetes, 82
Amphibians, 38, 40, 128–136
  behavior of, 18, 126–128, 262–263
Amphicyon, 307
Amphitragulus, 306

Amygdala, 73, 92
Amynodon, 306
Analysis of variance of *EQ,* 313–314, 462–465
Anatosaurus, 145
Anolis, 266
Anoplotheres, 293, 302, 317–318
Anoplotherium, 301–302, 306
Anthony's index, 56
Anthropoidea, definition of, 365
Anthropoids, 223, 387–395, 397–399, 418–420
Anticipation, 426
Aquatic
  mammals, 346–351
  vertebrates, 39
Archaeoceti, 346–348, 350
Archaeolemur, 375
Archaeopteryx, 177–194, 196–199, 275–277
Archaeotherium, 306
"Archaic"
  definition of, 204
  mammals, *see* name of order
Archosaurs, 137–140
Arctocyon, 231, 236, 240, 250, 305
Arctocyonides, 231, 236, 305
Arctocyonids as carrion feeders, 337
Argyrocetus, 348–351
Arminiheringia, 338
Arthrodira, 101
Artifacts, hominid, 432
Artiodactyls, 53, 287, 295, 300–302, 308–310
  archaic, 293, 306, 317–318
Association cortex, 402
Astrapotherium, 322, 325–326
Ateles, 392
Attention, 415
Auditory
  space-time, 22, 263–267, 413–415, 426

471

system, 9, 13, 19–20, 23–24, 269–271, 348–354, 415–417
and vestibular nerve (N.VIII), 90
Aulophyseter, 348–349
Australopithecines, 389–399
  adaptive zone, 421–423
  endocasts, 389–391
Australopithecus, 44, 367, 389–391, 394, 398–399, 420
Avian niches, 257, 276–278

Baboons, 362, 394
Baillarger's law, 297–298
Balaena, 347
Baluchitherium, 293, 344
Barylambda, 230, 232, 236, 305
"Basal" insectivores, 60, 218–219, 222–223
Bats, 40, 157, 352–354
Bettongia, 260
"Between species" defined, 4
Biological intelligence, 3, 5, 11, 16–25, *see also* Intelligence
Bipedalism, 403
Birds, 177–199, *see also Archaeopteryx, Numenius*
  adaptive radiation, 39–40, 178–181, 257–258, 276–281
  "primate" niches, 199, 277
  songs of, 274
  world of, 406, 412, 429
Blarina, 219
Body size
  estimation of, 50–54, 135, 144–145, 189–190, 209–211, 239–240, 378–380
  factor, 16, 42, 57, 73, 466–469
  formulas, 52–54, 102–104, 166–169, 210–211
  selection pressures and, 126, 202, 259
Body temperature, control of, 39, 259, 408, *see also* Homoiothermy
Boreosomus, 113–114, 147
Bothriolepis, 83
Brachiation, 403–404, 421–422
Brachiosaurus, 140, 143, 145
Brain, *see,* Endocast, names of animals,
  enlargement, evolution of, 386–387, 395, 411–415, 418
  evolution, "laws" of, 14–16

flexured, 283–284, 289, 300–303, 319
indices, 47–49, 55–63, 458–461
size of parts of, 235–238, 246–247, 457–461
structures and functions, 90–93
work of, 429
Brain:body
  computations, 102–104
  graphic data, *see also* Cumulative frequency
    amphibians 136, 147
    birds, 147, 169
    fish, 147
    mammals, 213, 243, 309, 310, 328, 360
    primates, 362, 382, 391, 394, 398
    reptiles, 147, 169
    vertebrates, 43, 44
  factor analysis, 72–75
  numerical data
    amphibians, 132, 135
    birds, 191, 194
    fossil mammals, 207, 212, 236–237, 305–307, 326–327, 334, 359, 381, 390
    living mammals, 212, 219, 247, 359, 381, 392–393, 467
    reptiles, 145, 153, 166
  relations in closely related species, 345–346, 359–362
  "space," 42–45
Brain:endocast relations, 28–32
  amphibians, 131, 134
  birds, 177, 189
  fish, 28, 84, 99, 104–108
  insectivores, 32, 353
  "lower" vertebrates, 28–30
  mammallike reptiles, 149–153, 205
  mammals, 30–32, 241, 295, 346, 388
  olfactory bulbs in mammals, 251
  reptiles, 28–30, 50, 142–146, 157, 162, 165
Brain:foramen magnum relations, 374–378, 466–469
Brain size
  as biological statistic, 14, 41, 55, 73, 388
  and brain functions, 74–81
  estimation, 49–51, 235–242
  evolutionary role of, 316–318
  meaning of, 63–74, 387–389

Index

parameters for *EQ*, 313, 331, 462–465
  theory of, 78–81, 360–362, 395–397, 399
Brontops, 293
Brontosaurus, 142
Brontotheres, 292, 317–318
Burrowing (fossorial) mammals, 354

Cainotheres, 293, 317
Cainotherium, 306
Callicebus, 392
Callithrix, 467
Camelids, 289–290, 293, 301
Camelops, 359
Camelus, 359
Camptosaurus, 145
Canids, 297
Canis, 234, 250–251, 359
Carnivore endocasts, morphology of, 295–299
Carnivore–herbivore balance, Neotropical, 338–339
Carnivores, 53, 290–291, 295–299, 304–315, 417–418
  archaic, 305, 308–315
Carnivorous condylarths, 231, 337
Carpiodes, 105
Cartilage, 28–30, 99–100, 109, 146, 149–150
Cat, domestic, 31
Cathartes, 268–269
Cebus, 392
Cell assemblies, 7, 10
Cenozoic
  birds, 194–197
  time scales, 37
Cephalaspids, 84–85, frontispiece
Cephalization, *see* Encephalization, Index of cephalization
Cephalized sensory systems, 21, 265, 416
Cephalogale, 307
Cercocebus, 393
Cercopithecoids, 367
Cercopithecus, 392–394
Cerebellar neuron counts, 81
Cerebellarization, 13, 249
Cerebellum, 13, 72, 85, 92–94, 247–249, 458–459, *see also* Hindbrain
  ablation of, 158
  Archaeopteryx, 186–189
  fish, 105

Iguana, 29
  mammallike reptiles, 152, 204–205
  prosimians, 370
  pterosaurs, 164
  whales, 348, 351
Cerebral hemispheres, *see* Forebrain
Cetacea, 346–351
Chimera, 109
Chirodipteras, 121–122
Chlorotalpa, 75, 219
Choanichthyes, 115
Chondrichthyes, 40, 100, 109–110, 191
Chrysochloris, 75, 219
C. I. (cerebellar index), 247, 458
Circadian rhythms, 163, 261
Climate in evolution, 35, 260–261, 424
Climatius, 83
Coelacanth, 28, 99, 119–121
Cognitive processes
  evolution of, 16–25, 160–161, 425, 430–432
  and imagery, 427
Colliculi, 92
Colonoceras, 294, 305
Color vision, 21, 24
Columba, 164
Communication and imagery, 427, 432
Competition, 336–338
Complexity of neuronal network, 64, 69–71
Computer and brain, 80, 125–126, 158
Condylarths, 227, 229–232, 236, 240, 290, 294, 305, 308, 321
Consciousness, 16–18, 160–161, 425–432, *see also* Perceptual worlds
Conservation principles, 12, 410
Conservatism of brain evolution, 136, 155, 410
"Constancy" principle, 20, 23, 278, 426
Continental drift, 35, 261
Convergent evolution, 278
Convolutions and brain advancement, 284–285, 289, 295–303
Cope's law, 293, 346, 355, 411
Corpus callosum:medulla ratio, 56
Cortex, *see* Association cortex, Neocortex, Paleocortex
Cortical
  neurons, number of, 64–71, 78–81
  surface area, 64–65

volume, 64–66, 69–70, 72–75
  activity in, 69–71
Corticalization, *see* Encephalization, Neocorticalization
Coryphodon, 32, 230–232, 236–237, 305
Craik, K. J. W., 17, 160, 415
Cranial
  capacity
    as a "parameter," 67–68
    as a "statistic," 14, 41, 55, 67–68, 73, 388
  nerves, 90–91
    in agnathans, 88
    in dinosaurs, 148
    in fish, 102
    in mammals, 89
Cranium in amphibians, 129
Creodonts, 227, 233–235, 237, 290, 305, 308–315, 417
Cretaceous birds, 192–194
Critical mass of brain, 411
Crocidura, 75, 219
Crocodilians, 138
Crocuta, 247
Crossopterygii, 99, 116
Crowding, 423
Cryptobranchus, 130–133, 136
Culture, 5, 428–430
Cumulative frequency
  of brain size in *Homo,* 401
  distributions of EQ, 218, 311, 312, 330, 383
  distributions of log EQ, 463
Cynelos ("*Amphicyon*"), 306
Cynodesmus, 307
Cynodonts, 150–154
Cynohyaenodon, 234, 237, 305

Dacrytherium, 301, 306
Daphoenus, 290, 306
Daubentonia, 384
Deer brain, 301
Deficiencies in human development, 428
Delayed reaction test, 420
Dendrites, 68–71
Desmana, 219
Diademodon, 153
Dicroceras, 300, 306
Dicynodon, 30
Dicynodonts, 149–153

Didelphids, 210–213, 217
Didelphis, 207, 212–213, 243
Diencephalon, 72, 74–75, 85, 92, 265
Diet, early hominid, 422
Dimensional analysis, 53, 60
Dinichthys, 83, 100
Dinictis, 290
"*Dinoceras*," 232
Dinocerata, 227, 232
Dinosaurs, 40, 138–149
  endocasts of, 142–146
  and homoiothermy, 259
Diplobune, 306
Diplodocus, 143, 145
Dipnoi, 116, 121–122
Displacement, 432
Diurnal habits, 35, 162, 220–221, 412
Diversification in EQ, 216–221, 310–319, 329–335
Diversity of specialization, 110–111, 221, 303, 321–323
Dolphin, bottlenose (*Tursiops*), 349–350
Domestication of animals, 424
Dorudon, 348
Dremotherium, 306
Dryopithecus, 366–367
Dubois, E., 55, 57, 62, 80, 360
Dunkleosteus, 83, 100

E, *see* Brain size
Ear, structure of, 270
Early environment and intelligence, 402
Echidna, 7, 260, 354
Echinops, 219
Echinorsorex, 207–208, 212–213, 223
Echolocation, 13, 40, 268, 351–354
Ecology of hominids, 403–405, 421–424
Ectosteorhachis, 116–119
Edentates, 321, 328, 334–335
Edinger, T., 15, 129–135, 164–165, 181–182, 192, 228, 299–300, 316, 352
Edops, 128–129, 133–136, 147
Elasmobranch brain, 108–110
Electric organs, 111–112
Elephants, 290, 342–345, 359
Elephantulus, 75, 219
Elephas, 247
Encephalization, 8, 11
  index of, (k), 57–60
Encephalization quotient(s) (EQ), 60–

Index

62, 211–218, 310–319, 331–339, 383–385
anthropoids, 392–393
carnivores, 305–307
cetaceans, 348
in closely related species, 345–346, 359–362
cumulative distributions, 218, 311, 312, 330, 383, 463
edentates, 334
Holarctic ungulates, 305–307
hominids, 390, 393
insectivores, 212, 219
marsupials, 212, 318, 339
"Mesozoic" niches, 212
monotremes, 354
Neotropical ungulates, 326–327
proboscids, 342, 344
prosimians, 381
Quaternary mammals, 299, 359
statistical analysis, 312–315, 331–335, 462–465
Endocast:foramen magnum relationship, 374–378, 467–469
Endocasts, 14, 26–32, *see also* Brain:endocast relations
earliest, 84–86, frontispiece
measurements on, 49–51, 235–242
morphology of
amphibians, 133–135
archaic ungulates, 240, 293–295
artiodactyls, 300–304
bats, 352–354
birds, 183–189, 194–196
carnivores, 295–299
dinosaurs, 141–144
mammallike reptiles, 149–153
Neotropical ungulates, 321–325
perissodactyls, 289, 293–295, 299–300
prosimians, 368–372
pterosaurs, 162–165
whales, 346–351
primer for, 88–94
Entelodonts, 293, 317
"*Eogyrinus*," 126
Eohippus, *see* Hyracotherium
Eotylopus, 306
Eozostrodon, 203
Epiphysis, *see* Pineal organ

EQ, *see* Encephalization quotient
Equids, 289, 300
Equipotentiality, 70
Equus, 51, 299–300
Erinaceus, 75, 219, 243, 458
Errors
in estimates of body size, 206, 233
of measurement, 31–32, 50–51, 56, 146–147, 241–242
Eryops, 124, 128–129, 133–136, 147
Esox, 112
Ethology, 17, 138
Euarctos, 231
Euparkeria, 164
Eusmilus, 290, 306
Evolution, 5–7, *see also* Adaptive radiation, Diversification, Rates, Selection pressures, Transitions
of behavior, theories of, 407–411
of birds, sketch of, 178–181
of diversity, 315–319
of horse brain, 299–300
of parts of brain, 245–253
Evolutionary trends in primate brains, 372–373
Explosive evolution, 38
"Extra neurons," *see* Theory of brain size

Facial nerve (N.VII), 90, 111
Factor analysis of brain measures, 72–75
Felids, 296–299
Figure:ground relations, 406, 415
Fire, hominid use of, 424
Fish
adaptive radiation of, 38, 100–101
body size in, 102–104
Fissipedes, 287
Fissurization, 32, 284–285, 295–303, 318–319, 345
Fixed action patterns, 279, 414
Flexured brain, 283–284, 289, 300–303, 319
Flight, number of neurons for, 157–159, 197
Flocculus of cerebellum, 89, 152, 189
Flying reptiles, *see* Pterosaurs
Foramen magnum, 236, 374–378, 466–469
Forebrain, 13, 73, 85, 87, 91–93
in birds, 184–189, 194–195

in bony fish, 105, 111
in dinosaurs, 146, 148
evolution of size of, 246–248
in *Iguana,* 29
in mammals, 236–237, 246–248, 459–461
olfaction and "time," 267–269
in pterosaurs, 163–165
in whales, 348–351
Forebrain:body map, 248
Fossa hypophyseos (pituitary), 89
Fossa sylvii, 89
Fossil brains, *see* Endocasts
Fossilization, 26–27
Fossorial (burrowing) habits, 354–355
Frogs, *see* Amphibians
Frontal lobes in prosimians, 370
"Functional equivalence" and "equipotentiality," 70

Galago, 75, 374, 385
Galemys, 75, 219
Genetic drift, 410
Geochronology, 34–37
Geographic dispersion, 336
Geological time scale, 36–37
Geomorphology, historical, 34–35
Gibbons, 362, 366
Gigantism, 293, 342–345, 358, 386, *see also* Cope's law
Gigantopithecus, 367
Giraffa, 247
Glaciation, 421, 423–424
Glia, 68–71
Glia/nerve cell index, 71
Glossopharyngeal nerve (N.IX), 90
Gnathostomes, definition of, 101
Gorilla, 362, 367, 392–394, 398–399
Graphic double integration, 50–51, 235–239
Ground sloths, 334
Growth rate of brain and body, 99, 242
Gyri, evolution of, 32, 293–303

Habitus, 54, 228–235, 239
Hapalops, 334
Haplolambda, 230, 232, 236
Hearing, *see* Auditory
Hedgehogs, *see* Erinaceus
Hegetotherium, 326
Helarctos, 359

Hemicentetes, 219
Hemicyclaspis, 83
Heptodon, 291–295, 303, 305
Herpestes, 306
Hesperocyon, 298, 306, 308
Hesperornis, 181, 192–193, 276
Heterostracids, 84–85
Hibernation, 259
"Higher" vertebrates, 43, 110
Hindbrain, 13, 73, 85, 91–93
in dinosaurs, 146
evolution of size, 249
in mammals, 236–237, 246–249, 459–460
Hindbrain:body map, 249
Hippocampus, 72
Homalodotherium, 322, 325, 327
Hominid "1470," 400
Hominid adaptive zone, 420–424
Hominids, 420–432
brain enlargement in, 396, 400–405
Hominoids, 367
Homo, evolution and adaptations of, 362, 368, 388–389, 398–405
brain, 400–402
H. erectus, 362, 368, 391, 398–402, 420, 423
H. habilis, 398, 400, 405, 420, 423
H. sapiens, 362, 388–391, 393, 398–402, 405, 420–423
"Sapient," criterion for, 388
Homoiothermy, 39, 259–261, 408–409
Hoplophoneus, 31–32, 239, 288–290, 298, 306, 359–360
Human bite, 404
Human preadaptation to zero gravity, 125–126
Hummingbird, 191
Hyaenodon, 234, 237, 290, 305
Hylobates, 393
Hyopsodus, 90–91, 230–232, 236, 240, 291, 305, 338, 352
Hypoglossal nerve (N.XII), 50–51, 91
Hypophysis, *see* Pituitary
Hyrachyus, 294, 305
Hyracotherium (eohippus), 51, 289, 291–295, 299–300, 303, 305

Ichthyomyzon, 86
Ichthyornis, 181, 192–193, 276

Ichthyostega, 124, 128
Iguana, 29
Iguanodon, 145
Imagery, 5, 425–427, 430
  and toolmaking, 430
Imprinting, 279
Index of cephalization (k), 3, 55–63, 288, 458–461
Inferior colliculi, 40, 92, 240, 265–266, 351–353, 416
Information-processing capacity, 41, 70, 76, 81
Inhibition, 415
Innate releaser mechanisms, 17, 279
Insect flight, 156–159
Insectivores, 210–224
Instincts, 263
Integrative functions, definition of, 20
Integrative systems, 25, 413–415
Intelligence, 411, 417–418, 424, 425, 429–433, *see also* Biological intelligence, Cognitive capacities, Imagery
Intrinsic control mechanisms, 149
Inverted perceptual space, 18–19
Isolation, geographical, 336

k, *see* Index of cephalization
Kiaeraspis, 85, frontispiece
Komba ("Progalgo"), 385
Kujdanowiaspis, 101–106

La Brea tar pits, 341, 356–358
Labyrinthine system, 102, 111, 133
Labyrinthodonts, 129
Lampreys, 82, 86–88
Language, 22–23, 405, 409–410, 425–432
Langurs (*Presbytis*), 393–394
Lartet, E., 14
Lashley, K. S., 3, 9, 11, 70, 76
Lateral geniculate bodies, 264–265, 416
Lateral-line system, 111
Latimeria, 28, 44, 99, 116, 119–121
Learning, 10–11, 280, 409, 419–420, 425
Learning sets, 24, 420
Lemur, 365, 379, 467
Lemuroid niches, 386, 403
Length:weight equations, 52–54, 102–104, 166–167, 210–211
Leontinia, 326
Lepilemur, 381
Lepospondyls, 129

Leptolambda, 232, 236, 305
Libypithecus, 367
Light-dark cycles, 35, 261
Limbs as pivot points, 127
Limnogale, 219
Linguistic models of "reality," 430
Lissamphibians, 129
Litopterna, 321
Living fossils, see Relicts
Lizards, 140
Localization
  of function, 7–9, 76, 296
  spatial, of objects, 267, 272
Localizing sound in space, 19–20, 92
Locomotion in fish and amphibians, 116, 126–128
Loris, 365, 467
"Lower" vertebrates, 43, 110
Loxodonta, 359
Lungfish, 116, 121–122
Lutra, 297
Lystrosaurus, 153

Macaca, 393
Macropetalicthys, 106–108
Madagascar lemurs, 373–376, 382, 385–387, 418–419
Mammallike reptiles, 40, 149–153, 200–205, 271
Mammals, 173–175, 283–285
Mandibular nerve (N.V$_3$), 89
Mandrillus, 393–394
Marsh, O. C., 14–15, 32, 179, 192–194, 288, 309, 316–318
Marsupials, 203–204, 210–213, 216–217, 290, 320–321, 338–339
Mass action, 8, 9
Mastodon, 342–343, 359
Medial geniculate, 265, 416
Medulla, 72, 88–94, 265–266
  in agnathans, 85, 87–88
  in bony fish, 105–107
  in *Iguana,* 29
  in mammals, 56, 376–377, 466–469
  in pterosaurs, 164–165
Megachiroptera, 352
Megaladapis, 375, 382, 385–386
Megalichthys, 116–119
Megalonyx, 334
Megaptera, 347

Megatherium, 344
Memory, 75, 415
Meniscotherium, 230–232, 236, 240, 305, 338
Menodus, 293, 300, 306
Mental deficiency, 427–428
Merychippus, 300, 306
Merycochoerus, 306
Merycoidodon, 306
Merycoidodonts, 308, 317–318
Mesatirhinus, 293, 306
Mesencephalon, *see* Midbrain
Mesocyon, 298, 307
Mesohippus, 289, 294, 299–300, 305
Mesonyx, 231, 237
Mesopithecus, 367
Mesotherium, 325, 327
Mesozoic
  birds, 179–181
  climate and geography, 260
  mammals, brains and bodies of, 202–208
  niches, 173–175, 201–204, 216
Miacid carnivores, 290, 338, 352
Microcebus, 381–383, 467
Microchiroptera, 352
Microgale, 219
Midbrain, 20, 85, 88, 91–93, *see also* Optic lobes, Tectum, Inferior colliculi, Superior colliculi
  birds, 184–189, 194–195
  dinosaurs, 146
  fish, 105
  frog, 130
  Iguana, 29
  pterosaurs, 163–165, 171
  whales, 351
Midbrain exposure, 239–240, 283–284, 352–354
Middle ear bones, 40, 270–271
Mind, 4
Miniature nervous systems, 9–10, 160
Minimum convex polygons, 44–48
Miocochilius, 327
Modeling of novel neural systems, 21–23, 410, 425–427, 429–430
Moeritherium, 334, 342–343
Monkeys, 362, 365–368, 392–395
Monotremes, 6–7, 226, 260, 354
Morganucodonts, 203

Mormyrus, 111–112
Motivation, 409
Multituberculates, 203–205, 228
Mustela, 233, 345
Mysticeti, 348

N.I–XII (cranial nerves), 90–91
N_c (extra neurons), *see* Theory of brain size
Neanderthal man, 368, 401
Necrolemur, 381, 384
Necturus, 130–133, 136
Neoceratodus, 121–122
Neocortex, 72, 91–93, 218, 289, 351, 370
Neocorticalization, 218, 319
Neogene, definition of, 36, 288
Neohipparion, 306
Neomys, 75, 219
Neotropical mammals, phylogeny of, 320–323
Nerve:muscle ratio, 159
Nesides, 119–120
Nesodon, 322, 327
Nesogale, 219
Neural capacity for memory, 269
Neurochemistry and brain size, 14, 71
Neurogymnurus, 207–208, 212–213, 220
Neuron(s)
  aggregates of, 69–71, 160
  evolutionary significance of, 9–10
  number in
    cerebellum, 81
    cerebral cortex, 66–67, 71, 79–81, 264, 390, 392–393
  number of,
    in auditory system, 264, 416
    for flight control, 157–159
    in visual system, 266, 416
Neuron density, 64–66, 68–71
Neuronal
  connectivity, 68–71
  packing, 20, 68, 202
N. I. (neopallial index), 247, 457
Niche and brain size, 255, 315–318
Niches
  mammalian, 200–202, 215, 217, 346–355, 402–405
  vertebrate, 37–41, 128, 276–278
Night vision system, 271–273
Nimraevus, 298

Nocturnal
  adaptive zone, 261–275
  niches, 20, 40, 201–202
Nocturnicity, 412
Notharctus, 364, 368, 379
Nothrotherium, 334
Notoryctes, 354
Notostylops, 324, 326
Notoungulates, 321
Numenius, 169–170, 181, 192, 194–197
Nyctosaurus, 163
Nythrosaurus, 204–205, 214

"Objects" as neural constructions, 17, 20, 22, 161, 413–414
Oculomotor nerve (N.III), 90
Oddity learning, 24
Odocoileus, 302
Odontoceti, 347
Odontognathae, 179–181, 192–194
Oldfieldthomasia, 326
Olfaction, 223–224, 251–253, 255, 267–269
Olfactory bulbs, 72–73, 85, 87, 89, 92
  in *Archaeopteryx,* 187
  and body size, 250–253
  in *Ectosteorachus,* 117, 119
  evolution of size of, 9, 221–224, 250–253, 255
  in *Iguana,* 29
  in mammals, 221–224, 236–237, 240, 250–253, 351
  in pterosaurs, 164
Olfactory nerve (N.I), 90, 148
Olfactory system, peripheral, 271
Optic lobes, 93–94
  in agnathans, 88
  in birds, 184–189, 194–195, 278
  in fish, 111–112, 115
  in *Iguana,* 29
  in pterosaurs, 163–165, 171
Optic nerve (N.II), 89–90, 416
Orbit of eye, 164
Oreodonts, 293, 317–318
Oriented evolution (orthogenesis), 340
Ornithorhynchus, 260, 354
Orohippus, 289
Orthogenetic evolution, 63, 125, 340
Orycteropus, 345
Oryzorictes, 219

Ossicles, auditory, 40, 270–271
Ossification, 100, 152
Osteichthyes, 40, 100, 110–115
Osteodontokeratic (bone-tool) culture
Ostracoderms, 82–84
Ostrich, 141–142
Otters, 296–297
Oxyaena, 337

P, see Body size
Pachycynodon, 306
Palaeosyops, 306
Paleocortex, 72–73, 91, 93, 268, 289, 370
Paleogene, definition of, 36, 288
Paleoniscid fish, 111–115
Pan, 367, 393, 398
Panthera, 359
Pantodonta, 227, 232
Pantolambda, 229–230, 232, 236, 305
Pantotheres, 203
Papio, 393, 397
Parameter, definition of, 14, 67
Paramylodon, 334
Parietal opening, see Pineal organ
Perceptual constancy, 20, 23, 278, 426
Perceptual world
  construction of, 17, 410–411, 415, 426–427, 431–432
  evolution of, 18–23
Perikarya (cell bodies), 68
Peripheral sensory apparatus, 13, 265, 269–271
Perissodactyls, 53, 289–293, 299–300, 308–310, 317–318
Petromyzon, 45
Phenacodus, 89–91, 230–232, 236, 238, 284, 291, 305
Phocaena, 348
Phyletic variate in factor analysis, 72–73
Phylogenetic lines, 139, 226, 291, 322, 366
Pike (*Esox*), 111–112
Piltdown forgery, 420
Pineal organ (epiphysis), 93, 106, 117, 122
  in agnathans, 85, 87–88
  in amphibians, 133
  in bony fish, 106, 117, 122
  in dinosaurs, 148
Pinnipeds, 346

Pituitary (hypophysis), 92–94, 106, 121, 142, 148
Placerias, 150
Placoderm fish, 40, 100–108
Plaina, 334
Plasticity, behavioral, 279, 402, 414–415, 428–433
Pleistocene radiation, 355–358
Plesiadapis, 368
Plesictis, 306
Plesiogulo, 307
Pleuraspidotherium, 231, 236, 305
Pliohippus, 299–300, 306
Poëbrotherium, 52, 289, 306
Poikilotherms, 39, 408
Pongids, 362, 397–399
Pongo, 367, 393, 398
Pons, 89, 93
Poraspis, 83
Potamogale, 75, 219, 467
Potamotherium, 296–297, 306
Preadaptations, 115–116, 125–128, 202
Predator-prey relationship, 320–327, 335–339, 417–419
Predators, social, 403–405, 422
Prefrontal areas, expansion in primates, 370
Presbytis, 393–394
Primate(s)
 olfactory bulbs in, 223
 phylogeny, 364–368
Primitive-progressive dimension, 293–295
Proadinotherium, 326
Proboscids, 290, 334, 342–345
Procamelus, 290, 306
Procavia, 247
Proconsul, 367, 387
Procyon, 247
Progalago, 385
Progressive evolution, 63, 76, 217
Progressiveness, measure of, 80
Promerycochoerus, 306
Promicrops, 119
Proper mass, principle of, 8, 16, 154, 284, 296
Propithecus, 467
Prosimians
 brain:body analysis in, 378–387
 diurnicity in, 418–419

olfactory bulbs in, 223
skulls and endocasts, 369, 371
Prosqualodon, 348
Proterotherium, 326
Protheosodon, 326
Protoceras, 306
Protoceratops, 145
Protohominids, 403–405, 420–422
Protolabis, 290, 306
Protylopus, 295, 301, 306
Protypotherium, 322, 327
Prozeuglodon, 347–348
Pseudaelurus, 298–307
"*Pseudocynodictis,*" 298
Psychic brain factor (E_c), 78
Psychobiology, fundamental problem of, 160
Pterodactyloids, 162, 164, 258
Pterodactylus, 162–170
Pterodon, 162–163, 166–170, 234–235, 237, 290, 305
Pterolepis, 83
Pteronisculus, 113
Pteroplax, 126–127
Pteropus, 167
Pterosaurs, 39, 40, 156, 161–174, 197, 276
Pterygolepis, 83
Ptilodus, 206–218, 222–223, 243
Pyriform lobe, 89, 91, 93, 351, 370

Quaternary fossil birds, 181, 197

Rabbit, 260
Race differences in brain size, 345, 362, 402
Radinsky, L. B., 32, 295–298, 374–378, 466–469
Raja, 105
Ramapithecus, 367, 389
Rana, 130–133, 136
Range, geographic, 404, 422
Rates of evolution, 6, 292–293, 335, 406–407
 in birds, 192, 195–197, 280
 in hominids, 400–401, 420–421, 424
 in primitive mammals, 213–222
 in progressive mammals, 299, 315–319
Rattus, 207, 211, 223
Ray (elasmobranch), 109

Index

Reality as construction of brain, 17, 22, 275, 410, 425–427, 429–430
Relay in evolution, 254, 292–293, 339
"Releasers" in birds, 279
Relicts ("living fossils"), 6, 28, 99, 119–121, 129, 140, 213
Reptiles, history of, 40, 138–141
Reptilian visual systems, 262–266
Retina, 262–266, 271–273, 416
Reversal learning, 11, 24
Rhamphorhynchids, 162, 164–170, 258
Rhamphorhynchus, 163, 166–170
Rhinal fissure, 89, 91, 93, 289, 370
Rhinencephalon, 73, 93
Rhinoceros, 316–317
Rhipidistians, 116
Rhynchippus, 326
Rhynchocephalia, 140
Rhynchocyon, 75, 219, 467
Rhyphodon, 324, 326
Rodents, 210–212, 217–221
Romer, A. S., 26, 33, 129–135, 417–418
Rooneyia, 366, 369–371, 375–376, 381–385, 403
"Rubicon" for sapiency, 388
Ruling reptiles, 137–140

Sabretooth, 31–32, 357–360
Sacher, G. A., 72–74, 374
Sacral "brain," 141
Saguinus, 392
Saimiri, 392
Samotherium, 306
Sarcopterygii, 115
Scalopus, 219
Scaphognathus, 16, 166–170
Sclerotic ring, 162
Selection pressures, 21, 201–202, 257–261, 400, 402–405, 421–424
Self, perception of, 429
Sensory adaptations for diurnicity, 413
Septum, 73
Setifer, 75, 207–208, 212, 219, 222
Sex differences, 362, 402
Sexual dimorphism, 78, 382, 392, 396
Sharks, 40, 109–110
Sign stimuli, 263, 279
Similitude, principle of, 379
Simpson, G. G., 33–34, 204–205
Sinoconodon, 203–205, 209, 214, 226

Sinopa, 233
Sirenians, 342, 346
Skeletal adaptations, early hominids, 422
Skull length, 210–211
Smilodectes, 366, 368–371, 374–376, 378–384, 469
Smilodon, 357–360
Snakes, 140
Social communication, 422–425
Social organization, 278, 404–405
Solenodon, 219, 467
Solipsistic dilemma, 22
Songbirds, 279
Sorex, 75, 219
Sounds, uses of, 279, 425–427, 430–431
Space, perceived, 17–22, 413–415
Species-specific traits, 23, 32, 428
Specific gravity of tissue, 30, 51, 171, 233
Speech, 372, 409, 425
Sphenodon, 28, 140, 144
Spinal accessory nerve (N.XI), 91
Spinal cord, 141–142
Squamata, 140
Statistical analysis of EQ, 312–315, 331–335, 462–465
Stegosaurus, 141–145
Stensiö, E., 84–85, 101–107, 110, 121–122
Stereotypy, 18, 262–263
St.I. (stem index), 247, 458
Subhyracodon, 306
Suncus, 219
Superior colliculi, 8–9, 92, 265, 275
Surface:volume relationship, 49
Sylvian fissure, 32, 295
Sylvisorex, 219
Symphalangus, 393
Synapsids, 149–155

Tachyglossus, 354
Talpa, 75, 219
Talpidae, 354
Tapirus, 247
Tarsiers, 365, 381, 384
Tarsius, 381
Taste receptors, 111
Taxis, 263
Tectum, 87–88, 93, 105
Telencephalon, 85, 88, 93, 265
Teleoceras, 306, 316–317

Temporal integration, 278, 414–415, 431, see also "Time"
Temporal lobe, 368, 370
Tenrec, 75, 219, 353, 467
Terrestrial adaptive zones, 38–39, 124–128
Territoriality, 269, 279
Tetheopsis, 232–233, 237, 305
Tetonius, 369–374, 380–384
Tetraclaenodon, 291
Tetrapods, 124–130
Thalamus, 93
Theory of brain size, 78–81, 360–362, 395–397, 399
Therapsid (mammallike) reptiles, 149–154, 201–205
Thinocyon, 233, 237, 305
Thought as model building, 160
Thrinaxodon, 150–154, 271
T. I. (total index), 247, 457
"Tillybat," 352
"Time," 17–23, 267, 269, 274–275, 414–415, 427, 429–431
Time binding, 17–23
Titanoides, 237
Titanotheres, 291–293, 300, 317–318
Tomarctus, 298, 307
Toolmaking, 403, 427, 430
Toxodon, 322, 325, 327
Trachytherus, 326
Transitions, evolutionary, 37–41
  australopithecine to pithecanthropine, 423–424
  fish to amphibian, 124–130
  insectivore to primate, 364–365
  monkeys to apes, 419
  prosimians to monkeys, 403, 418–419
  reptiles to mammals, 200–205, 213–214
Tremarcotohterium, 359
Triceratops, 140, 145
Trichosurus, 260
Triconodon, 203–210, 212–218, 223, 243
Trigeminal nerve (N.V.), 51, 90, 143, 148, 150–152, 324–325, 347

Trochlear nerve (N.IV), 90
Tuatara, 28, 140, 144
Tubulidentata, 354
Tupaia, 75
Tupaiids, 364–365
Tursiops, 349
Turtles, 140
Tylopods, 293
Typotheriopsis, 322, 325, 327
"Typhotherium" (Mesotherium), 325, 327
Tyrannosaurus, 30, 50–51, 140, 143, 145, 150

Uintatherium ("Dinoceras"), 230, 232–233, 237, 252, 305
Umwelt, 16–17
Ungulates, 287–293, 315, 318, 320–323
  archaic, 213, 215, 217–220, 290–291, 305, 307–308, 320–323
  Neotropical, 321–336, 338–339
Ursus, 247

Vagal lobe, 89, 105, 111
Vagus nerve (N.X), 51, 91, 148
Vertebrate history, 37–41
Vertebrates, "lower" and "higher," 28, 43
Vervets, 394
Vision in nocturnal mammals, 271–273
Visual
  system, 8–9, 18–21, 24, 262–266, 368, 415–417
  world, 171, 199, 275–278, 412–420

Whales, 342, 346–351
Weight:length formulas, 50–54, 102–104, 166–167, 210–211
Wing surface:body weight formula, 168
Wirz, K., 57, 246–247, 457–461
"Within species" defined, 4
Wolves as social predators, 409, 422

Zeuglodonts, 346–348
Zinjanthropus, 391, 400, 423